Metallic Resources 2

SCIENCES

Geoscience, Field Director – Yves Lagabrielle

Natural Resources: Applied Basic Research,
Subject Head – Philippe Boulvais

Metallic Resources 2

Geodynamic Framework and Remarkable Examples in the World

Coordinated by
Sophie Decrée

iSTE

WILEY

First published 2023 in Great Britain and the United States by ISTE Ltd and John Wiley & Sons, Inc.

ISTE Ltd
27-37 St George's Road
London SW19 4EU
UK

www.iste.co.uk

John Wiley & Sons, Inc.
111 River Street
Hoboken, NJ 07030
USA

www.wiley.com

Library of Congress Control Number: 2023935886

British Library Cataloguing-in-Publication Data
A CIP record for this book is available from the British Library
ISBN 978-1-78945-136-8

ERC code:
PE10 Earth System Science
PE10_10 Mineralogy, petrology, igneous petrology, metamorphic petrology

Contents

1

Lithium Mineralization, Contributions of Paleoclimates and Orogens

Éric GLOAGUEN[1,2], Jérémie MELLETON[1], Blandine GOURCEROL[1] and Romain MILLOT[1]
[1] *BRGM, Orléans, France*
[2] *ISTO, CNRS, University of Orléans, France*

1.1. Properties and distribution of lithium in minerals and Earth reservoirs

1.1.1. *Physicochemical properties of lithium*

Lithium is the third element in the periodic table, containing three protons, hence Z = 3, which defines its place in Mendeleev's periodic table of elements. It is a very light alkaline metal element with a density of 0.53, which is half the density of pure water. The average ionic radius of lithium is 0.76 Å. It is very similar to the ionic radius of magnesium (0.72 Å), which explains the frequent substitution of magnesium by lithium in minerals. When in pure state, it is a silvery metal, soft and breakable, with a melting point of only 180.5°C. Metallic lithium oxidizes under air and forms lithium oxide Li_2O. Given its very high reactivity, particularly with water and air, it does not exist in native state. Its boiling point at 1,342°C is high. It is the element with the highest specific heat capacity, 3,582 J/(kg·K). Lithium has three electrons, two on the first electron shell and a single electron on

---

For a color version of all figures in this chapter, see www.iste.co.uk/decree/metallic2.zip.

Metallic Resources 2,
coordinated by Sophie DECRÉE.

the second electron shell (Figure 1.1). The first electron shell reduces the attraction exerted by the protons in the nucleus on the electron of the second shell, which explains the low electronegativity of lithium, 0.98 on the Pauling scale, one of the lowest among all elements. It can thus very easily lose its electron to the benefit of a more electronegative element and form Li^+ cation. Lithium exists in the form of two natural and stable isotopes, 92.5% in the form of ^7Li and only 7.5 % in the form of ^6Li (Figure 1.1.). Lithium 7 (^7Li) is one of the rare nuclides that was formed during the Big Bang, while a very small amount of ^6Li was generated during this event. The large difference in relative mass (~16 %) between ^6Li and ^7Li may generate significant isotopic fractionation. The isotopic composition of lithium in a sample is denoted in relative deviation (per thousand, denoted by ‰) with respect to the value of a reference standard, which is lithium carbonate L-SVEC ($^7Li/^6Li$ = 12.02 ± 0.03‰) (Flesch et al. 1973), whose value of δ^7Li is 0‰ by definition:

$$\delta^7\text{Li}(‰) = \left(\frac{\left(\frac{^7\text{Li}}{^6\text{Li}}\right)_{\text{sample}}}{\left(\frac{^7\text{Li}}{^6\text{Li}}\right)_{\text{L-SVEC}}} - 1 \right) \times 10^3$$

Isotopic fractionation in the Earth's systems is very important, with a variation in δ^7Li of over 60‰ (Tomascak 2004).

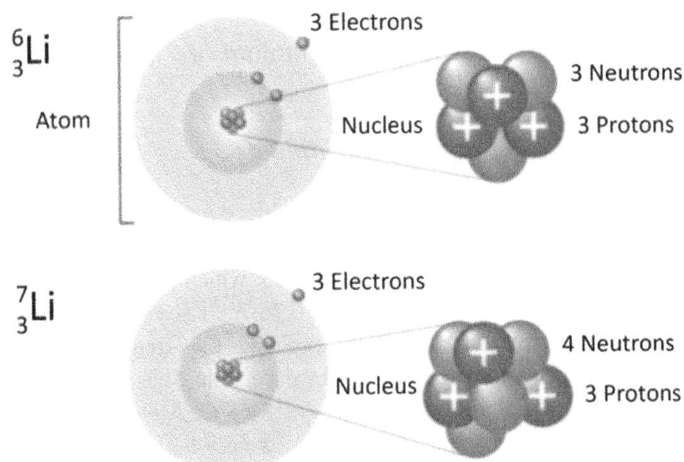

Figure 1.1. *Configuration of the two lithium isotopes*

At the first approximation, the more negative values of δ^7Li are measured in mantle rocks, with the majority of rocks having an average signature between 0 and 5‰, with the exception of carbonates (above +15‰). Waters have, on average, much higher signatures with larger variations than those measured for rocks. Finally, the present seawater has a value around +31‰.

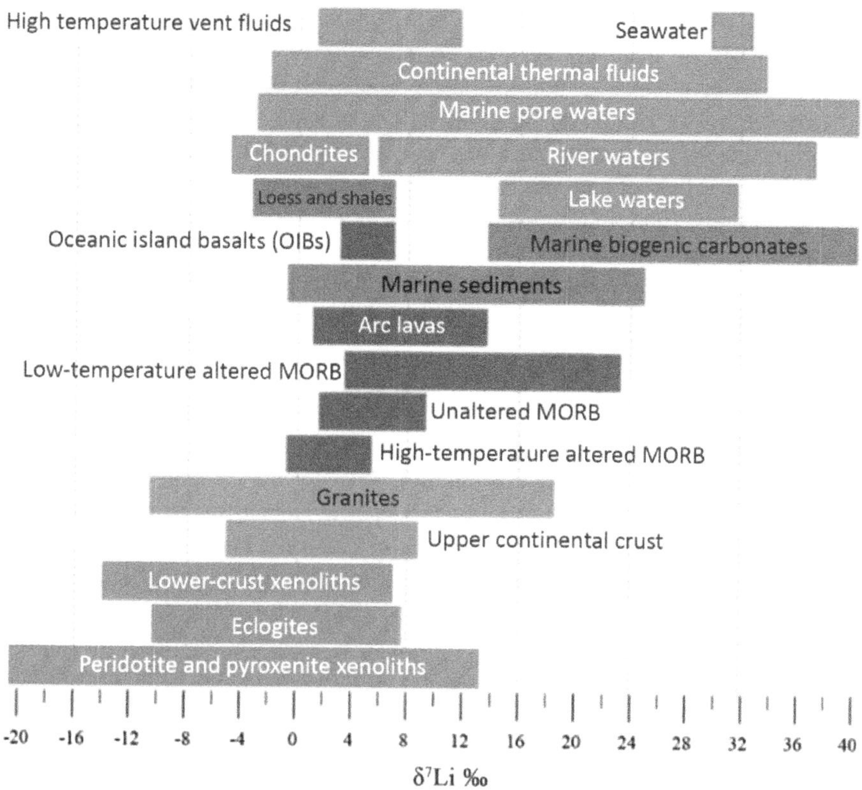

Figure 1.2. *Isotopic lithium composition in various Earth reservoirs. Modified after Tang et al. (2007)*

1.1.2. *Uses of lithium*

Lithium is used in various forms: minerals, metals, carbonates, hydroxides, various chemical compounds. In 2018, 50,750 tons of lithium (metal equivalent) was used in the world (Lefebvre and Tavignot 2020). These various forms are related to various purity degrees required for the uses. The field of ceramics does not generally require very pure lithium, which allows its direct use from natural minerals. On the other hand, the field of batteries and chemistry require very high purity lithium (≥99.5% Li), which involves complex treatment processes applied to specific minerals or brines. A very high level of purity is in fact required in order to reach the performances, longevity and also safety of batteries or chemical components. In 2018, the main uses of lithium were as follows (Lefebvre and Tavignot 2020): cathodes and electrolytes of batteries (58%), glass and ceramics (37%), lithium greases (11%), air treatment and air conditioning (5%), mold fluxes for steel production by continuous casting (5%), the production of pharmaceutical products and polymers (2%), aluminum metallurgy (2%) or other minority uses, including cement manufacturing, pyrotechnics or water treatment. Worldwide uses of lithium are driven by the field of batteries, whose lithium consumption has multiplied seven times in 10 years, from 4,260 tons in 2008 to 29,435 tons in 2018. This very high increase in lithium consumption for the field of batteries is expected to continue, leading to an annual consumption between 150,000 and 300,000 tons in 2030. This explains the large number of mineral exploration projects for lithium, conducted on all continents, for several years.

The various uses of lithium are presented in detail in Labbé and Daw (2012) and Christmann et al. (2015), with the following main uses being drawn and summarized from them. In lithium-ion batteries, lithium is used for the cathode, in the form of mixed oxide of lithium and another metal, such as cobalt, manganese, nickel or phosphorus. Hence, there are six main types of cathodes: (1) LCO (lithium cobalt oxide, $LiCoO_2$); (2) LMO (lithium manganese oxide spinel, $LiMn_2O_4$) ; (3) LNO ($LiNiO_2$); (4) NCA ($Li(Ni,Co,Al)O_2$); (5) LFP ($LiFePO_4$) and (6) NMC (Li(Ni,Mn,Co)O). For the electrolyte, lithium is used in the form of lithium hexafluorophosphate ($LiPF_6$), and also in the form of other various complex lithium salts ($LiBF_4$, $LiAsF_6$, LiI, $LiClO_4$, $LiCF_3SO_3$, LiN $(CF_3SO_2)_2$, $LiN(C_2F_5SO_2)_2$, $LiB(C_2O_4)_2$, etc.) in organic solvents or solid polymers. The anode is generally made from carbon (graphite), but sometimes from lithium titanate $Li_4Ti_5O_{12}$. In lithium-metal-polymer batteries, the anode is made from lithium metal. In the industry of glass and ceramics, lithium is used as an additive for manufacturing certain types of glass and ceramics, including vitro-ceramics. Due to the addition of lithium and titanium, the glass expansion is practically zero, because of the negative expansion coefficient of spodumene microcrystals that compensate for the expansion of the glass matrix. For the floor tiling industry, lithium lowers the melting temperature. Depending on the processes, lithium is used in the form of minerals, oxide or carbonate. Generally speaking, the presence of lithium makes it possible to lower the melting temperature of products (energy savings), decrease viscosity, reduce the amount of bubbles, reduce the thermal expansion coefficient and improve chemical resistance. In the industry of lubricating greases, for example for ball bearings, lithium is used in the form of metallic soap, resulting from the reaction of a fatty acid with lithium hydroxide. These greases contribute to maintaining good lubricating properties over a wide temperature range and good resistance to water and oxidation, while being compatible with other additives. The air treatment and purification sector uses about 1,000 tons of lithium per year. Lithium is used in the form of lithium bromide as refrigerant fluid for absorption climatization. Lithium is also used in dehumidification processes, particularly in the form of lithium chloride that can absorb nearly 10 times its weight in water. Finally, lithium aids air purification by eliminating CO_2 in enclosed spaces using lithium hydroxide or peroxide. Continuous casting of steel uses about 1,000 tons/year metallic Li equivalent in the form of spodumene, petalite or lithium carbonate. Lithium improves fluidity and accelerates casting. This process can be used to manufacture parts that are much larger than the mold, for example the iron tracks of railroads, as the solidified part is extracted simultaneously with molten metal casting in the mold. In the pharmaceutical industry, lithium is mainly used in the form of carbonate, for the treatment of bipolar disorder, in addition to the treatment of depression and other various neuropsychiatric disorders, and also for dermatological creams (lithium succinate). Lithium is widely used in aluminum metallurgy for the production of aluminum-lithium alloys for the aerospace industry. Indeed, the addition of lithium decreases the density of the alloy and increases its elasticity. Many alloys on the market have lithium contents ranging from 1.1 to 3.8%. These alloys are used in particular for parts of the structures of Airbus, Boeing and Bombardier aircraft, and also for fighter aircraft. NASA shuttles also use these alloys.

1.1.3. *Lithium minerals*

A minimum of 120 lithium-bearing mineral species are identified in the database of the International Mineralogical Association (IMA, see: https://mineralogy-ima.org/). This number is a minimum, as many mineral species containing low amounts of lithium are not recorded. These 120 species include 73% of silicates, 19% of phosphates and 8% of oxides, hydroxides, arsenates, chlorides or fluorides (Grew et al. 2019; Grew 2020). Among these, 12 species are the most common (Figure 1.3), with over 50 worldwide known occurrences (Grew 2020). Conversely, jadarite $LiNaSiB_3O_7(OH)$, a mineral from the family of zeolites, has only been known and identified in the mineralization of Jadar in Serbia (Stanley et al. 2007), but may represent a significant European deposit in the following years.

A ranking of these various minerals in descending order of their lithium oxide content (Figure 1.3) shows that the groups with the highest lithium content are phosphates, inosilicates and nesosilicates, followed by phyllosilicates. Lithium mineralization therefore depends on the lithium mineral species present in the rock and on their relative proportions in it. Finally, it can be noted that spodumene is the most abundant among lithium-rich minerals (8% weight, Figure 1.4), which partly explains the fact that lithium extraction processes for this mineral are several decades old.

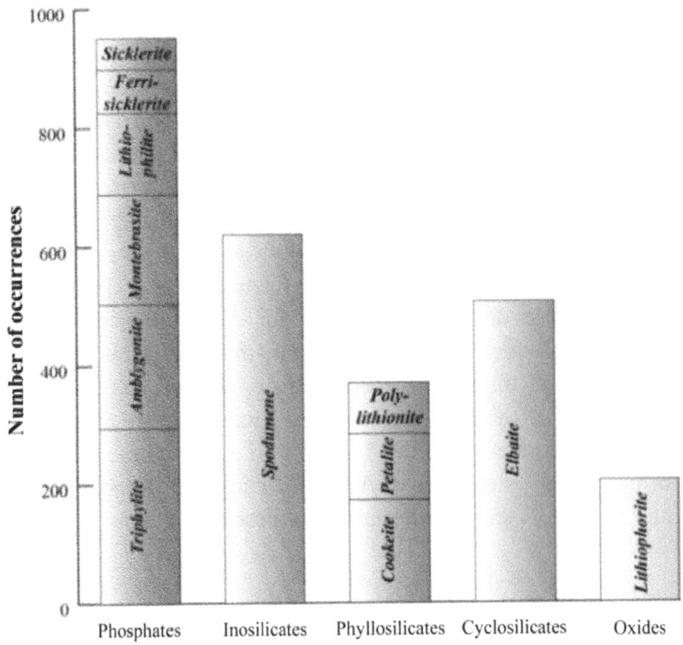

Figure 1.3. *Number of occurrences for the most widespread lithium minerals (Grew 2020)*

1.1.4. *Distribution of lithium in Earth's reservoirs*

Compared to other alkaline metals, such as sodium and potassium, which are among the major components of the upper continental crust, lithium is far less abundant, with an average concentration of only 0.002%, or 20 ppm, equivalent to 20 g of lithium per ton of rock (Taylor and McLennan 1985).

The average concentration of lithium in the upper continental crust is about 35 ± 11 ppm (2σ) (Teng et al. 2004), three times higher than in the ocean crust and basalts (Table 1.1). The concentration of granites is on average 30 ppm, but this average covers a very wide variability. Peraluminous granites are generally far richer in lithium.

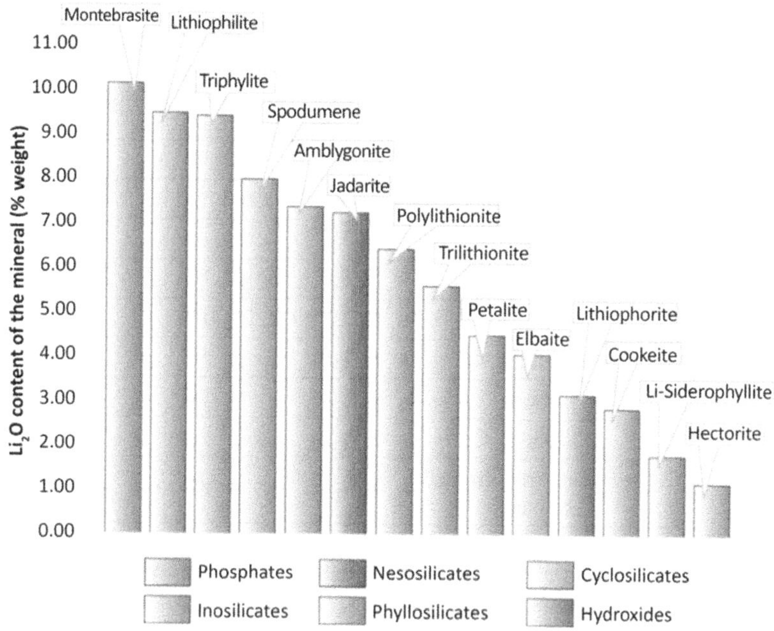

Figure 1.4. *Maximal theoretical concentration of lithium (expressed as Li_2O weight percentage) in lithium minerals that are most common or important in terms of mineralization*

COMMENT ON FIGURE 1.4.– *Source of data: International Mineralogical Association (https://mineralogy-ima.org). Lepidolite (not in the figure) is not a mineral species, but the name of a series representing the solid solution of polylithionite-trilithionite, whose general formula is* $K(Li,Al)_3(Si,Al)_4O_{10}(F,OH)_2$.

As an example, the Variscan prealuminous granites of Cornwall have an average content between 108 and 445 ppm (Simons et al. 2017), which varies according to the generations and excluding the rare-metal granites. Clay-rich marine sediments are relatively rich, with an average from 60 to 80 ppm, but once more this average covers a wide disparity. At the first approximation, the lithium content of aluminous detrital sediments seems, on average, to be above the average of the upper continental crust, as well as the crust types of granites resulting from the melting of sediments.

Lithology units	Lithium content (ppm)
Mantle	1,5[b]
Ocean crust (MORB)	9[a], 10[b]
Lower continental crust	13[b]
Upper continental crust	20[a]; 35–11[b]
Granites	30[a]
Micaschist	34[a]
Hemipelagic muds	79[a]; 57[c]
Deep-sea carbonates	5[c]
Clays	45[a]; 42–77[b]
Marine clays	66[a]; 28–109[b]
River clays	30[a]
Loess	35[a]; 29–16[b]
Soils	33[a]
Secondary (transported) bauxites	47–613[d]

Table 1.1. *Average lithium content in various Earth reservoirs, types of rocks, sediments and soils (source: (a) Li (2000); (b) Teng et al. (2004); (c) Taylor and McLennan (2008); (d) Ling et al. (2018))*

1.2. Lithium metallogeny and gitology

1.2.1. *Lithium behavior in geological processes*

1.2.1.1. *Lithium behavior in surface processes*

At the Earth's surface, several processes have an effect on the mobilization or immobilization of lithium, particularly processes of alteration, erosion and evaporation in the presence of variable amounts of water. Lithium has a strong affinity for the water in which it can be found as tetrahedral complex $[Li(H_2O)_4]^+$ because of its small ionic radius (Rudolph et al. 1995). By comparison, alkaline metals with larger ionic radius, such as rubidium and cesium, form octahedral complexes with water (Mähler and Persson 2012). In chemical alteration processes, rocks altered at the surface (alterites) may have their lithium content increase or decrease with respect to the content of the initial rock, irrespective of its lithium concentration. Indeed, lithium is shared between the neo-formed clay minerals and the surface waters related to alteration (Millot et al. 2010). In low-temperature aquifers, waters allow the formation of a relatively significant proportion of clays

(Pogge von Strandmann et al. 2016) that trap lithium, integrating it in the structure of clays and/or adsorbing it at their surface. Lithium may also be trapped by chlorites as a substitute for magnesium. Therefore, there is a relative proportion of lithium integrated in the structure of minerals that cannot be mobilized by waters unless the minerals are altered. Conversely, the proportion of interlayered lithium (between clay layers) can be mobilized via water circulation. Lithium retention in the structure of secondary clay minerals occurs because its ionic radius is well adapted to the octahedral cavity of the latter (Huh et al. 2004). Hence, during supergene alteration, lithium released by primary minerals is integrated proportionally with aluminum in neo-formed clays (Horstman 1957; Hirst 1962) and surface waters contain a small amount of dissolved lithium. This phenomenon is quite prominent in clay-rich geological formations, in which lithium concentration is correlated with aluminum oxide concentration (Teng et al. 2004) and also with the chemical alteration index (molar ratio of $Al_2O_3/[Al_2O_3 + CaO + Na_2O + K_2O]$) of Nesbitt and Young (1982). Supergene alteration processes therefore tend to concentrate lithium in alterites and in certain bauxites, even though there is not enough data to determine the averages. These processes can be observed through isotopic measurements of lithium. Indeed, surface waters (rivers, lakes, Figure 1.2) have a $\delta^7Li > 4‰$, which means that these waters are enriched in ^7Li compared to alterites and rocks. This enrichment illustrates the fact that the light ^6Li isotope is preferentially integrated in the site with the highest coordination degree (Oi et al. 1991), which corresponds to octahedral sites in secondary minerals (clays, chlorites). Minerals are thus enriched in ^6Li, while waters are enriched in ^7Li at the same time. The relative enrichment of continental waters in ^7Li largely explains the strongly enriched signature of $\delta^7Li = +31‰$ of seawater.

The dissolved salts composition of surface or near-surface waters depends mainly on the initial nature of the latter (rainwaters, etc.), on the nature of the rocks with which they interact and on the water residence time, which is the time of interaction between waters and rocks (e.g. Elango and Kannan 2007). Water gets loaded with lithium, provided that it interacts with a rock that contains it and can release it. This capacity depends on mineral phases that constitute the rock and on their stability in the presence of a given composition of water (Table 1.2).

Volcanic rocks contain variable proportions of glass, which correspond to fractions of magma that are not crystallized due to rapid cooling. This glass can be altered very rapidly, for example from 2,000 to 30,000 years in the case of ashes, into an assembly of aluminosilicates, allophane and imogolite (Hiradate and Wada 2005). The alteration of volcanic rocks by surface waters leads to an enrichment of the latter in potassium, lithium, magnesium, boron and, to a lesser extent, sodium and calcium (Warren 2010). The concentration of waters mineralized in interior (or semi-closed) basins and the enrichment in salts by evaporation transforms these waters into brines when the total concentration of dissolved salts exceeds 3.5%, with the concentration corresponding to that of seawater.

Phases	Formulas	Relative resistances	Chemical alteration processes
Halite	$NaCl$	Very low	Dissolving
Gypsum	$CaSO_4.2H_2O$	Very low	Dissolving
Pyrite	FeS_2	Low	Dissolving and oxidation
Calcite	$CaCO_3$	Low	Dissolving
Dolomite	$CaMg(CO_3)_2$	Low	Dissolving
Glass	Composition variable	Low	Hydrolysis
Olivine	$(Fe,Mg)_2SiO_4$	Moderately low	Oxidation, hydrolysis
Pyroxene	$(Ca,Na,Mg,Li)(Mg,Fe,Ti,Al)(Si,Al)_2O_6$	Moderate	Oxidation, hydrolysis
Plagioclases	$NaAlSi_3O_8$-$CaAl_2Si_2O_8$	Moderate	Hydrolysis
Amphiboles	$(Ca,Mg,Fe,Al,Na)_7\ Si_8O_{22}(OH)_2$	Moderate	Oxidation, hydrolysis
Biotite	$K(Mg,Fe)_3[AlSi_3O_{10}](OH,F)_2$	Moderate	Oxidation, hydrolysis
Potassium feldspar	$(K,Na)AlSi_3O_8$	Moderately high	Hydrolysis
Muscovite	$KAl_2AlSi_3O_{10}(OH)_2$	High	Hydrolysis
Quartz	SiO_2	Very high	Very slow dissolving
Clays	$K_{x+y}[(Si_{4-x}\ Al_x)O_{10}(Al_{2-y}Mg_y)(OH)_2]$	Very high	Hydrolysis

Table 1.2. *Behavior of glass and very common minerals under chemical alteration. Modified from Elango and Kannan (2007)*

1.2.1.2. *Behavior of lithium during metamorphism*

As already noted, due to a similar ionic radius between Li^+ (76 pm) and Mg^{2+} (72 pm), lithium, with 6 coordinance in phyllosilicates (Wenger and Armbruster 1991), substitutes for magnesium in octahedral sites, as well as in interlayer positions to balance electric charges in clays. At the Earth's surface, lithium is mainly carried by clays, particularly produced by supergene alteration. Lithium concentration in metapelites is thus higher when they result from continental domains with strong supergene alteration (Qiu 2011). From diagenesis to greenschist facies, lithium remains with magnesium successively in smectites, illites then chlorites during the increase in temperature. The latter has a negligible effect on the lithium concentration of metapelites at the approach of greenschist facies (~300°C).

Conversely, metamorphic dehydration from greenschist facies to granulite facies may lead to a significant, though not systematic, loss of lithium. Indeed, one of the key factors of lithium preservation in metapelites during metamorphism is the stability of hydrated magnesium phyllosilicates such as chlorite, biotite and staurolite. In particular, staurolite is by far the most lithium-rich mineral compared to other minerals in metapelites (Dutrow et al. 1986; Feenstra et al. 2003). Hence, following the pressure-temperature path and the mineralogical transformations, lithium is increasingly mobilized by the fluid (Chan et al. 1994; Chan and Kastner 2000; Chan et al. 2002) above the temperature of smectite-illite transformation.

If mineralogical carriers of lithium are destroyed, the amount of lithium lost by meta-sediment in contact with the fluid depends on the temperature at which the fluid loss occurs (Chan and Kastner 2000; Millot and Négrel 2007; Vigier et al. 2008; Scholz et al. 2009, 2010), but also on the nature of neo-formed minerals and their capacity to integrate lithium in their structure.

Consequently, a sedimentary formation may lose between 20 and 50% of its initial lithium content during metamorphic dehydration reactions between 250 and 650°C (Zack et al. 2003; Marschall et al. 2007; Teng et al. 2007; Qiu et al. 2011) and even beyond that temperature. Certain examples suggest that the passage from amphibolite facies to granulite facies may lead to a loss of 90% of the initial lithium content in metapelites (Qiu 2011). As an example, the whole-rock geochemistry analysis of the schists of Bündnerschiefer formation (marine metasediments of Western Alps) at various metamorphic grades by Garofalo (2012) shows, in this case, that the most significant loss of lithium occurs at the middle of facies of amphibolites, therefore above 500°C. This loss of lithium is associated with strong dehydration of rocks (Figure 1.5) and interpreted as being related to metamorphism, with only mobile elements in fluids (Li, B, Si) undergoing losses (Garofalo 2012). As a hypothesis, this loss may be related to the destruction of staurolite, chlorite and muscovite according to the following reaction (Spear et al. 1999): staurolite + chlorite + muscovite ⇨ andalusite + biotite + vapor.

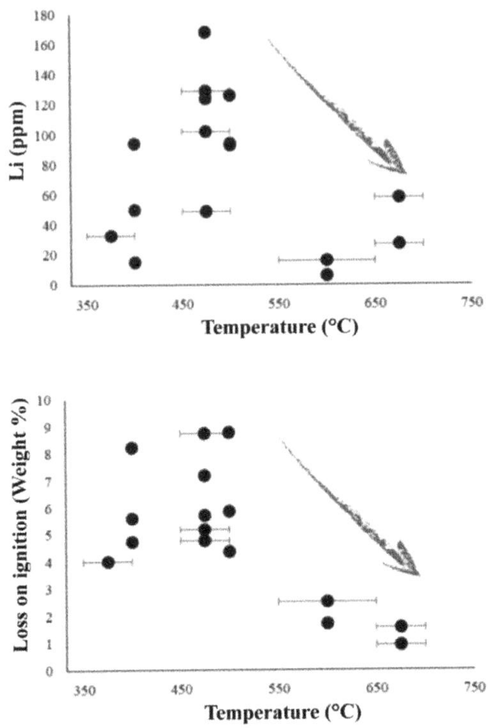

Figure 1.5. *Compared lithium and water content (approximated by loss on ignition) of schists and micaschists of marine origin of Bündnerschiefer formation (Western Alps) at various metamorphic grades (Garofalo 2012)*

1.2.1.3. *Behavior of lithium in magmatic processes*

During the crystallization of magmas of granitic composition, the crystallization of anhydrous minerals such as potassium feldspar (~40% vol.), quartz (~30% vol.) and plagioclases (~20% vol.) has the overall effect of increasing the relative concentration of residual magma in incompatible elements in aluminosilicates (H_2O, Li, P, LILE, HFSE, etc.) (Černý et al. 2005). A key to understanding this phenomenon is the mineral-magma partition coefficient (Kd), which reflects the preference of a given chemical element, denoted by i, to integrate the structure of a mineral or to remain in the residual liquid due to its incompatibility with the structure of the mineral. During a crystallization at equilibrium, this preference will be reflected by a difference between the concentration (denoted by C) of element i in the mineral and the concentration of the same element i in the magma where the mineral crystallizes. The ratio of concentrations of element i in the mineral and in the magma corresponds to this partition coefficient Kd:

$$\mathrm{Kd}_i = \frac{C^i_{\mathrm{mineral}}}{C^i_{\mathrm{magma}}}$$

Despite this simple definition, the mineral-magma partition coefficients for many elements still remain poorly constrained (Table 1.3), as many parameters such as the initial content of the considered chemical element $[C_0]_i$ in the initial magma, or still, the oxygen fugacity of magma, must influence the value of Kd, which may explain the significant variations between the published studies (Table 1.3).

	Biotite			Muscovite			Plagioclase	Orthose	Quartz
	Min	Preferred	Max	Min	Preferred	Max			
F	1.56	2.25	4.00	0.94	1.00	1.92	0.01	0.01	0.01
Li	0.41	0.55	1.67	0.12	0.19	0.82	0.10	0.05	0.01
Rb	1.11	1.90	2.13	0.96	1.60	1.75	0.06	0.83	0.01
Ba	5.63	14.40	15.67	3.41	4.50	8.14	0.19	6.70	0.01
Cs	0.27	0.30	0.57	0.16	0.20	0.36	0.44	0.13	0.01

Table 1.3. *Mineral-magma partition coefficients (Kd) for several elements, traces in granitic magmas. Modified from Villaros and Pichavant (2019) and included references*

Magma crystallizes in the form of several minerals: potassium feldspar, plagioclase feldspar, biotite, muscovite and quartz. The overall partition coefficient of the rock, denoted by D, corresponds to the modal fraction of each mineral, multiplied by the Kd and can be calculated as follows:

$$D = \sum_{k=1}^{n} (X_k \times Kd_{k_i})$$

where:

$$\sum_{k=1}^{n} X_k = 1$$

In the case of lithium, the partition coefficient D of a peraluminous[1] granite composed of 33% quartz, 27% potassium feldspar, 25% plagioclase, 9.5% muscovite and 5.5% biotite is therefore defined by:

$$D = 0.27 \times Kd_{feldspar\ K} + 0.33 \times Kd_{quartz} + 0.25 \times Kd_{plagioclase} + 0.095 \times Kd_{muscovite} + 0.055 \times Kd_{biotite}$$

According to the values of Kd employed (minimum, preferred or maximum, see Table 1.3), Kd of lithium varies significantly between 0.07 and 0.23. However, as Kd is below 1, residual magmatic liquids must get enriched in lithium (and also in Rb, Cs, etc.) during granite crystallization. In order to estimate the enrichment in lithium of a residual peraluminous magma related to the crystallization of peraluminous granite, various approaches are used. The first approach involves the use of Rayleigh's fractionation equation in the case of crystallization at equilibrium to model the enrichment in trace elements (Shearer et al. 1992):

$$C_i^l = C_i^0 \times F^{(D-1)}$$

where F is the fraction of residual magmatic liquid.

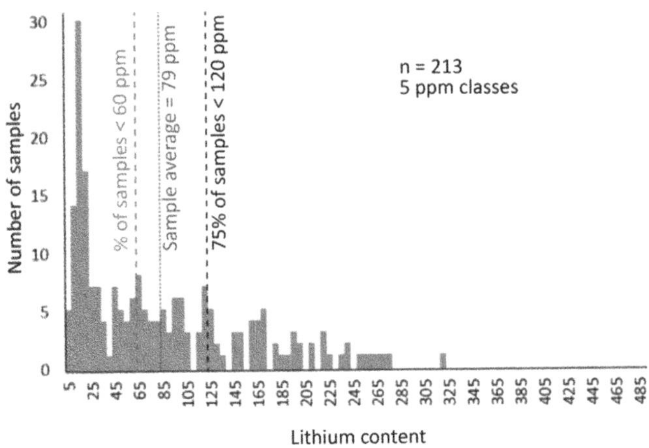

Figure 1.6. *Histogram of lithium content (in ppm) of a collection of 213 samples of granites of various origins according to whole-rock geochemistry analysis: 5 ppm class size (source: BRGM, the French Geological Survey)*

1 A granite is considered peraluminous when $Al_2O_3 > (CaO + K_2O + Na_2O)$, expressed in moles.

It is not easy to estimate the initial lithium content of granitic magmas. However, the initial lithium content of magma (C_0) is clearly one of the critical parameters. The higher it is, the richer the residual magma. Data available on granites generally indicate quite low values, 30 ppm Li on average for the granites (Table 1.1). It is higher for peraluminous crustal granites, of about 79 ppm Li, according to the data in Figure 1.6.

BRGM data on 213 whole-rock analyses performed on samples of granites of various origins range between 2.4 and 316 ppm. About 50% of these granites have a Li content ≤60 ppm and only 25% of them have a Li content >120 ppm. If the initial lithium content of the magma (C_0) is 120 ppm, therefore relatively high, this leads, in the case of an ideal crystallization at equilibrium, to a lithium concentration of the last percent of residual magmatic liquid between 4,100 and 8,466 ppm Li, depending on the Kd used.

This Rayleigh's fractionation model is however not realistic, as it assumes that 100% of the residual liquid is extracted between the crystals, which is not in agreement with the significant viscosities of these magmas (Clemens and Petford 1999) and the impossibility to separate crystals from liquids (Bea 2010).

The Langmuir model (1989) uses a more realistic configuration with the presence of a zone of solidification, mixture of crystals and magmas between the crystallized granite and the uncrystallized magma. In this model, the concentration of a trace element in the residual liquid is defined as follows:

$$C_i^l = C_i^0 \times F^{f(D-1)/[D(1-f)+f]}$$

where:

– f: fraction of fractionated liquid resulting from crystallization in the zone of solidification that returns to the non-fractionated magma in the magma chamber;

– F: total fraction of residual magma liquid.

If magma crystallizes very rapidly, then f is close to 0; if magma crystallizes very slowly, f tends to 1, and the residual magma has the time to escape from the zone of solidification, return and get mixed in the unfractionated magma.

This type of model seems to be applicable to magma chambers established deeply in the lower crust or for deep-seated plutons, but it is difficult to apply to plutons resulting from crustal melting, issued and established in the mid- to upper continental crust.

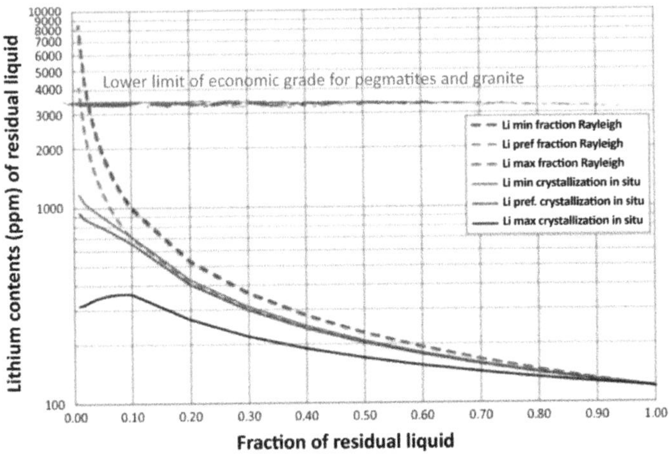

Figure 1.7. *Logarithmic scale graph of the evolution of Li content of residual liquid during the crystallization of a magma of granitic composition between f = 1 (liquid) and f = 0.01, composition of the last percent of residual liquid*

COMMENT ON FIGURE 1.7.– *The highest theoretical lithium concentrations (8,466 ppm) are obtained with Rayleigh fractionation in the ideal (unrealistic) case of crystallization at equilibrium, and the lowest (maximum 1,167 ppm) in the case of the Langmuir in situ crystallization model (1989), calculated for a fraction of fractionated liquid resulting from the crystallization in the zone of solidification that returns to the unfractionated magma in the magma chamber equal to 0.3 (f = 0.3) fixed up to 10 % of residual liquid, then linearly decreasing between 0.3 and 0.08 for a total fraction of residual magma liquid from 0.1 to 0.01 (F = 0.1 to F = 0.01), from a magma of identical initial lithium concentration of 120 ppm. It is also important to note the importance of different Kd resulting from Table 1.3.*

Based on the previous elements, it can be noted that, as the models get increasingly complex (and become more realistic), the enrichment of residual magma in incompatible elements becomes less significant (Figure 1.7). Moreover, the processes through which crustal granite intrusions are built are by themselves critical for the possibility to generate magmas rich in incompatible elements. Indeed, enrichments in incompatible elements by fractionated or in situ crystallization are possible in the case of magma chambers undergoing crystallization. However, the works of the last decade are increasingly calling into question the formation of crustal granite intrusions by the crystallization of magma chambers to the benefit of a construction by successive pulses of limited volumes of anatectic magmas within the same zone (Clemens et al. 2010; Clemens and Stevens 2016). The consequences of this mode of building crustal granite intrusions are dramatic for potential enrichments: the low volume of magma and its probable rapid crystallization very strongly limit the possible enrichments by fractionated or in situ crystallization, and must lead to a concentration in lithium and incompatible elements of residual liquids at the same order of magnitude as the initial liquids issued from partial melting. While crustal granites are probably capable of producing residual liquids at the origin of pegmatites-aplites, these are probably easily enriched in incompatible elements compared to the parent granite.

It is now important to examine the compositions in incompatible elements that can be generated by partial melting of metasediments (see Černý et al. (2005) and the references therein).

The enrichment in incompatible elements of the anatectic liquid during partial melting can be modeled by *batch melting* (Shaw 1970). In this model, the chemical equilibrium between restite and magma is maintained until the final extraction of magma, according to the equation:

$$C_i^l = \frac{C_i^0}{F + D(1 - F)}$$

This equation shows that the richer the melting metasediment is in micas, the lower the enrichment of the first liquid drops resulting from partial melting in incompatile elements. These elements have the tendency to partially remain in the micas, actually limiting the enrichment of the produced liquid.

Hence, for two metasediments with quartz-biotite-plagioclase-sillimanite-garnet-muscovite containing 120 ppm of lithium each, the first one containing 43% biotite and the second one 23% biotite, the first percentage of liquid produced will have a lithium concentration of 556 ppm and 777 ppm, respectively.

Figure 1.8 shows the essential role of initial concentration (C_0) in obtaining liquids very rich in incompatible elements. In order to obtain, with a metasediment composed of 43% biotite, 1–5% anatectic liquids with lithium contents of the order of magnitude of whole rock contents of economic pegmatites, or of the order of 9,000–10,000 ppm, these meta-sediments should contain a minima about 1,800 to 2,500 ppm lithium before starting to melt.

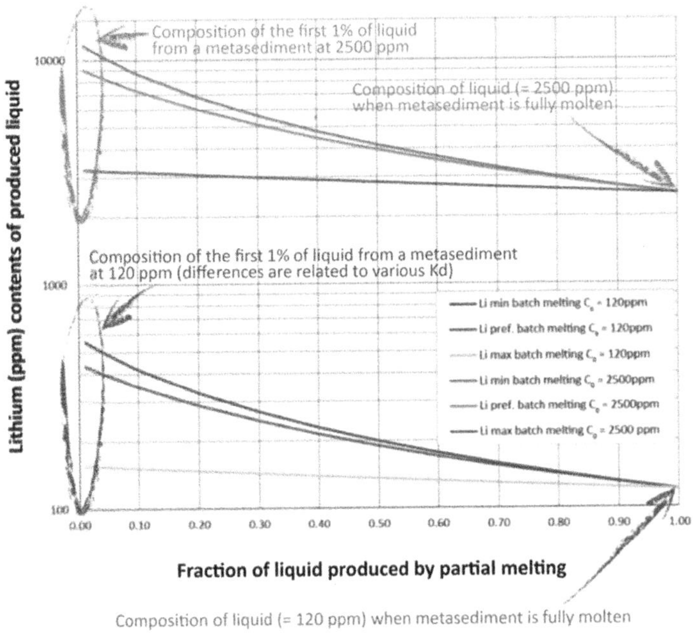

Figure 1.8. *Logarithmic scale graph of the evolution of the content of liquids formed by batch melting of two metasediments composed of 43% biotite, 25% plagioclase, 15% sillimanite, 11% quartz, 3% muscovite and 3% garnet*

COMMENT ON FIGURE 1.8.– *The first meta-sediment has a total content of 120 ppm and the second of 2,500 ppm. For each sample, there are three curves, depending on the value of Kd used for the micas that lead to a variation of the global partition coefficient D of the rock. It can be seen here that in order to obtain magmas with lithium contents of the order of magnitude of economic pegmatites, a significant pre-enrichment is required, in the absence of which this composition cannot be obtained by partial melting.*

1.2.1.4. *Behavior of lithium in hydrothermal processes*

1.2.1.4.1. Magmatic fluids

At the end of crystallization of rare-metal granites, high-temperature hydrothermal fluids, mainly rich in silica, phosphorus, lithium and fluorine are sometimes produced (Roda-Robles et al. 2016). When these fluids are trapped, they form veins or lodes with lithium quartz-phosphates. These are generally phosphates of the amblygonite-montebrasite series whose general formula is $(Li_{1-x}Na_x)Al(PO_4)(OH,F)$. These veins are mainly observed inside or in the proximity of rare-metal granites (Roda-Robles et al. 2016). The best known example of a quartz-amblygonite lode system is represented by the rare-metal granite of Argemela and by the quartz-amblygonite mine of Argemela in Portugal (Michaud et al. 2020). These hydrothermal fluids can also be diffused in formations hosting rare-metal granites and form significant diffusion halos. These halos can be evidenced by geochemistry campaigns on river sediments or soil. The research conducted by Aubert (1969) on the rare-metal granite of Beauvoir (France) is a good illustration.

1.2.1.4.2. Geothermal fluids

In hydrothermal systems, the partitioning of lithium between fluids and rocks interacting with fluids depends on many parameters. The main parameters are the temperature of fluid-rock interaction, the typology of fluids (salt content, pH, etc.), the lithium content of host rocks, the mineralogical carriers structural position of lithium (exchangeable versus non-exchangeable), the granulometry and the porosity-permeability (Romer et al. 2014). Among all these parameters, the fluid-rock interaction temperature is one of the most important and this importance is illustrated, on the one hand, with the global increase in lithium concentration in various types of fluids with temperature (Figure 1.9) (Fouillac and Michard 1981), and, on the other hand, with the existence of geothermometers based on the Na/Li ratio (see Sanjuan et al. (2014) and the references therein) according to an equation of the type:

$$T(°K) = \frac{x}{log\frac{Na}{Li} + y}$$

with $855 < x < 2\,002$ and $-1.275 < y < 1.322$, for values of x and y varying depending on geological contexts of geothermal environments (sedimentary, volcanic, basement, etc., see Figure 1.9).

As already noted in section 1.2.1.1, during the alteration of lithium-bearing rocks by the fluids, there is a competition between the dissolution of primary minerals and the formation of secondary minerals (Pogge von Strandmann et al. 2020). If the primary minerals are dissolved, and there is no formation of secondary minerals, mainly clays, the lithium content of fluids will be significant and the isotopic signature of lithium will preserve the signature of the primary rock. Conversely, if there is a formation of secondary minerals, these will capture a part of lithium and induce a modification of the isotopic signature in lithium of the fluid (equilibrium fractionation). The lithium content of fluids will therefore depend on the lithium concentration of primary rocks (and of the mineralogical carriers of lithium), on the nature of the fluids present and on their temperature. Lithium can be integrated in smectites, smectite-chlorites mixed-layers, illites and chlorites (see Millot et al. (2010) and the references therein). Between 200 and 300°C, lithium is partitioned between the fluid and the newly formed clay minerals (Pogge von Strandmann et al. 2016). In experiments on the crystallization of lithium-bearing smectites, either dioctahedral smectites (Li-montmorillonites) or trioctahedral smectites (hectorites), the amount of lithium integrated in the smectites increases between 25 and 250°C (Calvet and Prost 1971; Vigier et al. 2008).

As an example, laboratory experiments show that the lithium concentration of hectorites increases from 120 to 3,400 ppm between 25 and 250°C (Vigier et al. 2008). Conversely, in high-temperature geothermal systems, above 300°C, the formation of clay minerals is inhibited and lithium is partitioned in the geothermal fluid (Pogge von Strandmann et al. 2016), which therefore has its lithium content increased.

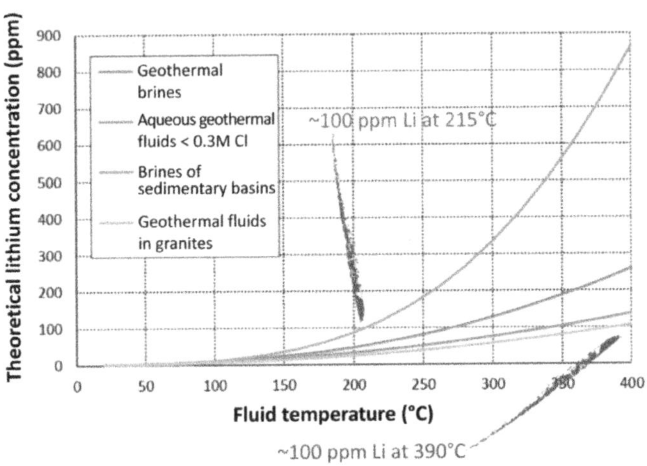

Figure 1.9. *Theoretical evolution of lithium concentration (in ppm or mg/L) of fluids in various geological environments with temperature increase*

COMMENT ON FIGURE 1.9.– *Strong lithium concentration requires strong sodium concentration, which is here arbitrarily fixed at 10 g/L, which is geologically impossible for fluids in all these environments. This concentration is arbitrarily fixed in order to show that geothermal fluids of basements require a much higher temperature than geothermal brines of sedimentary basins in order to reach the same lithium concentration. Sodium concentration also evolves with temperature, and it is the Na/Li ratio that is the geothermometer and not the lithium concentration in itself. The equations employed are: Fouillac and Michard (1981) for aqueous or salted geothermal fluids; Michard (1990) for geothermal fluids in granites; Kharaka et al. (1982) and Kharaka and Mariner (1989) for the brines of sedimentary basins.*

1.2.2. *Lithium gitology and metallogeny: geological settings and types of mineralization*

1.2.2.1. *Overall view*

Lithium mineralization can be grouped in two families. A first family covers sedimentary-hydrothermal mineralization formed at the surface or relatively near the surface. The genesis conditions are strongly controlled by climate, topography and geological setting. A second family covers magmatic to hydrothermal mineralization, whose genesis and location spread from middle crust to the surface. There are frequently more or less significant relations between these two families, with the genesis of magmatic-hydrothermal mineralization having a direct contribution to the stock of lithium of sedimentary-hydrothermal mineralization directly via magmatic fluids or indirectly via volcanic products or rocks.

1.2.2.2. *Sedimentary-hydrothermal mineralization*

Sedimentary-hydrothermal mineralization with lithium takes several distinct forms: *brines*, *clays*, *zeolites* (jadarite) and *bauxites*. Only the first three types can actually constitute deposits. The main environments (or paleo-environments) in which these sedimentary-hydrothermal mineralization occur are mainly intra-mountain or continental basins, rifts and calderas with acid magmatism. The climate associated with sedimentation is mostly dry, or with a closed hydrological system, which seems to be a prerequisite for avoiding the dilution of the lithium contained in surface waters. Stratiform sedimentary-hydrothermal mineralization and lithium bauxites are thus testifying for environments or paleo-environments that have a large number of common characteristics. Indeed, the formation of stratiform mineralization with lithium requires the trapping of a flow of lithium in a fixed place for a certain period of time in order to have an accumulation. The source and the trap may be very close to one another or coincident in the case of clays. The trap may also take the form of a physical depression that may be a closed (endoreic), semi-closed or open continental sedimentary basin, a graben, a rift or still a volcanic caldera. The trap may be a bauxite formation itself. The flow of lithium is provided by the surface water and underground waters that come from drainage where there are volcanic rocks and/or basement regionally surrounding the trap. A part of the lithium flow can also be provided by a volcano-sedimentary detrital fraction deposited in the sedimentary sequence or by hydrothermal and/or in situ supergene alteration of volcano-sediments that are generally of rhyolitic composition. Magmatic brines can also contribute to the lithium stock (Benson et al. 2017; Castor and Henry 2020). The entrapment of lithium-bearing waters occurs in a variable manner by the competition between the evaporation rate and the opening degree of the basin (or of the trap). The most significant lithium mineralization is then encountered under dry climate in an interior basin surrounded by rocks whose lithium is significantly leachable.

1.2.2.2.1. Lithium-bearing brines

According to Bowell et al. (2020), lithium brines are generally defined as hypersaline, with a salinity ranging between 1.7 and 24 times that of seawater, that is, 35 g of dissolved salt per liter. Lithium brines tend to be characterized by concentrations of dissolved ions near saturation for many salts, particularly halite in mature salars. Brines are composed of several ions, including sodium, chlorine, potassium, boron, sulfate and lithium. Nevertheless, lithium enrichment of continental brines is not systematic, highlighting the fact that lithium may come from certain appropriate geological sources (volcanic glass, favorable basement rocks, etc.). Lithium-bearing brines are mainly encountered in three types of environments: in *evaporite basins* (salar type) mainly in South America and Tibet, in certain *geothermal fields* and certain *oilfields*. In these various environments, lithium brines are pumped to the surface then stored in basins where salts can be concentrated by evaporation. Processes have also been elaborated to directly extract lithium from the brine and thus avoid the long evaporation phase.

Lithium brines of evaporite basins are located in recent continental basins containing lacustrine evaporites generated by high evaporation rates with respect to precipitations (Kesler et al. 2012). At the surface of these basins, there are frequently salt crusts, including surface lakes known as salars, whose sedimentary filling is essentially constituted of evaporites. Brines are located in various aquifers inside these porous salt-bearing formations. These brines are encountered in various types of ecological environments defined by two poles, *mature salars* with chemical dominant and *immature salars* with detrital dominant (Houston et al. 2011) (Figure 1.10). Depending on climate and on the detrital contribution, the filling of the sedimentary trap varies from mainly lacustrine fluviatile detrital sequences (clays, sand, etc.) to rare evaporites to salars with an important evaporite nucleus. There is a continuum of possibilities between these two poles. The circulations of fluids with a remobilization of lithium are significant in these basins. Moreover, the circulations of fluids are often amplified by significant thermal gradients related to proximal volcanism. Stratiform lithium mineralization is mainly represented by evaporites containing Li, clays containing Li, and brines circulating in the basin. Mature salars are observed in drier regions (e.g. West Puna, Argentine), while immature salars are located in more humid regions and/or in more humid climate periods (NE Puna) (Houston et al. 2011; Bowell et al. 2020). Mature salars are dominated by a massive halite nucleus, with alternations and lenses of detrital sediments (silt/clay, sand). The halite nucleus consists of a thin layer of upper porous halite and a thick lower sequence of massive halite (Figure 1.10).

In-depth porous spaces are completely filled by diagenetic cementation, leading to thick sequences (>60–70 m) of massive halite (Warren 2010). Immature salars are dominated by detrital sediments such as clay, silt and sand with several intercalated evaporites (halite and gypsum). In these salars, brines do not generally attain saturation in halite because of more humid conditions. Brine volumes are generally high due to the high porosity of detrital sedimentary layers and several brine aquifers may be encountered in-depth. The lithium content of brines present in aquifers of salars are not at all homogeneous and vary strongly from one zone to another, illustrating the control of lithologies situated around and under the salar (Figure 1.11). This is also consistent with the existence of salar that does not contain lithium-bearing brines.

Clayton Valley (Nevada, United States) is an example of mineralization in the form of lithium-bearing brines hosted by a basin with clastic dominance where lithium-bearing clays are also located (Figure 1.12). Until now, only the brines are exploited. Nevertheless, several exploration companies are evaluating the potential of clays as a lithium resource. Clayton Valley is located in the Basin and Range province of the Western United States, which is characterized by a dry climate. This zone corresponds to a semi-closed basin, lined by high mountains comprising a complex metamorphic nucleus and a system of normal faults (Davis et al. 1986) (Figure 1.12). The sediments deposited in the basin are mainly silts, sands and gravels intercalated with clays in the form of illite, smectite and kaolinite (see Munk et al. (2016) and the references therein). The basin is also filled with tuffs and silicic rhyolites from Miocene to Pliocene, and tuffaceous lacustrine facies containing a significant amount of lithium, 215 to 490 ppm Li_2O on average (see Munk et al. (2016) and the references therein). The alternation of dry and humid periods during Pleistocene accounts for the formation of layers of clay, evaporites and associated brines.

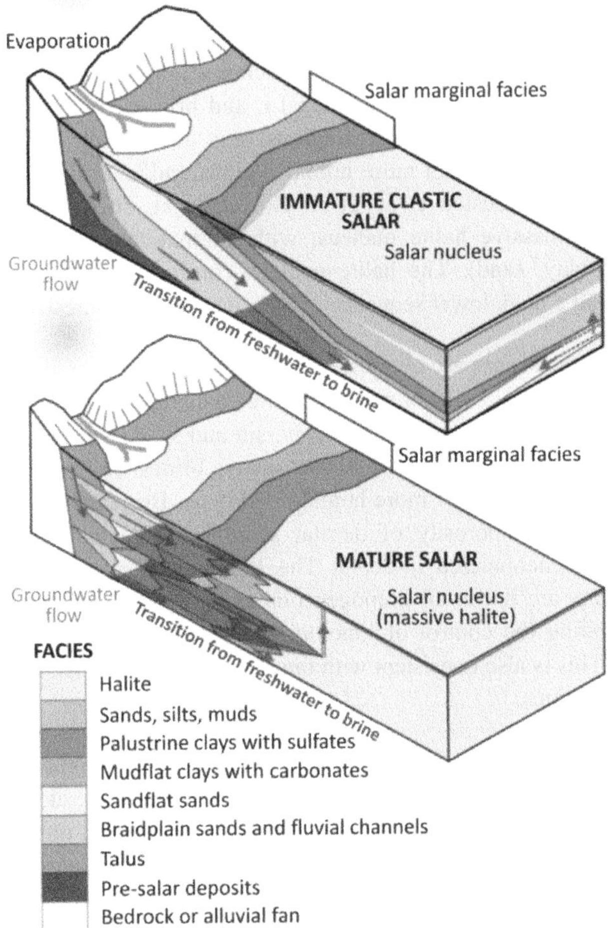

Figure 1.10. *Block models of mature and immature salars showing the distribution of facies and of the main hydrological components. Simplified from Houston et al. (2011)*

COMMENT ON FIGURE 1.10.– *In the mature model, the extension and recession of marginal facies resulting from tectonics and from the climate variation lead to the possibility to transfer diluted waters into the salar nucleus, which is extremely developed. In the immature model, the transmission of diluted waters into the nucleus, much more limited, is also possible.*

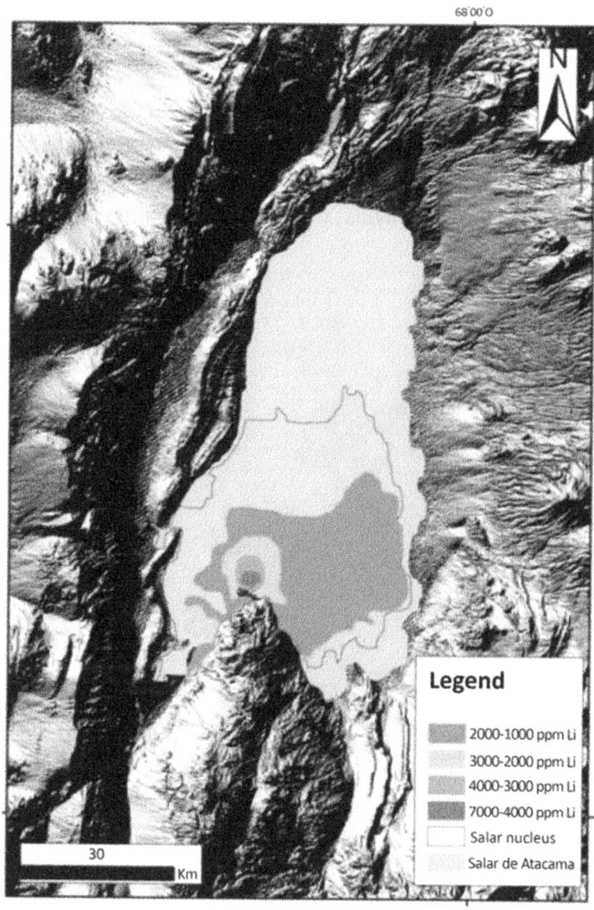

Figure 1.11. *Variations of lithium concentration in brines in the salt-bearing nucleus of the Salar de Atacama (Chile)*

COMMENT ON FIGURE 1.11.– *This map suggests a privileged lithium contribution of surface and underground waters from drainages south of the Atacama Salar. Outside the rich zones, the salt-bearing nucleus surrounded by a black line contains brines whose lithium concentration ranges between 900 and 1,000 ppm. Gruber and Medina (2010) data on the numerical model of shaded terrain Shuttle Radar Topography Mission 90 m of NASA.*

The variations in the quantity of inputs of detrital sediments with respect to evaporite deposits, combined with the play of faults, have allowed the formation of at least six aquifers containing brines (Figure 1.12). A combination of hydrothermal activity and leaching of volcanic ashes and clays throughout time is probably at the origin of the presence of lithium in the aquifers of Clayton Valley (Munk et al. 2016). Lithium brines are located within 6 aquifer units of various natures (layers of volcanic ashes, halite layers, tuff layers, travertine layers and gravel layers).

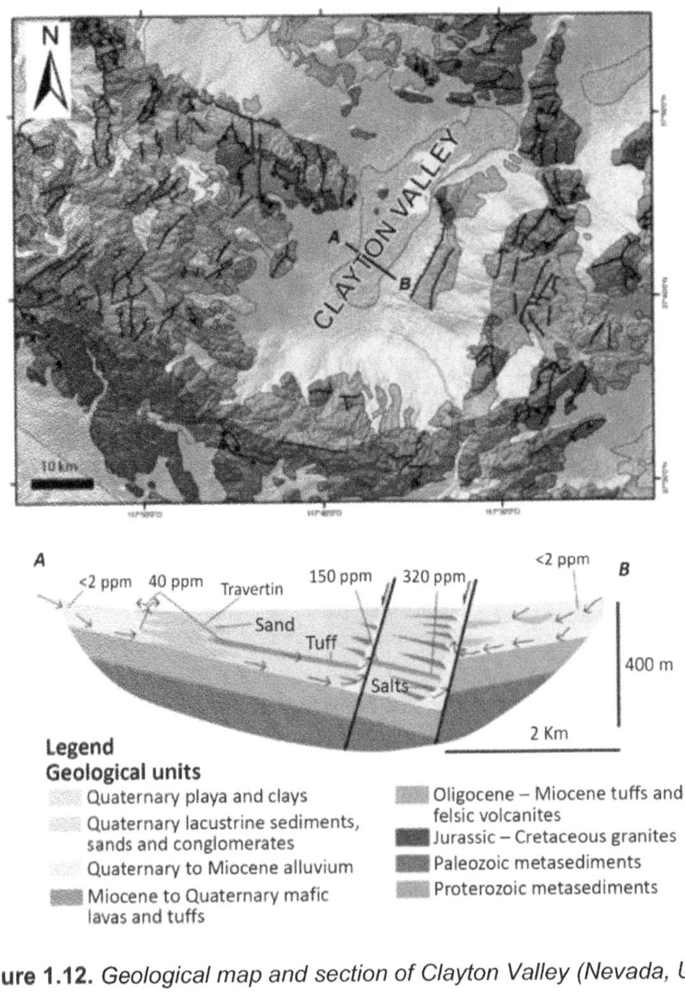

Figure 1.12. *Geological map and section of Clayton Valley (Nevada, USA)*

COMMENT ON FIGURE 1.12.– *Simplified geology according to the 1:500,000 geological map of Nevada of the Nevada Bureau of Mines and Geology, encased in the numerical model of shaded terrain Shuttle Radar Topography Mission 30 m of NASA. According to the shaded DTM, the Clayton Valley is practically a closed basin. Section from Davis et al. (1986). Lithium-bearing brines are present in six aquifers, and lithium-bearing clays under exploration are also present. The values in ppm indicate the lithium concentrations of the circulating fluids. Values from Davis et al. (1986).*

Other geological environments are also sometimes favorable to the presence of lithium brines. This is the case for certain geothermal fields and certain oilfields. For most of the geothermal fluids in contact with volcanic rocks or granites in the in-depth reservoir, rock leaching and hydrothermal alteration (precipitation of illite or smectite) are probably the main processes controlling the lithium concentration in brines. Na/Li ratios depend not only on temperature, but also on other parameters such as the salinity and origin of fluids, or the nature of reservoir rocks in contact with geothermal fluids (Sanjuan et al. 2014). European examples include the geothermal field of Cesano (Italy, Li = 350 mg/L, volcano-sedimentary), the geothermal field of the Upper Rhine Valley (France and Germany, Li = 220 mg/L, Triassic evaporites and Variscan granite) and the geothermal field of the South Crofty mine (South-West England, Li = 125 mg/L, Variscan granite).

1.2.2.2.2. Lithium-bearing clays

Lithium-bearing clays correspond to several species of clays in which lithium is, either integrated in the mineral structure, or exchangeable in interlayer position or adsorbed in the form of ions at the surface of clay. Depending on lithologies and alteration conditions, various types of minerals can be formed (Table 1.4): trioctahedral smectites (e.g. hectorite), dioctahedral smectites (e.g. swinefordite), chlorite-smectite mixed-layer (e.g. lithium-tosudite), chlorites (e.g. cookeite) or dioctahedral deficient micas (illite). Trioctahedral smectites neo-formed during the dry periods in lagoon-lacustrine environments are frequently rich in lithium, but with highly variable concentrations ranging between 400 and 6,000 ppm Li (Tardy et al. 1972). The main known deposits are located in the Basin and Range province (Western United States), and include the deposits of McDermitt caldera, Fish Lake Valley, Hector, Mexico (Sonora) and Peru (Falchani-Macusani). Other locations are also known in the Miocene basins in Serbia, Turkey, Morocco, but there are probably in many other locations worldwide. In these deposits, the source of lithium is frequently a rhyolitic volcanism (Benson et al. 2017) whose volcanic products (tuffs, glass, etc.) located in a depression (caldera, graben, lake, etc.) were altered by hydrothermal fluids contemporary with the neighboring volcanic activity, as volcanic glasses are easily alterable (Table 1.2). Mineralization takes the form of clay layers separated by barren layers. Lithium-bearing clays are also encountered in evaporitic basins, where their role in the enrichment or depletion of coexisting lithium-bearing brines is not yet clearly defined.

	Formulas	Types	Li (%)
Swinefordite	$(Li,Ca_{0,5},Na)_{0,72}(Li,Al,Mg)_{2,66}(Si,Al)_4O_{10}(OH,F)_2 \cdot 2(H_2O)$	Dioctahedral Smectite TOT	1.75
Hectorite	$Na_{0,3}(Mg,Li)_3Si_4O_{10}(OH)_2$	Trioctahedral Smectite TOT	0.54
Li-Tainiolite	$KLiMg_2Si_4O_{10}F_2$	Dioctahedral deficient mica (illite) TOT	1.13
Li-Tosudite	$(Ca_{0,01}Na_{0,15}K_{0,18})_{0,34}(Al_{2,95}Li_{0,22}Fe_{0,01}Ti_{0,01})_{3,19}(Si_{3,5}Al_{0,5})_4O_{10}(OH)_5$	Chlorite-smectite interlayered	0.33
Li-Donbassite	$(Na_{0,07}K_{0,04})_{0,11}(Al_{3,8}1Li_{0,52}Fe_{0,01}Ca_{0,02}Mg_{0,01}Mn_{0,01})_{4,38}(Si_{3,81}Al_{0,19})_4O_{10}$	Chlorite TOTO	0.74
Borocookeite	$Li_{1,6}Al_{3,8}B_{0,6}Si_3O_{10}F_{0,3}(OH)_{7,8}$	Chlorite TOTO	2.21
Cookeite	$LiAl_4(Si_3Al)O_{10}(OH)_8$	Chlorite TOTO	1.33

Table 1.4. *TOT clays, illite, chlorite-smectite mixed-layer and lithium-bearing chlorites (source: webmineral.com; Merceron et al. 1987; Castor and Henry 2020). The weight percentage of metal lithium is given as an illustration, as it is variable. TOT signifies a phyllosilicate whose structure is composed of a three-layers pattern, an octahedral layer (O) surrounded by two tetrahedral layers (T), hence TOT*

The lithium-bearing clay deposits of the Miocene caldera of McDermitt (40 × 25 km at the Nevada-Oregon border, United States, Figure 1.13) are a good example (see Benson et al. (2017) and Castor and Henry (2020), and the references therein). According to Henry et al. (2017), this caldera was formed at 16.39 ± 0.02 Ma during the eruption of about 1,000 km^3 of tuffs. The formation of the caldera was preceded by basaltic flows at 16.69 ± 0.02 Ma, then by many rhyolitic lavas. From 16.45 to 15.66 Ma, a minimum of 200 m tuffaceous sediments accumulate in the caldera. The sedimentary series starts with a 15 m – thick coarse silicified conglomerate followed by tuffaceous sandstones, argilites, pyroclastics, rare conglomerates and fossiliferous limestones. The deposit of this series is partly contemporary with hydrothermal circulations dated at 16.3 ± 0.1 Ma that form uranium deposits in the volcanic flows under sediments, in the western part of the caldera and mercury deposits in the north-western part. Hydrothermal lithium mineralization occured later, dated at 14.87 ± 0.05 Ma on the Thacker Pass deposit.

Lithium mineralization is located in clay layers within 200 m of the lacustrine sedimentary series that constitutes the trap and probably the main source of lithium (Castor and Henry 2020). The lower part seems the most mineralized (>2,000 ppm Li). The drillings of the first 100 m of the base of the sedimentary series at the Thacker Pass deposit show that mineralogy evolves with depth. At the base of the series there is lithium-bearing illite, then about 50 m higher up there are illite and smectite, and the following 40 m contain smectite (hectorite), which indicates that a strong temperature gradient influences the base of these sediments. Illite sediments are very rich (3,000–7,000 ppm) and the part with smectite is far less rich (1,000–3,000 ppm). Lithium enrichments are associated with enrichments in K, Rb, Mo, As, Mg, F and Sb, compatible with hydrothermal alterations. These fluid circulations are distinct from the circulations of uranium, which is not correlated with lithium. Before alteration, these sediments were mainly composed of rhyolitic volcanic glass rich in Li (Benson et al. 2017), ashes, pumices, fragments of minerals and rocks (Castor et Henry 2020). This sedimentary sequence has been modified by a closed hydrological system within Caldera, with a circulation of fluids probably generated by the cooling of the underlying magma chamber.

According to Benson et al. (2017) and Castor and Henry (2020), mineralization had presumably been formed by the alteration of rhyolites by meteoric waters, as well as by hydrothermal fluids rich in lithium issued from magmatic degassing of rhyolitic magmas initially containing 1,100 to 1,500 ppm of lithium (Benson et al. 2017). This mixture of fluids circulating preferentially in boundary faults of the caldera had presumably altered the tuffaceous sediments deposited in it. In the alteration zones, lithium-bearing illites and smectites were formed, leading to mineralized zones. Therefore, there are three sources of lithium: rhyolites, intra-caldera ignimbrites and magmatic degassing.

Lithium mineralization of Sonora in Mexico is also located in the Basin and Range province. The basement is mainly composed by the upper Oligocene-Miocene volcanic complex of Western Sierra Madre (andesitic basalt flows, rhyolitic tuffs, ignimbrites and breccia). This complex was subjected to an extension reflected by the development of a series of half-grabens during Miocene. Fluvial-lacustrine sediments with intercalated tuffs were deposited in the Miocene in the half-grabens. By contrast, Miocene basalt flows cover the basement and the basins. The whole formation was subjected to a subsequent fall that led to a dominant eastward dipping.

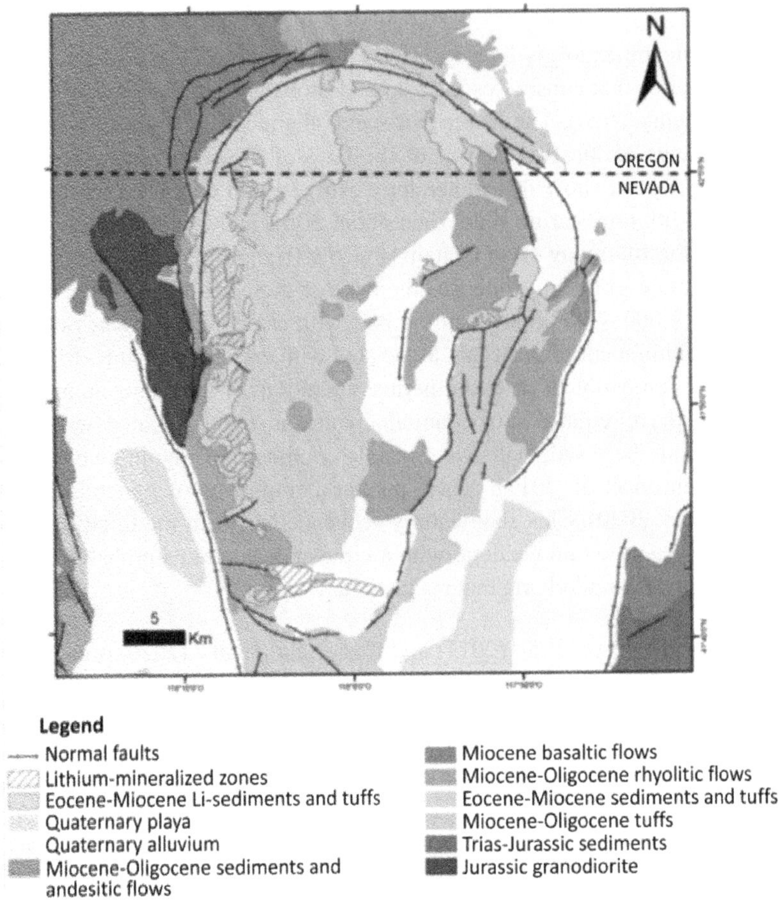

Figure 1.13. *Geological map of McDermitt caldera (at the Nevada-Oregon limit, USA)*

COMMENT ON FIGURE 1.13.– *Simplified geology after the 1:500 000 geological map of Nevada of the Nevada Bureau of Mines and Geology. Detailed geological contours of mineralized lacustrine sediments and mineralized zones according to Castor and Henry (2020). The main mineralized zones are located on the western (Kings Valley) and southern (Thacker Pass) flanks.*

According to Pittuck and Rawbone (2015) and Pittuck et al. (2018), mineralization is contained in a stratiform assembly dominated by pyroclastic formations comprising two distinct stratiform tuffaceous layers containing clays. The tuffaceous units are separated by an ignimbrite layer and the upper tuffacoeus unit is covered by Miocene basalt flows. The lower tuffaceous unit lies on basalt flows, breccias and Oligocene-Miocene tuffs and is covered by an ignimbrite sheet. The average thickness of the lower tuffaceous unit is about 20 m and reaches 40 m in places. The intercalated ignimbrite layer generally has a thickness of 6 m and is covered by the upper tuffaceous unit that measures on average 22 m and reaches up to 70 m in thickness. The mineralized tuffaceous units consist of several mineralized subunits. The lower tuffaceous unit is composed of a basal unit of red siltite-sandstone-conglomerate, tuffaceous sediments, thin layers of lapilli tuff and reworked tuff layers intercalated with lithium-rich clay layers. The upper tuffaceous unit consists of several poorly sorted rough sandstones subunits with clay or limestone matrix and altered clay volcano-sediments.

According to the data of Pittuck et al. (2018), the mineralized tuffaceous levels are mainly composed of volcanic glass (~29%), potassium feldspar (~19%), quartz (~20%), calcite (~18%), clays (~9%) and plagioclase (~4%), with minor quantities of analcime and fluorite. The proportions of clays vary between 3.5 and 18.8%. Clays are essentially illites with local swinefordites (smectites) instead of illites, which suggests a vertical zonality of the mineralogy. Nevertheless, depending on the extent of alteration, the proportions of clays must strongly vary. The mineralized zone has Li contents ranging from 1,000 to over 3,000 ppm Li. In these zones, lithium is correlated with calcium, cesium, magnesium, potassium, rubidium and strontium, marking a global potassium alteration. The origin of the mineralization remains open, but an hydrothermalism can be assumed within a lacustrine basin with an enrichment of volcano-sediments by hydrothermal alteration, as well as a possible contribution of underlying magma fluids or a hydrothermal alteration of the underlying basement.

1.2.2.2.3. Zeolites

In 2004, Rio Sava (subsidiary of Rio Tinto mining company) discovered the Jadar deposit in a lacustrine basin in Serbia. This deposit, located in the Neogene sedimentary basin of Lopare, is now considered a world class "unconventional[2]" lithium deposit thanks to the occurrence of jadarite ($LiNaSiB_3O_7(OH)$; Stanley et al. 2007), a zeolite. In 2022, the sum of the indicated resources (85.4 Mt at 1.76% Li2O) and the inferred resources (58.1 Mt at 1.87% Li2O) (Rio Tinto 2022) represented a giant deposit (2,589,510 tons Li2O).

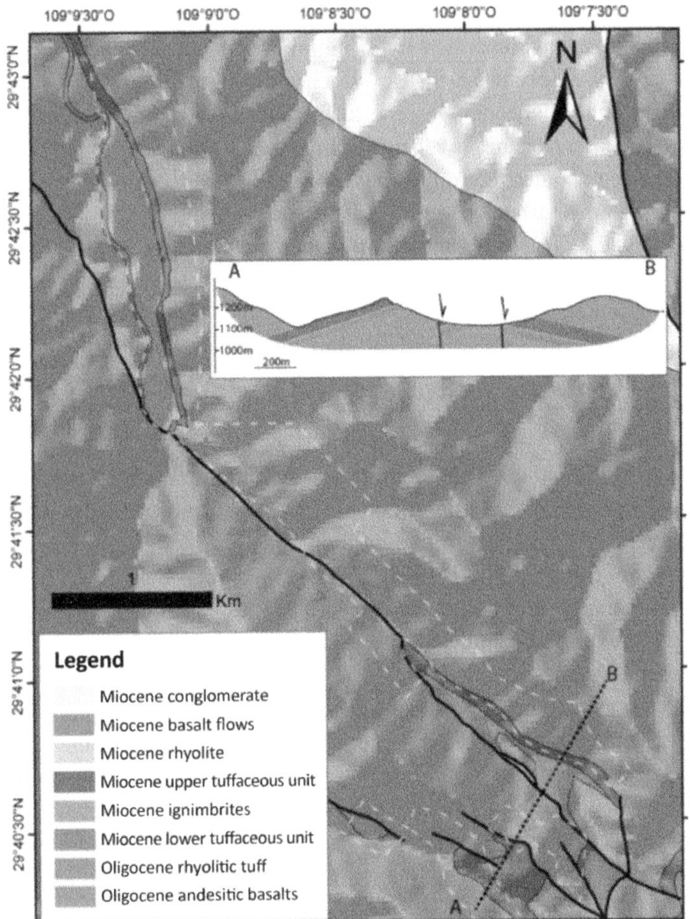

Figure 1.14. *Geological map and section of the Miocene deposit of Sonora (Mexico)*

2 "Unconventional": by contrast with "conventional" deposits (spodumene-bearing pegmatites and lithium-bearing brines), which are historically the main source of high purity lithium.

COMMENT ON FIGURE 1.14.– *Geology and section modified after Verley (2014). The layout of the A-B section is indicated on the map. The mineralized tuffaceous units are in green, see the legend. The right-of-way of the awash and nonawash mineralized zone is marked by the yellow dotted line. Shaded Digital Terrain Model Shuttle Radar Topography Mission 30 m of NASA. Projection NAD 1927 UTM zone 11N.*

Basins are distributed in the form of intra-mountain basins formed during Oligocene and Miocene. They were formed at the end of Dinarides orogen, contemporary with the post-orogen breakdown and the back-arc extension (Marović et al. 2002). These basins generally consist of clay-carbonate rocks, sandstone and clay rocks formed in a volcano-tectonic depression (basin/subaqueous caldera). The early stage of formation of the basin is marked by a lacustrine sedimentation comprising tuffs from fractionated contemporary calc-alkaline volcanic rocks. Lopare consists of an intra-mountain evaporitic lacustrine basin that is relatively large (>20 km long), composed of dolomite, marble, various siliciclastic sedimentary rocks, pyroclastic units and bituminous schists (Obradović et al. 1997). An interesting fact is that the basement of these basins includes LCT pegmatites and granitic intrusions from Cretaceous to Miocene, suggesting a local enrichment in lithium in the basement (Stojadinovic et al. 2017). The mineralization (Figure 1.15) is hosted in a Miocene sedimentary unit whose thickness ranges between 400 and 500 m dominated by calcareous clay, siltites, sandstone and clastic rocks, covering in discordance a Cretaceous basement composed of various metasedimentary rocks, limestone, sandstone and granite, including Miocene intrusions. Lithium and borate mineralization takes the form of stratiform lenses whose thickness ranges between 1.5 and 35 m of three slightly tilted tabular zones covering a surface area of 3–2.5 km (Figure 1.15). The ore is composed of silts and clays containing jadarite with locally intercalated borate lenses (ezcurrite, kernite, borax). Jadarite is present in the form of grains, nodules or white and round concretions from 1 to 10 mm in the matrix of silts or clay that contains various amounts of calcite, dolomite, potassium feldspar, rutile, albite, pyrite, muscovite and ilmenite (Stanley et al. 2007; Rio Tinto 2017). There are few constraints on the formation of jadarite. Certain authors have suggested an interaction between clastic sedimentary rocks and evaporitic brines, potentially involving a hydrothermal alteration of the andesitic-dacitic pyroclastic material, or an alteration of clay minerals may have contributed to its precipitation (Stanley et al. 2007; Stojadinovic et al. 2017).

1.2.2.2.4. Lithium-bearing bauxites

Bauxite deposits may contain various amounts of lithiophorite, cookeite and tosudite (Nishiyama et al. 1975). It seems that secondary bauxites localized in karsts are the richest, but this remains to be confirmed. Lithiophorite is the Li-bearing mineral of manganese oxide that is the most common in karstic bauxites. Aluminum and manganese are of special interest in this type of deposit. The presence of lithiophorite and local cookeite reflects Li-enrichment, though it is not systematically evaluated. However, values up to 0.53% of Li_2O in bauxite are known in China (Wang et al. 2013) and in the alterites of Carboniferous rocks in the states of Alabama, Ohio, Pennsylvania, Kentucky and Missouri (Tourtelot and Brenner-Tourtelot 1977).

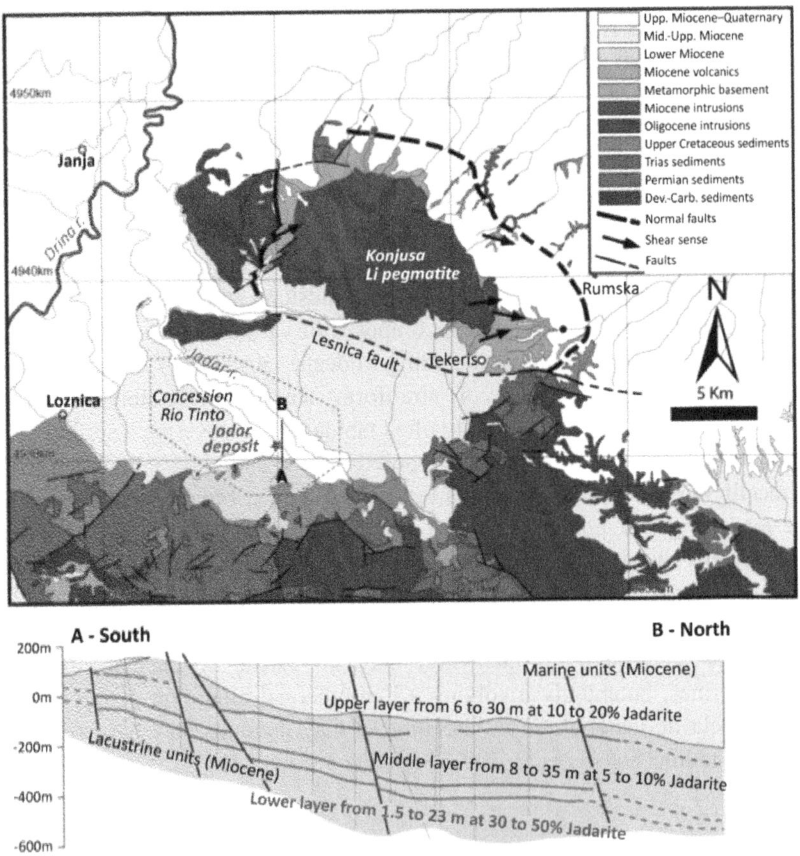

Figure 1.15. *Geological map and section of Miocene basin of Jadar (Serbia)*

COMMENT ON FIGURE 1.15.– *Geology modified after Stojadinovic et al. (2017) and north-west section after Rio Tinto (2017). The line of the A-B section is indicated on the map. The location of the Jadar deposit is indicated by a red star and the contour of the Rio Tinto concession by a dotted line. Many pegmatites, among which spodumene-bearing pegmatites are identified in the metamorphic basement of Cer at the north-east of the basin.*

Lithium is mainly carried by lithium chlorites (cookeite), with 5% of the disseminated cookeite being responsible for a whole-rock content of 1,000 ppm Li (Tourtelot and Brenner-Tourtelot 1977). Bauxite deposits from Jurassic to Cretaceous with lithiophorite are present in Hungary and Greece, suggesting a tropical climate during this period. The minerals containing Li in the bauxite deposits are from Cretaceous in Hungary (D'Argenio and Mindszenty 1986) and from Jurassic to Cretaceous in Greece, where they are located along the north shores of the Mediterranean (Bárdossy 1982).

In Hungary, lithium is present in several locations, such as the deposits from Halimba, Fenyöfö and Kincsesbanya, where the aluminum and gallium in the deposits of bauxite have been extracted since 1950 until recently (Anderson 2015). These deposits are stratiform, where the basement is carbonated and karstic, in which the bauxite horizons are relatively large. The mining district of Halimba is one of the largest, with a thickness of bauxite varying from 1 to 40 m. Here, lithiophorite, mineral containing Li, comes from the circulation of secondary fluid through the host rock (Bárdossy 1982) forming Mn-rich layers in the supergene levels. Research should still be conducted in order to evaluate the potential of bauxites as a possible economic source of lithium.

1.2.2.3. *Mineralization related to magmatism and to high-temperature hydrothermalism*

Lithium mineralization related to magmatism and to high temperature hydrothermalism have significant connections. Pegmatites and rare-metal granites (Figure 1.16) are probably the result of partial melting of lithium-rich metasediments (>2,500 ppm Li), among which many types were previously mentioned and which can allow the genesis of small volumes of magmas exceeding 7,000 ppm Li by partial melting, themselves.

Pegmatites are not present with rare-metal granites. The latter are generally near large crustal faults, which suggests that these faults play a role in the extraction of larger volumes of magmas issued from the partial melting to form larger volumes.

A metallogenic model according to these concepts was proposed by Gourcerol et al. (2019) (see Figure 1.17).

Hydrothermal lithium mineralization is related to fluids resulting from the end of crystallization of magmas. Either the alteration of granite (or of pegmatite) by its own fluids will create a mineralization essentially constituted of micas, known as greisens (Gourcerol et al. 2019) (Figure 1.17), or the fluids that escape the granite will form veins of quartz with lithium phosphates (Michaud et al. 2020). In certain cases, the diffusion of fluids in the hosts leads to a geochemical anomaly without forming mineralization.

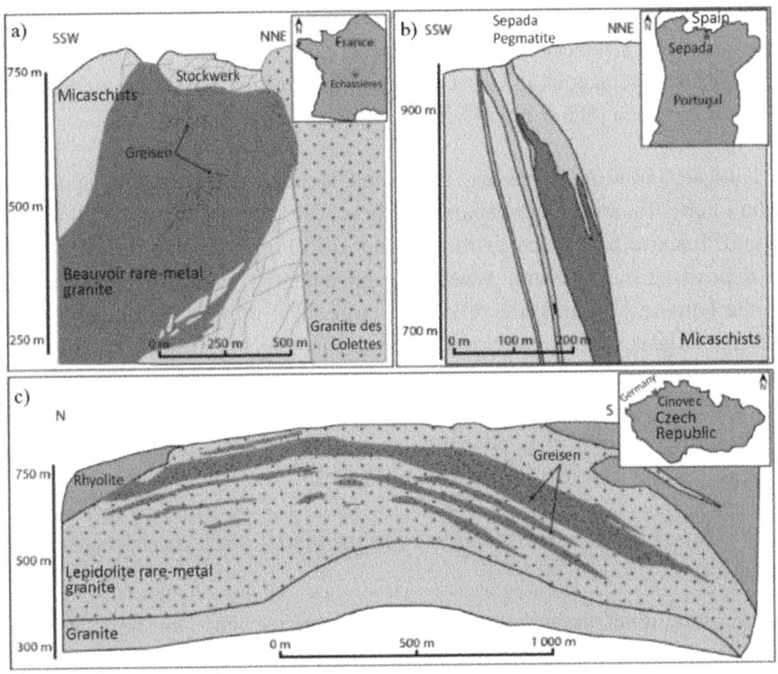

Figure 1.16. *Cross sections of various lithium deposits*

COMMENT ON FIGURE 1.16.– *Lithium units are identified in red in each section. (a) Beauvoir rare-metal granite (France) and associated greisens; (b) Sepeda pegmatite in Portugal, the orange zone represents sterile pegmatite; (c) Cinovec greisen in a rare-metal granite in the Czech Republic. Modified from Gourcerol et al. (2019).*

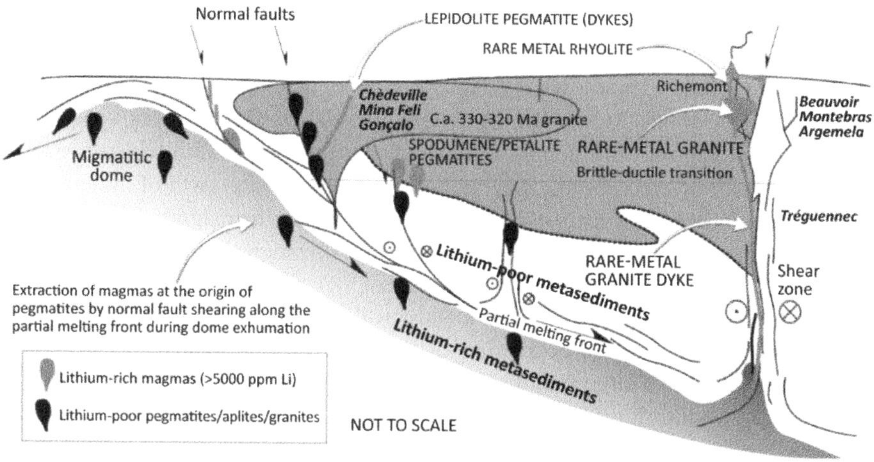

Figure 1.17. *Anatectic model for the generation of pegmatites (barren and Li-rich) and rare-metal granites. Several examples of European mineralization are indicated. Modified from Gourcerol et al. (2019)*

1.2.2.3.1. Lithium-rich rare-metal pegmatites

According to London (2008), a pegmatite is defined as follows: *an essentially igneous rock, commonly of granitic composition, that is distinguished from other igneous rocks by its extremely coarse but variable grain-size, or by an abundance of crystals with skeletal, graphic or other strongly directional growth-habits. Pegmatites occur as sharply bounded homogeneous to zoned bodies within igneous or metamorphic host-rocks*. In addition to this definition, it is important to note that pegmatites are bodies whose volume ranges from several cm^3 to less than 1 km^3. Furthermore, a pegmatite is never isolated, but in groups of several bodies of variable geochemistry up to several thousand individuals (Černý et al. 2005), distributed in the form of clusters of pegmatites (Deveaud et al. 2013). The several studies that focus on the emplacement of pegmatites (Demartis et al. 2011; Deveaud et al. 2013; Silva et al. 2018) show a significant structural control, highlighting the role of faults (drain) and the maximal compression stress (tension gashes), as well as the pressure of fluids associated with the crystallization of pegmatites (Brisbin 1986).

The large size of crystals in the pegmatites seems paradoxical for small intrusions of that crystallize rapidly. Experimental crystallizations of pegmatitic textures conducted by Devineau et al. (2020) contributed to understanding this paradox. The texture cannot be produced unless the pegmatitic magma is suddenly cooled by over 100°C below its *liquidus* temperature. This generates an extremely rapid increase of acicular skeletal crystals that share the same crystallographic axis. A maturation phase then allows crystallization among acicular crystals, which thus produces large crystals. Moreover, the dynamic crystallization produces a sequenced growth of minerals according to the following sequence: potassium feldspar > albite > quartz. As for layered aplites, they seem to be linked to a growth of minerals that is faster than the rate of diffusion of chemical elements in the magma, which leads to the formation of a limit layer and a sequenced crystallization (London 2008).

There is a large chemical and mineralogical variety in pegmatites. The most widely used classification of pegmatites is that of Černý and Ercit (2005), according to which there are two large families (section A1.1), lithium-cesium-tantalum (LCT) and niobium-yttrium-fluorine (NYF). Each family is then subdivided into class, subclass, type and subtype. Economic lithium pegmatites (>0.5 % Li_2O, for ceramic and battery industries) correspond only to certain types and subtypes of pegmatites in the LCT family. In other types and subtypes, lithium-bearing minerals can be present in the last stages of crystallization, but the volume of lithium-bearing minerals is very low. This is particularly the case for micas of the lepidolite series and lithium phosphates (lithiophillite-triphyllite) found in the albite zones of pegmatites of the beryl-columbite-phosphate subtype.

Pegmatites of the LCT family are of crustal origin, with their composition being similar to granites of type S that are linked to the partial melting of aluminous metasediments (Chappell and White 1992). They are therefore mainly encountered, similar to crustal granites, in environments of continental collisions where partial melting of metasedimentary sources is spread. LCT pegmatites are peraluminous rocks, whose mineralogy consists mainly of quartz, potassium feldspar, albite, ± biotite, ± muscovite, ± garnet, ± cordierite, ± sillimanite/andalusite, ± tourmaline. Lithium, cesium and tantalum enrichments – even relative – are one of the most prominent characteristics of LCT pegmatites (Černý et al. 2012). In LCT pegmatites, the main lithium carriers are spodumene, petalite, elbaite, the series of lepidolites and phosphates of the amblygonite-montebrasite and lithiophilite-triphylite series. Cesium is generally present in micas and beryl, more rarely in the most geochemically evolved pegmatites in the form of pollucite ($CsAlSi_2O_6$) (Černý et al. 2005; London 2008).

Tantalum is generally contained in colombo-tantalite oxides with niobium and also in cassiterite (Tindle and Breaks 1998), but also in tantalum oxides by the end of crystallization of the most advanced pegmatites. Besides lithium, cesium and tantalum, these LCT pegmatites are remarkable because of their content in various fluid elements such as phosphorus, boron or fluorine.

At the global scale, the number of economic lithium pegmatites remains extremely low compared to the total number of identified pegmatites. These rich pegmatites are always associated with fields of barren pegmatites, such as in the Ambazac mountains (north of Massif central, France, Figure 1.18; Gloaguen et al. 2018), but they may also be located on the periphery of fields, due to their ability to develop at higher structural levels. These pegmatites have been observed from the Archean to the present day.

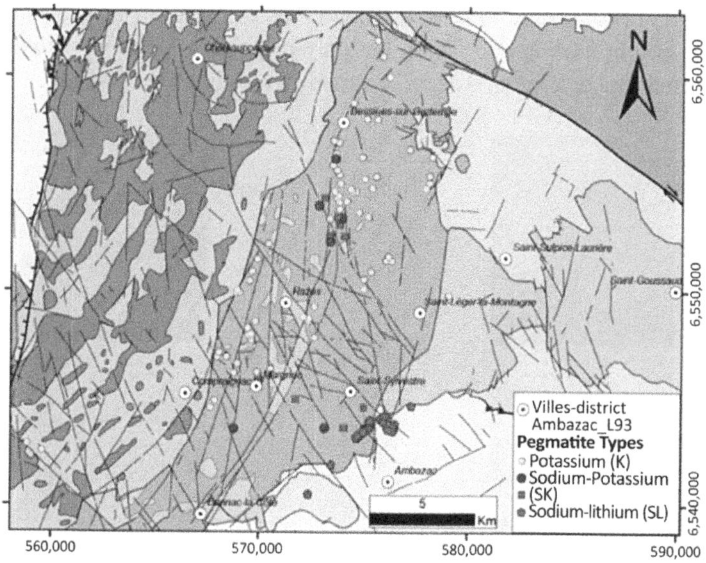

Figure 1.18. *Example of lithium pegmatites at the border of a field of sterile pegmatites. Modified from Gloaguen et al. (2018)*

COMMENT ON FIGURE 1.18.– *Pegmatites of Ambazac mountains and pegmatites intersecting the granite of Saint-Sylvestre (France). Lithium mineralized pegmatites are sodolithic pegmatites (Na and Li-rich pegmatites, in purple). The units in green correspond to mica schists and migmatitic gneiss. The facies in purple, pink and orange are various granites that are older than the lithium pegmatites intersecting them.*

Some of the Archean to Meso-proterozoic lithium pegmatites have the specificity of being giant pegmatites (e.g. Greenbushes in Australia, Manono in

DRC), meaning that their volume is similar to or slightly exceeds that of a rare-metal granite. Lithium pegmatites have spodumene, petalite, micas of the lepidolite series and various lithium phosphates as main carrier minerals.

At the pegmatite scale, lithium concentrations are never homogeneous and vary strongly from one zone to another. These variations are linked to chemical heterogeneities during the crystallization of pegmatites and phenomena of hydrothermal alterations (greisens).

Certain lithium pegmatites are mainly constituted of aplites, such as the lepidolite-petalite pegmatites at Chèdeville-La Chaise (Ambazac mountains, Figures 1.18 and 1.19) (Raimbault 1998; Gloaguen et al. 2018). Conversely, other lithium pegmatites do not have aplitic facies, but only pegmatitic facies with rough granite (e.g. spodumene pegmatite of Adagoï, Barroso-Alvão fields in Portugal).

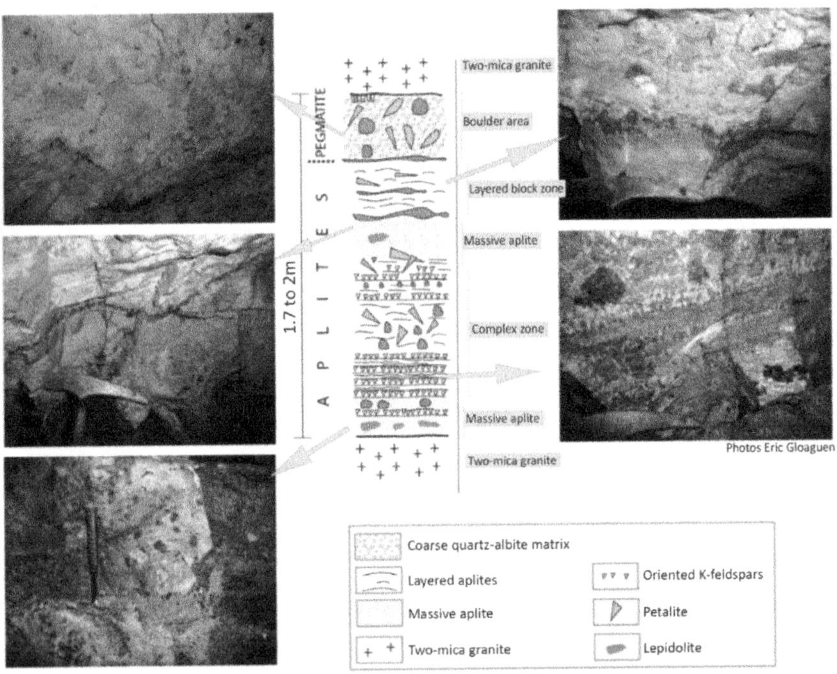

Figure 1.19. *Schematic vertical section of a lithium mineralized pegmatite-aplite dyke at Chèdeville (Ambazac mountains, France). This dyke is composed of 75% layered aplites and 25% pegmatites at the top of the dyke*

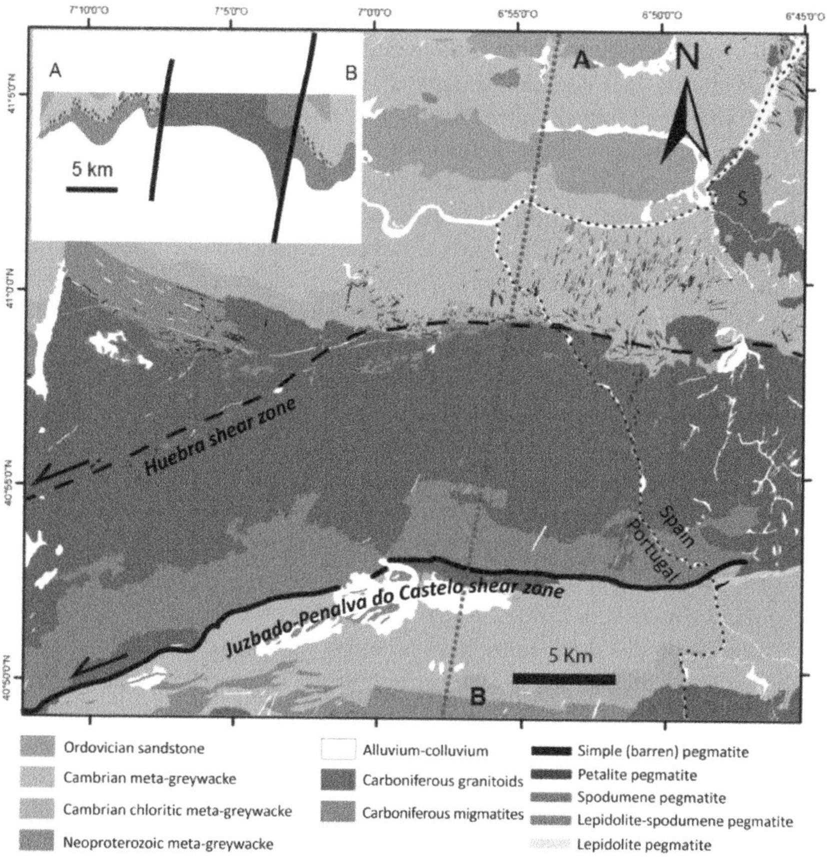

Figure 1.20. *Simplified geological map and section of the Almendra-La Fregeneda pegmatites field at the Portugal-Spain border*

COMMENT ON FIGURE 1.20.– *Simplified geology after the geological maps at 1:50 000ᵉ scale, section AB modified and simplified from Díez Fernández and Pereira (2016). Pegmatites from Roda-Robles et al. (2010). FCRL: granitic complex of Figueira de Castelo Rodrigo-Lumbrales; S: Saucelle granite.*

The field of pegmatites of La Fregeneda-Almendra at the border between Portugal and Spain is a good example of a field of mineralized pegmatites (Figure 1.20). This field is located in the central-Iberian zone that is constituted of

Neo-proterozoic sediments overlain by a Paleozoic greso-pelitic cover, mainly Cambrian to Silurian in age. This sedimentary pile was strongly deformed and transformed during the Variscan Orogen. It is interpreted as an ancient passive margin deformed and buried under thrust sheets (Arenas et al. 2016; Mateus et al. 2016). Metasediments have been affected by three successive phases of deformation and metamorphosed. Phase D1 is related to crustal thickening between 365 and 340 Ma and is accompanied by prograde metamorphism (Dallmeyer et al. 1997). Phase D2, between 340 and 320 Ma, is a syn-convergence phase of extension with the formation of large detachments (Martínez Catalán et al. 1997) and the emplacement of a first generation of granitoids (Castro et al. 2002; Gutiérrez-Alonso et al. 2018). Phase D3, between 320 and 295 Ma, is a phase of final necking associated with large strike-slip shear zones that accommodate the motions between thickened exhumed domains and external rigid domains (Martínez Catalán et al. 2014). During D3, these metasedimentary formations are affected by a significant partial melting that leads to the emplacement of many types of granites and granitoids (Valle Aguado et al. 2017). These various elements are found in the La Fregenda-Almendra zone, where the meta-sediments are strongly folded by east-west folds axes. The zone is intersected by two large D3 sinistral strike-slip shear zones (Juzbado-Penalva do Castelo and Huebra) (Pereira et al. 2017) (Figure 1.20). These strike-slip shear zones have a normal fault component whose effect is the exhumation of migmatites that lie beneath them. The emplacement of the large granitic complex of Figueira de Castelo Rodrigo-Lumbrales is controlled by those shear zones (in red, Figure 1.20) between the migmatites and the unmolten metasediments from 317 ± 2.1 to 313 ± 1.9 Ma (Ferreira et al. 2019). The granitic complex results from the partial melting of underlying neoproterozoic metasediments (Ferreira et al. 2020) whose metamorphic peak was recorded at a temperature of 761 ± 50°C and a pressure of 5 ± 1 kbar (Pereira et al. 2017). A late granitic injection occurs at 300 ± 2.2 Ma (Ferreira et al. 2019), in the south, along the sinistral strike-slip shear zone of Juzbado-Penalva do Castelo.

The field of pegmatites of Fregeneda-Almendra is located at the north of the granitic complex of Figueira de Castelo Rodrigo-Lumbrales (Figure 1.20). The very wide majority of pegmatites are located along the northern border of the granitic complex and between the complex and Saucelle granite (denoted by S in Figure 1.20), mainly in metasediments metamorphosed in the facies of sillimanite to amphibolite, less frequently in the greenschist facies. The characteristics of these pegmatites were presented in detail by Roda Robles et al. (2009, 2010) and Vieira et al. (2011), which are the sources of the data below. This field of pegmatites is composed of several hundred pegmatitic bodies that are essentially barren. Simple (barren) pegmatites are located within metamorphic foliation and are mainly composed of potassium feldspar, albite, quartz, muscovite, tourmaline ± andalusite ± garnet. The more evolved barren pegmatites intersect the metamorphic foliation. These pegmatites contain a bit of lithium in the form of montebrasite and many iron-manganese phosphates. The rubidium and cesium content of micas and feldspars is higher than that of barren pegmatites. Lithium mineralized pegmatites are less frequent. They are discordant on foliation. Four types of lithium mineralized pegmatites have been identified: (1) petalite pegmatites, (2) spodumene pegmatites, (3) lepidolite-spodumene pegmatites and (4) lepidolite pegmatites (Figure 1.20). *Petalite pegmatites* are generally located between 600 m and 1.5 km north of the granitic complex, in the andalusite-sillimanite isograd, only in the Portuguese part. These pegmatites have no obvious internal zonation. Their maximal thickness is 30 m (Mina Bajoca, Portugal, Figure 1.21), but 5-m-thick bodies are more common (Almeida 2003). Feldspars, quartz and petalite are the main minerals, with some muscovite, montebrasite and cassiterite (SnO_2). Fe-Mn phosphates and apatite are accessory minerals. *Spodumene pegmatites* are located between 1 and 4 km north of the granitic complex, near the biotite/chlorite isograd boundary. These pegmatites are not zoned and exhibit thickness ranging between 4 and 15 m (Mina Alberto, Spain, Figure 1.21). Their mineralogy is simple, with feldspars, quartz and spodumene as main minerals and, in lesser proportions, muscovite, montebrasite and petalite. Accessory minerals are essentially beryl ($Be_3Al_2Si_6O_{18}$), Fe-Mn phosphates and cassiterite. *Lepidolite-spodumene pegmatites* are only noted on the Spanish side (Mina Feli, Figure 1.21), between 4 and 4.5 km north of the granitic complex contact, at the proximity of chlorite/biotite isograd. These pegmatites generally have a thickness of less than 3 m, but they may reach 15 m. They have a layered zonation where quartz-lepidolite layers alternate with potassium feldspar and albite layers. The main minerals are quartz, lepidolite, albite, potassium feldspar and spodumene. Montebrasite and cassiterite are minor minerals, while beryl, apatite, ferrocolumbite ($FeNb_2O_6$), stannite (Cu_2FeSnS_4) and eucryptite ($LiAlSiO_4$) are accessory minerals. *Lepidolite pegmatites* appear between 2 and 4 km north of the granitic complex, on both Spanish and Portuguese sides of the field, in the less metamorphic zones with biotite and chlorite. Thickness is generally below 3 m. Their internal structure is zoned, in layers, similar to that of the previous type. The main minerals are quartz, lepidolite, albite and potassium feldspar, and the minor minerals are montebrasite and cassiterite.

Pegmatites were dated using the U-Pb method on colombite-tantalite (Melleton and Gloaguen, not published) and the $^{40}Ar/^{39}Ar$ method on micas (Roda Robles et al. 2009). The location of simple pegmatites is dated using the argon method between 305.0 ± 3.3 and 304.8 ± 4.7. Three types of lithium mineralized pegmatites have been dated using U/Pb to 305 ± 4 Ma (petalite pegmatite of Mina Bajoca), 301 ± 13 Ma (spodumene pegmatite of Mina Alberto) and 307 ± 5 Ma (lepidolite-spodumene pegmatite of Mina Feli). Argon-based dating of mineralized pegmatites yields ages 15–20 Ma younger, between 295.1 ± 4.2 and 296.4 ± 3.5 Ma, which indicates that metamorphic hosts remained above the blocking temperature of the Argon isotopic system (about 300°C) during this time. This phenomenon is related to the underlying presence of migmatites. Pegmatites ages form after the emplacement of the granitic complex, but are synchronous with a proximal partial melting of metasediments that produces late granites (Saucelle, Feli, 305.0 ± 3.3 to 300.0 ± 3.1 Ma, Figure 1.20) and the pegmatites that constitute this field. The partial melting of levels of lithium-rich metasediments (Gourcerol et al. 2019) is probably at the origin of lithium-rich mineralized pegmatites.

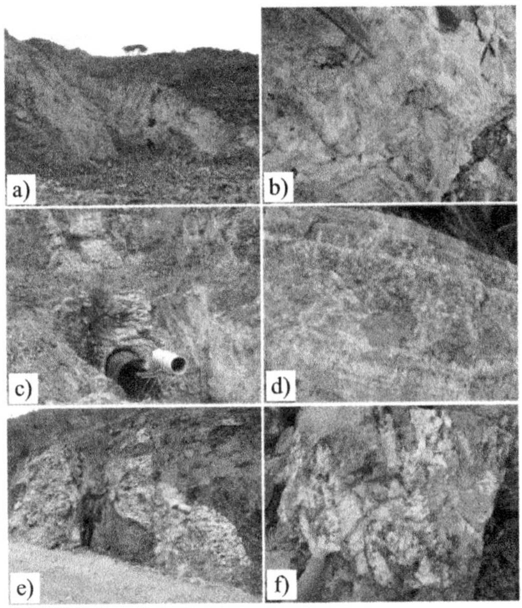

Figure 1.21. *Photos of lithium-bearing pegmatites of the La Fregeneda-Almendra field*

COMMENT ON FIGURE 1.21.– *(a) Mina Bajoca; (b) petalite of Mina Bajoca; (c) vertical pegmatite dyke of Mina Feli underground mining; (d) layered lepidolite (purple) facies of Mina Feli; (e) mineralized pegmatite dykes at Mina Alberto; (f) spodumenes (creamy white) of Mina Alberto (source: E. Gloaguen).*

1.2.2.3.2. Lithium-rich rare-metal granites and their volcanic equivalents

Rare-metal (or rare-element) granites, as defined by Černý et al. (2005) and Linnen and Cuney (2005), are quartz-poor, albite-rich granites, of peralkaline to peraluminous composition, exhibiting a significant enrichment in various elements, with extremely low Zr/Hf and Nb/Ta ratios (<18 and <5, respectively; Ballouard et al. 2016), and with a spectrum of elements related to their geochemistry. This metallic enrichment is reflected by a mineralization disseminated throughout the granite mass (Cuney et al. 2002; Černý et al. 2005; Linnen and Cuney 2005). These granites are mainly found in late orogenic contexts, but also in subduction contexts (in the form of porphyries) or even in anorogenic contexts. They are emplaced in the vicinity of paleo-surfaces in the form of isolated intrusive bodies (Figure 1.22), very small in size, less than or equal to 1 km^3 (Beauvoir, France) (Cuney et al. 1992), microgranite dykes (Tréguennec, France) (Lulzac and Apolinarski 1986), rhyolite (Richemont, France) (Raimbault and Burnol 1998) or of volcanic rocks such as Spor Mountain (USA), which is the largest beryllium deposit in the world (see Dailey et al. (2018) and the references therein). Rare-metal granites are divided into three groups (Černý et al. 2005; Linnen and Cuney 2005): rare-metal peralkaline granites, phosphorus-free metaluminous to peraluminous granites and phosphorus-rich peraluminous granites. Only the latter represent lithium deposits, mainly used to date for the glass and ceramic industry. Phosphorus and albite-rich peraluminous granites are highly enriched in tantalum, tin, fluorine and lithium, and are formed in contexts of continental collision. In these types of granites, the average lithium concentrations vary strongly between 0.5 and 1% Li_2O. These concentrations are carried by various minerals, such as phosphates (amblygonite-montebrasite) and phyllosilicates (polylithionite-trillithionite, Li-muscovite, and rarely zinnwaldite).

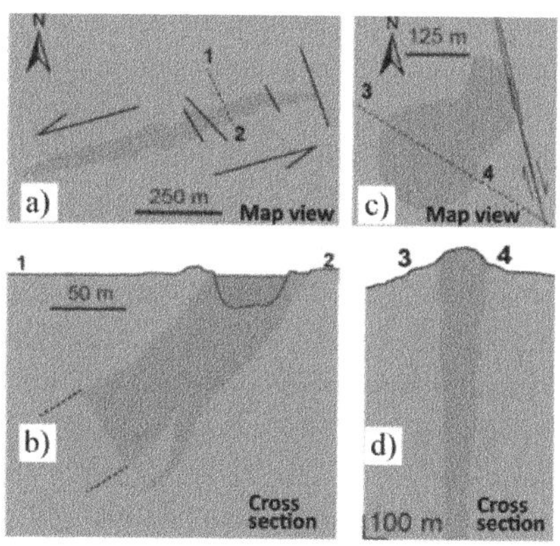

Figure 1.22. *Two examples of rare-metal granites in dyke, and in pipe map view (upper image) and cross-section view (lower image)*

COMMENT ON FIGURE 1.22.– *The section lines are dotted on the maps. (a) Main Tréguennec dyke of rare-metal granite (Finistère, France); (b) simplified section of main Tréguennec dyke, intersected by test drilling, simplified from Lulzac and Apolinarski (1986); (c) map of rare-metal granite of Argemela (Portugal), simplified from Michaud et al. (2020); (d) section of rare-metal granite of Argemela stressed by gravimetry, showing the pipe form of granite, modified from Michaud et al. (2020).*

Representative examples are the north intrusions of the French Massif central, such as Beauvoir (Allier) and Montebras (Creuse), and their volcanic equivalent, the rhyolite of Richemont (Haute-Vienne), or the dykes of Tréguennec (Finistère, Armorican massif) or the intrusion of Argemela in Portugal (Figure 1.23). The geochemistry and the enrichments of these granites are similar to those of LCT pegmatites that are lithium-rich and also phosphorus-rich. The main differences between rich and phosphorus-bearing rare-metal granites and LCT pegmatites are as follows: the volume difference (granite >> pegmatite), the difference in the number of intrusions (rare-metal granites as isolated intrusions or as several scarce dykes to be compared with fields of pegmatites with several hundred individuals) and the difference in structural level of emplacement, as rare-metal granites are not observed in fields of pegmatites, and are generally emplaced near the surface (volcanic or subvolcanic facies). These three points suggest that magma generation processes for granites and pegmatites are similar, but the extraction and ascension of these magmas may be facilitated by the presence of a large structure of ductile strike-slip faults (Gourcerol et al. 2019).

The privileged hypothesis for the genesis of these phosphorus-rich rare-metal granites is that of partial melting of meta-sediments (Černý et al. 2005). Nevertheless, these rare-metal granites could also represent the last injection, and the most fractionnated one, within vertically differentiated plutons (Kovalenko et al. 1970; Cuney et al. 1992; Yin et al. 1995), though this hypothesis may seem unlikely as the parent granite should initially contain of the order of 2,500 ppm lithium, which seems unrealistic. Therefore, there is a strong similarity between these phosphorus-rich rare-metal granites and LCT pegmatites. An important point seems to be the spatial link between rare-metal granites and the large areas of ductile shearing at crustal scale (Figure 1.17) (Gourcerol et al. 2019).

1.2.2.3.3. Lithium phosphates-bearing quartz veins

At the end of the crystallization of rare-metal granites, high-temperature hydrothermal fluids, mainly rich in silica, phosphorus, lithium and fluorine are sometimes produced (Roda-Robles et al. 2016). When these fluids are trapped, they form veins or lodes of quartz and lithium phosphates. These are generally phosphates of the amblygonite-montebrasite series of general formula $(Li_{1-x}Na_x)Al(PO_4)(OH,F)$.

These veins are mainly observed within or in the vicinity of rare-metal granites (Roda-Robles et al. 2016). One of the best examples is the quartz-amblygonite vein system located in the rare-metal granite of Argemela (Figure 1.23) and at the level of the quartz-amblygonite vein system of the Argemela mine (Portugal) hosted by hornfels (Michaud et al. 2020). These hydrothermal fluids can also diffuse within host formations surrounding rare-metal granites and form large diffusion halos. These halos can be evidenced by campaigns of stream sediments geochemistry and/or soil geochemistry surveys. The geochemical halo in the micaschists hosting the rare-metal granite of Beauvoir (France) is a good example (Aubert 1969).

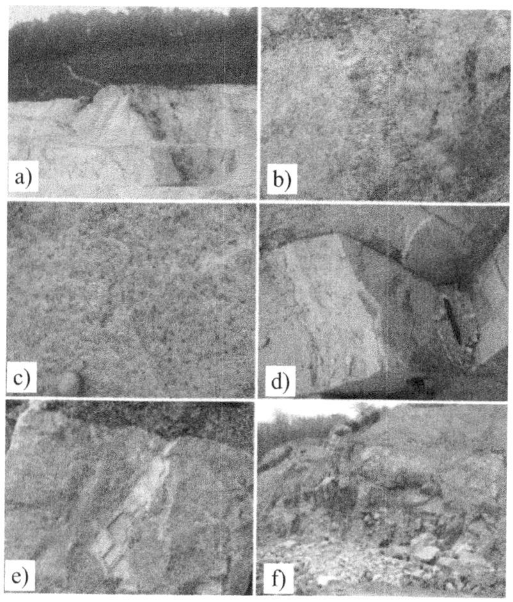

Figure 1.23. *Examples of magmatic and hydrothermal facies associated with rare-metal granites*

COMMENT ON FIGURE 1.23.– *(a) Top of Beauvoir granite (white) with micaschists above (brown). The granite is intersected by a greisen associated with quartz veins (brown-ochre area in the granite), Beauvoir, Échassières, Allier; (b) quartz-feldspar-montebrasite veins intersecting the rare-metal granite of Argemela (Portugal); (c) texture of rare-metal granite of Argemela, note the very light color of the granite and the presence of quartz snow balls related to an emplacement near the surface; (d) detailed view of rare-metal granite of Tréguennec (Finistère), the bedding is a mylonitic foliation related to the shearing of the granite dyke during the sinistral variscan deformation; (e) detailed view of the stockscheider at the top of the rare-metal granite of Montebras (Creuse), showing feldspars of 20 cm pushing from the cut-off contact with the host (dark gray). Feldspars are surrounded by a granite matrix; (f) Montebras stockscheider (pink) between the Chanon granite (at the top, altered) and the rare-metal granite of Montebras under the stockscheider (source: E. Gloaguen).*

1.2.2.3.4. Lithium-rich greisens

Greisens correspond to certain parts of granitic intrusions transformed by hot hydrothermal fluids (Figure 1.23) in an mineral assemblage mainly composed of muscovite and quartz (Černý et al. 2005). Greisens are porous rocks and in these pores there are apatite, cassiterite and sulfides. Greisens are located on the tops of granitic intrusions, and along the faults crosscutting the granites. The existence of a granitic apex where hydrothermal fluids are concentrated (Launay 2018; Launay et al. 2018, 2019) may lead to a total greisenization of the apex with a thickness of greisens that may exceed 100 m. The formation of greisens is linked to the circulation of acid and hot magmatic fluids resulting from the end of the crystallization of granitic magma. These fluids hydrolyze the feldspars and biotites and eliminate the solubilized elements, which leads to the formation of porosity in the greisen despite the development of hydrothermal muscovite. As proven by Launay (2018), greisens are active alterations that, by creating porosity, greatly improve hydrothermal flows, leading to the formation of mineralization.

Greisens developed in crustal peraluminous granites classically associated with Sn-W deposits are sometimes tin (cassiterite) mineralized greisens, but do not seem to lead to the formation of lithium enrichment. For example, the world-class W-Sn-(Cu) deposit of Panasqueira (Portugal) contains greisens averaging about 732 ppm of Li_2O (Bussink 1984). Conversely, the greisens developped in phosphorus-free peraluminous rare-metal granites can be mineralized in lithium, while the host granite is mineralized to a little extent. For example, the rare-metal granite of Cinovec (at the Germany-Czech Republic border) contains an indicated resource of 372.4 Mt of mineralized greisens with a 0.4% Li_2O content mainly in the form of zinnwaldite. Other coproducts are associated with lithium in these greisens, particularly tin, tungsten and other small metals (In, Sc, etc.), as appropriate. These greisens are to be explored on the tops of phosphorus-free peraluminous rare-metal granites and along the zones of faults or high temperature hydrothermal veins intersecting them.

1.2.3. *Geochemical cycle of lithium*

As a result of the study of the geochemical behavior of lithium in various processes operating at the Earth's surface and in its depths, completed by the study of various types of lithium deposits, surface processes seem to be essential for obtaining a sufficiently lithium-rich sedimentary-hydrothermal source (>1,800–2,400 ppm Li), which seems necessary for producing magmatic liquids exceeding the 4,000 ppm Li. Nevertheless, the systematic presence of volcanic glasses of rhyolitic composition suggests that the first contribution is itself magmatic, followed by a lithium concentration by climate-controlled alteration and evaporation processes. Moreover, in the endoreic basins or calderas in dry environment, the direct magmatic contribution in the mineralization via magmatic brines is frequently invoked.

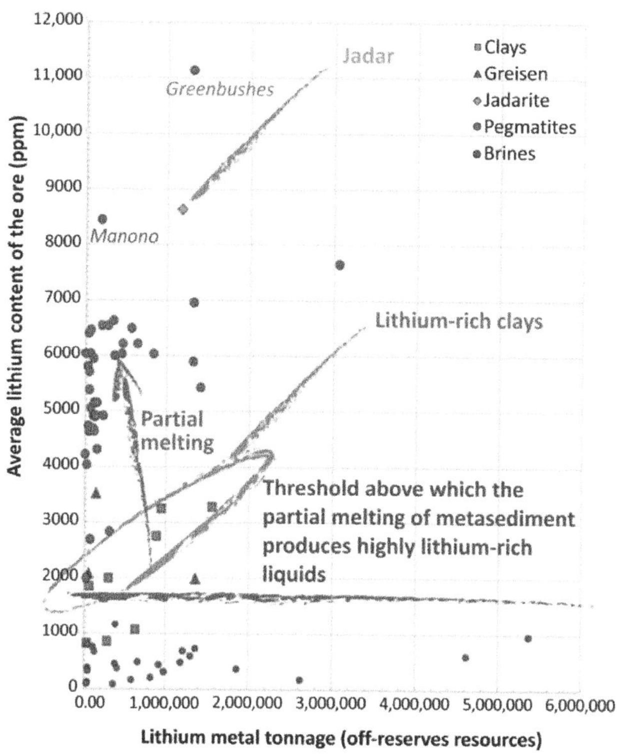

Figure 1.24. *Metal content tonnage depending on the ore content for lithium-bearing mineralization such as brines, clays, greisens, pegmatites and jadarite*

COMMENT ON FIGURE 1.24.– *Data from Lefebvre and Tavignot (2020). These data result from mining exploration projects integrating a large number of analyses over large volumes, which yields representative averages. These data clearly show here the existence of lithium-bearing clays whose metamorphic equivalents may exhibit possible sources of rare-metal magmatism. The presence of rare pegmatites with very high content (>7,000 ppm Li) suggests the existence of even richer metasediments, Jadar being a potential representative.*

Lithium preservation in metasediments during burial and metamorphism is not systematic. The small number of lithium rich pegmatites and granites at global scale rather suggests that this preservation is more exceptional than systematic. The partial melting of lithium rich metasediments seems to mainly produce pegmatites with 4,000–6,500 ppm Li (Figure 1.24). The origin of this interval should be sought in the percentage of partial melting (weak or strong) and in the average lithium concentration of metasediments (all types included) at the global scale.

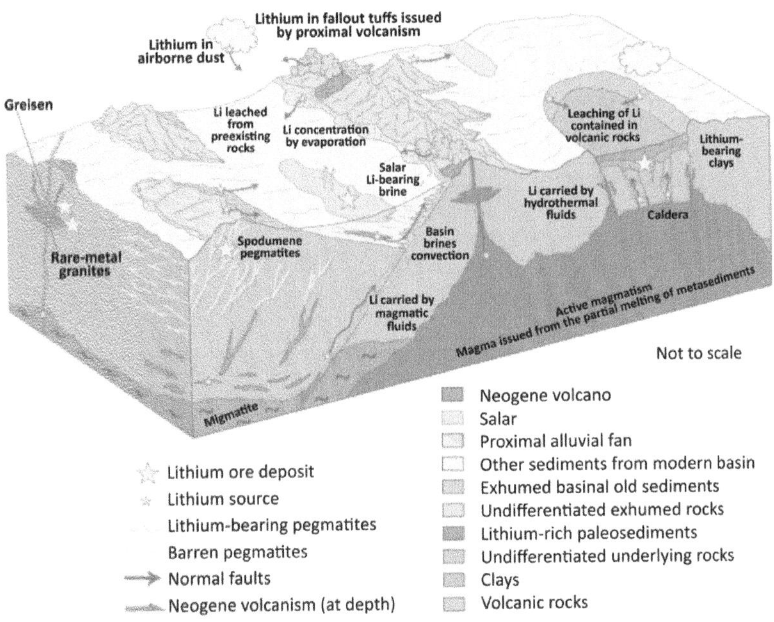

Figure 1.25. *Geochemic cycle of lithium and lithium mineralization. Modified from Lefebvre and Tavignot (2020)*

Certain exceptional pegmatites, particularly certain giant pegmatites, have extremely high average lithium concentrations, above 7,000 ppm Li (Figure 1.24).

These pegmatites suggest the existence of geological formations that are very rich in lithium, with concentrations probably exceeding 5,000 ppm Li (or 1.08% Li_2O). The hydrothermal-sedimentary deposit of Jadar whose average concentration of resources is 8,400 ppm Li (or 1.81% Li_2O) obviously shows that geological paleoformations very rich in lithium can exist.

All of these elements form a geological cycle of lithium, presented in Figure 1.25, showing lithium mineralization as evidence of contributions from paleoclimates and orogens.

1.3. Acknowledgments

This chapter includes many scientific advances in lithium metallogeny resulting from research projects conducted by BRGM, Labex Voltaire (ANR-10-LABX-100-01), ANR Varpeg (ANR-15-CE01-0001) and TRANSFAIR (ANR-21-CE01-0022), ERA-MIN (NewOres project, ANR-14-EMIN-0001) and by the European Institute of lithium (EIT-RM eLi 18136 project, see: https://www.lithium-institute.eu). We would like to thank our colleagues from UMR 7327 ISTO (University of Orléans, CNRS, BRGM), and particularly Charles Gumiaux, Fabrice Gaillard, Yannick Branquet, Stanislas Sizaret, Laurent Guillou-Frottier, Alexis Plunder and Michel Pichavant for our fruitful discussions, and also our collaborators Alexandre Lima (University of Porto), Milan Novák (Masaryk University, Brno) and Karel Breiter (Institute of Geology of the Czech Academy of Sciences). Special thanks go to our former students Sarah Deveaud, Josselyn Garde, Mélanie Lete, Julie Michaud, Floriane Léopold dit Offite, Maxime Picault, David Silva and Bárbara Salgado Da Silva Moreira. Finally, we would like to thank Sophie Decrée (Royal Belgian Institute of Natural Sciences) and Philippe Boulvais (Geosciences Rennes) for their enthusiasm, their words of encouragement, their enduring patience and all their editorial support.

1.4. References

Anderson, S.T. (2015). 2015 Minerals Yearbook: The Mineral Industry of Hungary. Report, USGS.

Arenas, R., Sánchez Martínez, S., Díez Fernández, R., Gerdes, A., Abati, J., Fernández-Suárez, J., Andonaegui, P., González Cuadra, P., López Carmona, A., Albert, R. et al. (2016). Allochthonous terranes involved in the Variscan suture of NW Iberia: A review of their origin and tectonothermal evolution. *Earth-Science Reviews*, 161, 140–178.

Aubert, G. (1969). Les coupoles granitiques de Montebras et d'Échassières (Massif central français) et la genèse de leurs minéralisations en étain, lithium, tungstène et béryllium. Thesis, Mémoires BRGM 46, Orléans.

Ballouard, C., Poujol, M., Boulvais, P., Branquet, Y., Tartèse, R., Vigneresse, J.L. (2016). Nb-Ta fractionation in peraluminous granites: A marker of the magmatic-hydrothermal transition. *Geology*, 44(3), 231–234.

Bárdossy, G. (1982). *Karst Bauxites, Bauxite Deposits on Carbonate Rocks.* Elsevier, Amsterdam.

Bea, F. (2010). Crystallization dynamics of granite magma chambers in the absence of regional stress: Multiphysics modeling with natural examples. *Journal of Petrology*, 51(7), 1541–1569.

Benson, T.R., Coble, M.A., Rytuba, J.J., Mahood, G.A. (2017). Lithium enrichment in intracon-tinental rhyolite magmas leads to Li deposits in caldera basins. *Nature Communications*, 8(1), 270.

Bowell, R.J., Lagos, L., de los Hoyos, C., Declercq, J. (2020). Classification and characteristics of natural lithium resources. *Elements*, 16(4), 259–264.

Bussink, R.W. (1984). Geochemistry of the Panasqueira tungsten-tin deposit, Portugal. *Geologica Ultraiectina*, 33, 1–170.

Calvet, R. and Prost, R. (1971). Cation migration into empty octahedral sites and surface properties of clays. *Clays and Clay Minerals*, 19(3), 175–186.

Castor, S.B. and Henry, C.D. (2020). Lithium-rich claystone in the McDermitt Caldera, Nevada, USA: Geologic, mineralogical, and geochemical characteristics and possible origin. *Minerals*, 10(1), 68.

Castro, A., Corretgé, G.L., De La Rosa, J., Enrique, P., Martínez, F.J., Pascual, E., Lago, M., Arranz, E., Galé, C., Fernández, C. et al. (eds). *The Geology of Spain.* Geological Society, London.

Černý, P. and Ercit, S.T. (2005). The classification of granitic pegmatites revisited. *The Canadian Mineralogist*, 43, 2005–2026.

Černý, P., Blevin, P.L., Cuney, M., London, D. (2005). Granite-related ore deposits. In *Economic Geology - One Hundredth Anniversary Volume 1905-2005,* Hedenquist, J.W., Thompson, J.F.H., Goldfarb, R.J., Richards, J.P. (eds). Society of Economic Geologists, Littleton.

Černý, P., London, D., Novák, M. (2012). Granitic pegmatites as reflections of their sources. *Elements*, 8(4), 289–294.

Chan, L.H. and Kastner, M. (2000). Lithium isotopic compositions of pore fluids and sediments in the Costa Rica subduction zone: Implications for fluid processes and sediment contribution to the arc volcanoes. *Earth and Planetary Science Letters*, 183(1), 275–290.

Chan, L.H., Gieskes, J.M., You, C.F., Edmond, J.M. (1994). Lithium isotope geochemistry of sediments and hydrothermal fluids of the Guaymas Basin, Gulf of California. *Geochimica et Cosmochimica Acta*, 58(20), 4443–4454.

Chan, L.H., Leeman, W.P., You, C.F. (2002). Lithium isotopic composition of Central American volcanic arc lavas: Implications for modification of subarc mantle by slab-derived fluids: Correction. *Chemical Geology*, 182(2), 293–300.

Chappell, B.W. and White, A.J.R. (1992). I- and S-type granites in the Lachlan Fold Belt. *Transactions of the Royal Society of Edinburgh*, 83, 1–26.

Christmann, P., Gloaguen, E., Labbé, J.F., Melleton, J., Piantone, P. (2015). Global lithium resources and sustainability issues. In *Lithium Process Chemistry*, Chagnes, A. and Światowska, J. (eds). Elsevier, Amsterdam.

Clemens, J.D. and Petford, N. (1999). Granitic melt viscosity and silicic magma dynamics in contrasting tectonic settings. *Journal of the Geological Society*, 156(6), 1057–1060.

Clemens, J.D. and Stevens, G. (2016). Melt segregation and magma interactions during crustal melting: Breaking out of the matrix. *Earth-Science Reviews*, 160, 333–349.

Clemens, J.D., Helps, P.A., Stevens, G. (2010). Chemical structure in granitic magmas – A signal from the source? *Earth and Environmental Science Transactions of the Royal Society of Edinburgh*, 100(1–2), 159–172.

Cuney, M., Marignac, C., Weisbrod, A. (1992). The Beauvoir topaz-lepidolite albitic granite (Massif Central France). A highly specialized granite with disseminated Sn-Li-Ta-Nb-Be mineralization of magmatic origin. *Economic Geology*, 87, 1766–1794.

Cuney, M., Alexandrov, P., Le Carlier de Veslud, C., Cheilletz, A., Raimbault, L., Ruffet, G., Scaillet, S. (2002). The timing of W-Sn-rare metals mineral deposit formation in the Western Variscan chain in their orogenic setting: The case of the Limousin area (Massif Central, France). In *The Timing and Location of Major Ore Deposits in an Evolving Orogen*, Blundell, D.J., Neubauer, F., von Quadt, A. (eds). The Geological Society of London.

D'Argenio, B. and Mindszenty, A. (1986). Cretaceous bauxites in the tectonic framework of the Mediterranean. *Rendiconti della Societa Geologica Italiana di Mineralogia e Petrologia*, 9, 257–262.

Dailey, S.R., Christiansen, E.H., Dorais, M.J., Kowallis, B.J., Fernandez, D.P., Johnson, D.M. (2018). Origin of the fluorine- and beryllium-rich rhyolites of the Spor Mountain Formation, Western Utah. *American Mineralogist*, 103(8), 1228–1252.

Dallmeyer, R.D., Martinez Catalan, J.R., Arenas, R., Gil Ibarguchi, J.I., Gutiérrez Alonso, G., Farias, P., Aller, J. (1997). Diachronous Variscan tectonothermal activity in the NW Iberian Massif: Evidence from $^{40}Ar/^{39}Ar$ dating of regional fabrics. *Tectonophysics*, 277, 307–337.

Davis, J.R., Friedman, I., Gleason, J.D. (1986). Origin of the lithium-rich brine, Clayton Valley, Nevada. *U.S. Geological Survey Bulletin*, 1622, 131–138.

Demartis, M., Pinotti, L.P., Coniglio, J.E., D'Eramo, F.J., Tubía, J.M., Aragón, E., Agulleiro Insúa, L.A. (2011). Ascent and emplacement of pegmatitic melts in a major reverse shear zone (Sierras de Córdoba, Argentina). *Journal of Structural Geology*, 33(9), 1334–1346.

Deveaud, S., Gumiaux, C., Gloaguen, E., Branquet, Y., Deveaud, S., Gumiaux, C., Gloaguen, E., Branquet, Y. (2013). Spatial statistical analysis applied to rare-element LCT-type pegmatite fields: An original approach to constrain faults-pegmatites-granites relationships. *Journal of Geosciences*, 58(2), 163–182.

Devineau, K., Champallier, R., Pichavant, M. (2020). Dynamic crystallization of a haplogranitic melt: Application to pegmatites. *Journal of Petrology*, 61(5). doi: 10.1093/petrology/egaa054.

Díez Fernández, R. and Pereira, M.F. (2016). Extensional orogenic collapse captured by strike-slip tectonics: Constraints from structural geology and UPb geochronology of the Pinhel shear zone (Variscan orogen, Iberian Massif). *Tectonophysics*, 691, 290–310.

Dutrow, B.L., Holdaway, M.J., Hinton, R.W. (1986). Lithium in staurolite and its petrologic significance. *Contributions to Mineralogy and Petrology*, 94(4), 496–506.

Elango, L. and Kannan, R. (2007). Rock–water interaction and its control on chemical composition of groundwater. In *Concepts and Applications in Environmental Geochemistry*, Sarkar, D., Datta, R., Hannigan, R. (eds). Elsevier, Amsterdam.

Feenstra, A., Ockenga, E., Rhede, D., Wiedenbeck, M. (2003). Li-rich zincostaurolite and its decompression-related breakdown products in a diaspore-bearing metabauxite from East Samos (Greece): An EMP and SIMS study. *American Mineralogist*, 88(5–6), 789–805.

Ferreira, J.A., Bento dos Santos, T., Pereira, I., Mata, J. (2019). Tectonically assisted exhumation and cooling of Variscan granites in an anatectic complex of the Central Iberian Zone, Portugal: Constraints from LA-ICP-MS zircon and apatite U–Pb ages. *International Journal of Earth Sciences*, 108(7), 2153–2175.

Ferreira, J.A., Mata, J., Bento dos Santos, T., Pereira, I. (2020). The role of melting on the geochemical evolution and isotopic variability of an anatectic complex in the Iberian Variscides. *Lithos*, 378–379, 105769.

Flesch, G.D., Anderson, A.R., Svec, H.J. (1973). A secondary isotopic standard for 6Li/7Li determinations. *International Journal of Mass Spectrometry and Ion Physics*, 12(3), 265–272.

Fouillac, C. and Michard, G. (1981). Sodium/lithium ratio in water applied to geothermometry of geothermal reservoirs. *Geothermics*, 10(1), 55–70.

Garofalo, P.S. (2012). The composition of Alpine marine sediments (Bündnerschiefer Formation, W Alps) and the mobility of their chemical components during orogenic metamorphism. *Lithos*, 128–131, 55–72.

Gloaguen, E., Melleton, J., Lefebvre, G., Tourlière, B., Yart, S. (2018). Ressources métropolitaines en lithium et analyse du potentiel par méthodes de prédictivité. Public report BRGM/RP-68321-FR, BRGM, Orléans.

Gourcerol, B., Gloaguen, E., Melleton, J., Tuduri, J., Galiegue, X. (2019). Re-assessing the European lithium resource potential – A review of hard-rock resources and metallogeny. *Ore Geology Reviews*, 109, 494–519.

Grew, E.S. (2020). The minerals of lithium. *Elements*, 16(4), 235–240.

Grew, E.S., Hystad, G., Toapanta, M.P.C., Eleish, A., Ostroverkhova, A., Golden, J., Hazen, R.M. (2019). Lithium mineral evolution and ecology: Comparison with boron and beryllium. *European Journal of Mineralogy*, 31(4), 755–774.

Gruber, P.W. and Medina, P.A. (2010). Global lithium availability: A constraint for electric vehicles? PhD Thesis, University of Michigan, Ann Arbor.

Gutiérrez-Alonso, G., Fernández-Suárez, J., López-Carmona, A., Gärtner, A. (2018). Exhuming a cold case: The early granodiorites of the northwest Iberian Variscan belt – A Visean magmatic flare-up? *Lithosphere*, 10(2), 194–216.

Henry, C.D., Castor, S.B., Starkel, W.A., Ellis, B.S., Wolff, J.A., Laravie, J.A., McIntosh, W.C., Heizler, M.T. (2017). Geology and evolution of the McDermitt caldera, northern Nevada and southeastern Oregon, western USA. *Geosphere*, 13(4), 1066–1112.

Hiradate, S. and Wada, S.I. (2005). Weathering process of volcanic glass to allophane determined by ^{27}Al and ^{29}Si solid-state NMR. *Clays and Clay Minerals*, 53(4), 401–408.

Hirst, D.M. (1962). The geochemistry of modern sediments from the Gulf of Paria II: The location and distribution of trace elements. *Geochimica et Cosmochimica Acta*, 26(11), 1147–1187.

Horstman, E.L. (1957). The distribution of lithium, rubidium and caesium in igneous and sedimentary rocks. *Geochimica et Cosmochimica Acta*, 12(1), 1–28.

Houston, J., Butcher, A., Ehren, P., Evans, K., Godfrey, L. (2011). The evaluation of brine prospects and the requirement for modifications to filing standards. *Economic Geology*, 106(7), 1225–1239.

Huh, Y., Chan, L.H., Chadwick, O.A. (2004). Behavior of lithium and its isotopes during weathering of Hawaiian basalt. *Geochemistry, Geophysics, Geosystems*, 5, Q09002. doi: 10.1029/2004GC000729.

Kesler, S.E., Gruber, P.W., Medina, P.A., Keoleian, G.A., Everson, M.P., Wallington, T.J. (2012). Global lithium resources: Relative importance of pegmatite, brine and other deposits. *Ore Geology Reviews*, 48, 55–69.

Kharaka, Y.K. and Mariner, R.H. (1989). Chemical geothermometers and their application to formation waters from sedimentary basins. In *Thermal History of Sedimentary Basins: Methods and Case Histories*, Naeser, N.D. and McCulloch T.H. (eds). Springer-Verlag, New York.

Kharaka, Y.K., Lico, M.S., Law, L.M. (1982). Chemical geothermometers applied to formation waters, Gulf of Mexico and California Basins. *Am. Ass. Petrol. Geol. Bull.*, 66, 588.

Kovalenko, V.I., Kuz'min, M.I., Letnikov, F.A. (1970). Magmatic origin of lithium and fluorine bearing rare metal granites. *Doklady Akademii Nauk SSSR, Earth Sciences Section*, 190, 189–192.

Labbé, J.F. and Daw, G. (2012). Panorama 2011 du marché du lithium. Public report BRGM/RP-61340-FR, BRGM, Orléans.

Langmuir, C.H. (1989). Geochemical consequences of in situ crystallization. *Nature*, 340(6230), 199–205.

Launay, G. (2018). Hydrodynamique des systèmes minéralisés péri-granitiques : étude du gisement à W-Sn-(Cu) de Panasqueira (Portugal). PhD Thesis, Université d'Orléans.

Launay, G., Sizaret, S., Guillou-Frottier, L., Gloaguen, E., Pinto, F. (2018). Deciphering fluid flow at the magmatic-hydrothermal transition: A case study from the world-class Panasqueira W–Sn–(Cu) ore deposit (Portugal). *Earth and Planetary Science Letters*, 499, 1–12.

Launay, G., Sizaret, S., Guillou-Frottier, L., Fauguerolles, C., Champallier, R., Gloaguen, E. (2019). Dynamic permeability related to greisenization reactions in Sn-W ore deposits: Quantitative petrophysical and experimental evidence. *Geofluids*, 5976545.

Lefebvre, G. and Tavignot, D. (2020). Le marché du lithium en 2020 : enjeux et paradoxes. *Bulletin Ecomine, 2020-2* [Online]. Available at: https://www.mineralinfo.fr/fr/ecomine/marche-du-lithium-2020-enjeux-paradoxes.

Li, Y.H. (2000). *A Compendium of Geochemistry: From Solar Nebula to the Human Brain.* Princeton University Press.

Ling, K.Y., Zhu, X.Q., Tang, H.S., Du, S.J., Gu, J. (2018). Geology and geochemistry of the Xiaoshanba bauxite deposit, Central Guizhou Province, SW China: Implications for the behavior of trace and rare earth elements. *Journal of Geochemical Exploration*, 190, 170–186.

Linnen, R.L. and Cuney, M. (2005). Granite-related rare-element deposits and experimental constraints on Ta-Nb-W-Sn-Zr-Hf mineralization. In *Rare-Element Geochemistry and Mineral Deposits*, Linnen, R.L. and Samson, I.M. (eds). Geological Association of Canada Short Course Notes.

London, D. (2008). Pegmatites. *The Canadian Mineralogist*, Special Publication 10, Mineralogical Association of Canada, Quebec.

Lulzac, Y. and Apolinarski, F. (1986). Inventaire du territoire métropolitain. Les minéralisations à étain, tantale et lithium de Tréguennec (Finistère). État des connaissances au 31 mars 1986. Public report 86 DAM 011 OP4, BRGM, Orléans.

Mähler, J. and Persson, I. (2012). A study of the hydration of the alkali metal ions in aqueous solution. *Inorganic Chemistry*, 51(1), 425 438.

Marović, M., Djoković, I., Pešić, L., Radovanović, S., Toljić, M., Gerzina, N. (2002). Neotectonics and seismicity of the southern margin of the Pannonian basin in Serbia. In *Neotectonics and Surface Processes: The Pannonian Basin and Alpine/Carpathian System*, Cloetingh, S.A.P.L., Horváth, F., Bada, G., Lankreijer, A.C.A.C. (eds). EGU, Munich.

Marschall, H.R., Pogge von Strandmann, P.A.E., Seitz, H.M., Elliott, T., Niu, Y. (2007). The lithium isotopic composition of orogenic eclogites and deep subducted slabs. *Earth and Planetary Science Letters*, 262(3), 563–580.

Martínez Catalán, J.R., Arenas, R., Díaz García, F., Abati, J. (1997). Variscan accretionary complex of northwest Iberia: Terrane correlation and succession of a tectonothermal events. *Geology*, 25(12), 1103–1106.

Martínez Catalán, J.R., Rubio Pascual, F.J., Díez Montes, A., Díez Fernández, R., Gómez Barreiro, J., Dias da Silva, Í., González Clavijo, E., Ayarza, P., Alcock, J.E. (2014). The late Variscan HT/LP metamorphic event in NW and Central Iberia: Relationships to crustal thickening, extension, orocline development and crustal evolution. *Geological Society Special Publications*, 405, SP405.1.

Mateus, A., Munhá, J., Ribeiro, A., Tassinari, C.C.G., Sato, K., Pereira, E., Santos, J.F. (2016). U–Pb SHRIMP zircon dating of high-grade rocks from the Upper Allochthonous Terrane of Bragança and Morais Massifs (NE Portugal); geodynamic consequences. *Tectonophysics*, 675, 23–49.

Merceron, T., Inoue, A., Bouchet, A., Meunier, A. (1987). Donbassite et tosudite lithinifères à Echassières. *Géologie de la France*, 2–3, 271–278.

Michard, G. (1990). Behaviour of major elements and some trace elements (Li, Rb, Cs, Sr, Fe, Mn, W, F) in deep hot waters from granitic areas. *Chemical Geology*, 89(1), 117–134.

Michaud, J.A.S., Gumiaux, C., Pichavant, M., Gloaguen, E., Marcoux, E. (2020). From magmatic to hydrothermal Sn-Li-(Nb-Ta-W) mineralization: The Argemela area (central Portugal). *Ore Geology Reviews*, 116, 103215.

Millot, R. and Négrel, P. (2007). Multi-isotopic tracing (δ^7Li, δ^{11}B, ^{87}Sr/^{86}Sr) and chemical geothermometry: Evidence from hydro-geothermal systems in France. *Chemical Geology*, 244(3), 664–678.

Millot, R., Vigier, N., Gaillardet, J. (2010). Behaviour of lithium and its isotopes during weathering in the Mackenzie Basin, Canada. *Geochimica et Cosmochimica Acta*, 74(14), 3897–3912.

Munk, L.A., Hynek, S.A., Bradley, D.C., Boutt, D., Labay, K., Jochens, H. (2016). Lithium brines: A global perspective. In *Rare Earth and Critical Elements in Ore Deposits*, Verplanck, P.L. and Hitzman, M.W. (eds). Society of Economic Geologists, Littleton.

Nesbitt, H.W. and Young, G.M. (1982). Early Proterozoic climates and plate motions inferred from major element chemistry of lutites. *Nature*, 299(5885), 715–717.

Nishiyama, T., Shimoda, S., Shimosaka, K., Kanaoka, S. (1975). Lithium-bearing tosudite. *Clays and Clay Minerals*, 23(5), 337–342.

Oi, T., Kawada, K., Hosoe, M., Kakihana, H. (1991). Fractionation of lithium isotopes in cation-exchange chromatography. *Separation Science and Technology*, 26(10–11), 1353–1375.

Pereira, I., Dias, R., dos Santos, T.B., Mata, J. (2017). Exhumation of a migmatite complex along a transpressive shear zone: Inferences from the Variscan Juzbado–Penalva do Castelo Shear Zone (Central Iberian Zone). *Journal of the Geological Society*, 174(6), 1004–1018.

Pittuck, M.F. and Rawbone, C. (2015). NI 43-101 mineral resources estimate for the Sonora lithium project, Mexico. Report UK6355, SRK Consulting Limited, London.

Pittuck, M.F., Lane, G.S., Welhener, H.E., Carrasco, J.A. (2018). Technical report on the feasibility study for the Sonora lithium project, Mexico. Report 101304-FS-0004, Ausenco Services Pty, Brisbane.

Pogge von Strandmann, P.A.E., Burton, K.W., Opfergelt, S., Eiríksdóttir, E.S., Murphy, M.J., Einarsson, A., Gislason, S.R. (2016). The effect of hydrothermal spring weathering processes and primary productivity on lithium isotopes: Lake Myvatn, Iceland. *Chemical Geology*, 445, 4–13.

Pogge von Strandmann, P.A.E., Kasemann, S.A., Wimpenny, J.B. (2020). Lithium and lithium isotopes in Earth's surface cycles. *Elements*, 16(4), 253–258.

Qiu, L. (2011a). Lithium and δ^7Li behavior during metamorphic dehydration processes and crustal evolution. PhD Thesis, University of Maryland, College Park.

Qiu, L., Rudnick, R.L., McDonough, W.F., Bea, F. (2011b). The behavior of lithium in amphibolite- to granulite-facies rocks of the Ivrea–Verbano Zone, NW Italy. *Chemical Geology*, 289(1–2), 76–85.

Raimbault, L. (1998). Composition of complex lepidolite-type granitic pegmatites and of constituent columbite-tantalite, Chèdeville, Massif Central, France. *Canadian Mineralogist*, 36, 563–583.

Raimbault, L. and Burnol, L. (1998). The Richemont rhyolite dyke, Massif Central, France; A subvolcanic equivalent of rare-metal granites. *The Canadian Mineralogist*, 36(2), 265–282.

Rio Tinto (2017). Increase to Jadar project mineral resources. Notice to ASX.

Rio Tinto (2022). Rio Tinto updates ore reserves and mineral resources at Jadar. Notice to ASX [Online]. Available at: https://www.riotinto.com/-/media/Content/Documents/Invest/Reserves-and-resources/2021/RT-Jadar-reserves-resources-2021.pdf

Roda-Robles, E., Vieira, R., Lima, A., Pesquera Pérez, A. (2009). Petrogenetic links between granites and pegmatites in the Fregeneda-Almendra area (Salamanca, Spain and Guard, Portugal): New insights from ^{40}Ar/^{39}Ar dating in micas. *Estudos Geológicos*, 19(2), 305–310.

Roda-Robles, E., Vieira, R., Pesquera, A., Lima, A. (2010). Chemical variations and significance of phosphates from the Fregeneda-Almendra pegmatite field, Central Iberian Zone (Spain and Portugal). *Mineralogy and Petrology*, 100(1), 23–34.

Roda-Robles, E., Pesquera, A., Gil-Crespo, P.P., Vieira, R., Lima, A., Garate-Olave, I., Martins, T., Torres-Ruiz, J. (2016). Geology and mineralogy of Li mineralization in the Central Iberian Zone (Spain and Portugal). *Mineralogical Magazine*, 80(1), 103–126.

Romer, R.L., Meixner, A., Hahne, K. (2014). Lithium and boron isotopic composition of sedimentary rocks – The role of source history and depositional environment: A 250Ma record from the Cadomian orogeny to the Variscan orogeny. *Gondwana Research*, 26(3), 1093–1110.

Rudolph, W., Brooker, M.H., Pye, C.C. (1995). Hydration of lithium ion in aqueous solutions. *The Journal of Physical Chemistry*, 99(11), 3793–3797.

Sanjuan, B., Millot, R., Ásmundsson, R., Brach, M., Giroud, N. (2014). Use of two new Na/Li geothermometric relationships for geothermal fluids in volcanic environments. *Chemical Geology*, 389, 60–81.

Scholz, F., Hensen, C., Reitz, A., Romer, R.L., Liebetrau, V., Meixner, A., Weise, S.M., Haeckel, M. (2009). Isotopic evidence (87Sr/86Sr, δ7Li) for alteration of the oceanic crust at deep-rooted mud volcanoes in the Gulf of Cadiz, NE Atlantic Ocean. *Geochimica et Cosmochimica Acta*, 73(18), 5444–5459.

Scholz, F., Hensen, C., De Lange, G.J., Haeckel, M., Liebetrau, V., Meixner, A., Reitz, A., Romer, R.L. (2010). Lithium isotope geochemistry of marine pore waters – Insights from cold seep fluids. *Geochimica et Cosmochimica Acta*, 74(12), 3459–3475.

Shaw, D.M. (1970). Trace element fractionation during anatexis. *Geochimica et Cosmochimica Acta*, 34(2), 237–243.

Shearer, C.K., Papike, J.J., Jolliff, B.L. (1992). Petrogenetic links among granites and pegmatites in the Harney Peak rare-element granite-pegmatite system, Black Hills, South Dakota. *The Canadian Mineralogist*, 30(3), 785–809.

Silva, D., Lima, A., Gloaguen, E., Gumiaux, C., Noronha, F., Deveaud, S. (2018). Spatial geostatistical analysis applied to the Barroso-Alvao rare-elements pegmatite field (Northern Portugal). In *GIS – An Overview of Applications*, Teodoro, A.C. (ed.). Bentham Science Publishers, Sharjah.

Simons, B., Andersen, J.C.Ø., Shail, R.K., Jenner, F.E. (2017). Fractionation of Li, Be, Ga, Nb, Ta, In, Sn, Sb, W and Bi in the peraluminous Early Permian Variscan granites of the Cornubian Batholith: Precursor processes to magmatic-hydrothermal mineralisation. *Lithos*, 278–281, 491–512.

Spear, F.S., Kohn, M.J., Cheney, J.T. (1999). P-T paths from anatectic pelites. *Contributions to Mineralogy and Petrology*, 134(1), 17–32.

Stanley, C.J., Jones, G.C., Rumsey, M.S., Blake, C., Roberts, A.C., Stirling, J.A.R., Carpenter, G.J.C., Whitfield, P.S., Grice, J.D., Lepage, Y. (2007). Jadarite, LiNaSiB3O7(OH), a new mineral species from the Jadar Basin, Serbia. *European Journal of Mineralogy*, 19, 575–580.

Stojadinovic, U., Matenco, L., Andriessen, P., Toljić, M., Rundić, L., Ducea, M.N. (2017). Structure and provenance of Late Cretaceous–Miocene sediments located near the NE Dinarides margin: Inferences from kinematics of orogenic building and subsequent extensional collapse. *Tectonophysics*, 710–711, 184–204.

Tang, Y.J., Zhang, H.F., Ying, J.F. (2007). Review of the lithium isotope system as a geochemical tracer. *International Geology Review*, 49(4), 374–388.

Tardy, Y., Krempp, G., Trauth, N. (1972). Le lithium dans les minéraux argileux des sédiments et des sols. *Geochimica et Cosmochimica Acta*, 36(4), 397–412.

Taylor, S.R. and McLennan, S. (2008). *Planetary Crusts: Their Composition, Origin and Evolution*. Cambridge University Press.

Teng, F.Z., McDonough, W.F., Rudnick, R.L., Dalpé, C., Tomascak, P.B., Chappell, B.W., Gao, S. (2004). Lithium isotopic composition and concentration of the upper continental crust. *Geochimica et Cosmochimica Acta*, 68(20), 4167–4178.

Teng, F.Z., McDonough, W.F., Rudnick, R.L., Wing, B.A. (2007). Limited lithium isotopic fractionation during progressive metamorphic dehydration in metapelites: A case study from the Onawa contact aureole, Maine. *Chemical Geology*, 239(1), 1–12.

Tindle, A.G. and Breaks, F.W. (1998). Oxide minerals of the separation rapids rare-element granitic pegmatite group, northwestern Ontario. *The Canadian Mineralogist*, 36(2), 609–635.

Tomascak, P.B. (2004). Developments in the understanding and application of lithium isotopes in the earth and planetary sciences. *Reviews in Mineralogy and Geochemistry*, 55(1), 153–195.

Tourtelot, H.A. and Brenner-Tourtelot, E.F. (1977). Lithium in flint clay, bauxite, related high-alumina materials and sedimentary rocks in the United States: A preliminary survey. Open file report 77-786, U.S. Geological Survey.

Valle Aguado, B., Azevedo, M.R., Nolan, J., Medina, J., Costa, M.M., Corfu, F., Martínez Catalán, J.R. (2017). Granite emplacement at the termination of a major Variscan transcurrent shear zone: The late collisional Viseu batholith. *Journal of Structural Geology*, 98, 15–37.

Verley, C.G. (2014). Updated and reclassified lithium resources, Sonora lithium project for Bacanora Minerals Ltd. Report, NI 43-101, Amerlin Exploration Services Ltd., Richmond.

Vieira, R., Roda-Robles, E., Pesquera, A., Lima, A. (2011). Chemical variation and significance of micas from the Fregeneda-Almendra pegmatitic field (Central-Iberian Zone, Spain and Portugal). *American Mineralogist*, 96(4), 637–645.

Vigier, N., Decarreau, A., Millot, R., Carignan, J., Petit, S., France-Lanord, C. (2008). Quantifying Li isotope fractionation during smectite formation and implications for the Li cycle. *Geochimica et Cosmochimica Acta*, 72(3), 780–792.

Villaros, A. and Pichavant, M. (2019). Mica-liquid trace elements partitioning and the granite-pegmatite connection: The St-Sylvestre complex (Western French Massif Central). *Chemical Geology*, 528, 119265.

Wang, D., Li, P., Qu, W., Yin, L., Zhao, Z., Lei, Z., Wen, S. (2013). Discovery and preliminary study of the high tungsten and lithium contents in the Dazhuyuan bauxite deposit, Guizhou, China. *Science China Earth Sciences*, 56(1), 145–152.

Warren, J.K. (2010). Evaporites through time: Tectonic, climatic and eustatic controls in marine and nonmarine deposits. *Earth-Science Reviews*, 98(3), 217–268.

Wenger, M. and Armbruster, T. (1991). Crystal chemistry of lithium; oxygen coordination and bonding. *European Journal of Mineralogy*, 3(2), 387–399.

Yin, L., Pollard, P.J., Shouxi, H., Taylor, R.G. (1995). Geologic and geochemical characteristics of the Yichun Ta-Nb-Li deposit, Jiangxi Province, South China. *Economic Geology*, 90(3), 577–585.

Zack, T., Tomascak, P.B., Rudnick, R.L., Dalpé, C., McDonough, W.F. (2003). Extremely light Li in orogenic eclogites: The role of isotope fractionation during dehydration in subducted oceanic crust. *Earth and Planetary Science Letters*, 208(3), 279–290.

2

Metallogeny of the Abitibi Greenstone Belt, Canada

Patrick MERCIER-LANGEVIN, Benoît DUBÉ, Michel G. HOULÉ, Valérie BÉCU, Anne-Aurélie SAPPIN, Jean-Luc PILOTE and Sébastien CASTONGUAY

Geological Survey of Canada, Natural Resources Canada, Quebec, Canada

2.1. Introduction

Greenstone belt is a term that designates elongated or linear assemblages of *green rocks* present in Precambrian shields and that are primarily composed of igneous metamorphosed and/or altered rocks of intermediate to mafic composition, whose color is due to the presence of greenish minerals such as chlorite, actinote and epidote, characteristic of the greenschist metamorphic facies. Individual belts may comprise more than one metamorphosed and deformed pile of volcano-sedimentary rocks in which the composition of volcanic rocks typically evolves from mafic to felsic. Most of these belts have a strong mineral potential.

The Abitibi greenstone belt, extending from northeastern Ontario to northwestern Quebec in Eastern Canada (section A2.1), is the largest belt of Neoarchean supracrustal rocks on the planet, and also one of the richest in terms of base and precious metals, including several critical and strategic metals. This belt, being among the best preserved from the Archean period, during which various types of deposits were preseved, has a complex geological history, spanning over nearly 200 million years (ca. 2.8–2.6 Ga). The mining history is also very rich,

---

For a color version of all figures in this chapter, see www.iste.co.uk/decree/metallic2.zip.

Metallic Resources 2,
coordinated by Sophie DECRÉE.

though very recent, since prospection only started at the end of the 19th century and the first mining started at the beginning of the 20th century. Despite this relatively short mining history, the Abitibi has been one of the largest poles of mining activity in Canada and in the world for several decades. The Abitibi also remains a first-level exploration target, and a driving force for scientific and technical developments both in the understanding of metallogenic processes, and in the exploration, exploitation and processing of ores, as well as for the responsible development and management of resources.

This chapter presents an overview of the geology and metallogeny of the Abitibi greenstone belt. It does not aim to provide a detailed description of the mechanisms and processes of deposit formation, as this information is available in many high-quality publications, but to highlight the exceptional mineral wealth of this belt and illustrate the typology of mineral systems and their distribution in space and time.

2.2. Mining history

The Abitibi region, located far inland, was not the object of significant prospection until the end of the 19th century, despite the presence of indigenous peoples and the discovery of North America by the Europeans in the 16th century.

The Canadian territory was first accessed through the network of lakes and rivers. Afterwards, the colonization efforts and the abundance of logging resources led to railway development in the early 1900s. Several major discoveries rapidly followed, among others in the sector of Témiscamingue Lake with the Ag and Co ore deposits of the Cobalt district. The auriferous potential of the region was revealed at the end of the 19th century by government works. In 1906, the first auriferous deposits were discovered in the sector of what will later become the town of Rouyn-Noranda. However, the huge potential of this belt was only truly revealed in 1909 by the discovery, in the Timmins sector, of the Hollinger, McIntyre and Dome ore deposits, which are still three of the largest auriferous ore deposits in North America. Shortly afterwards, other major discoveries followed in the regions of Timmins, Kirkland Lake and Matachewan in Ontario and Val-d'Or in Quebec (Dubé and Mercier-Langevin 2020a). These first auriferous discoveries in Ontario and Quebec revealed the existence of two prolific lineaments, generally oriented east-west and approximately parallel, and which will be known as the Larder Lake-Cadillac and Destor-Porcupine fault zone or *breaks* (Poulsen 2017). The potential of these fault zones has then been confirmed by several discoveries in the sectors of Val-d'Or and Malartic in Quebec, in 1920 and 1930 (Dubé and Mercier-Langevin 2020a). The rail and road network development in the period from 1940 to 1960 allowed the discovery of many additional and farther deposits, as well as the development of previously discovered deposits.

Although considered a “mature” mining region, new discoveries are still being made in the Abitibi, and the optimization of the operating procedures makes it possible to exploit deposits that were previously considered non-economic, thus generating new development opportunities.

2.3. Geological context

The Abitibi greenstone belt was the subject of numerous mapping projects and specific studies at various scales that were conducted by prospectors and by geologists from both the industry and the governmental bodies and academia. It is however impossible to mention here all the works and cite all the publications that contributed to this effort.

For the sake of simplicity, the most recent references or those considered particularly authoritative are preferentially cited below. Knowledge on the geology and metallogeny of this exceptional belt is constantly evolving, and only a summary of this knowledge is presented here.

The Superior Province consists of Mesoarchean to Neoarchean terranes of east-west dominant orientation assembled in the Neoarchean, forming the largest Archean craton of the planet (Percival et al. 2012). The Abitibi greenstone belt, located in the southern part of the Superior Province, corresponds to one of the belts of the Abitibi-Wawa terrane (Figure 2.1). With a surface area of approximately 100,000 km^2 (maximal dimensions of about 515 km E-W by 300 km N-S: Figure 2.2), the Abitibi greenstone belt represents the largest extent of Neoarchean supracrustal rocks of the planet. The belt is bordered to the north by high grade metamorphic rocks of the Opatica Subprovince, to the west by the tectonic zone of Kapuskasing, and to the south and south-east by the sedimentary rocks of the Pontiac Subprovince and Grenville Province (Figures 2.1 and 2.2).

The southwestern and northeastern parts of the belt are locally covered by Proterozoic sedimentary rocks of the Huronian Supergroup (south-west) and Chibougamau Formation (north-east) that lie in angular unconformity on the Archean rocks (Figure 2.2).

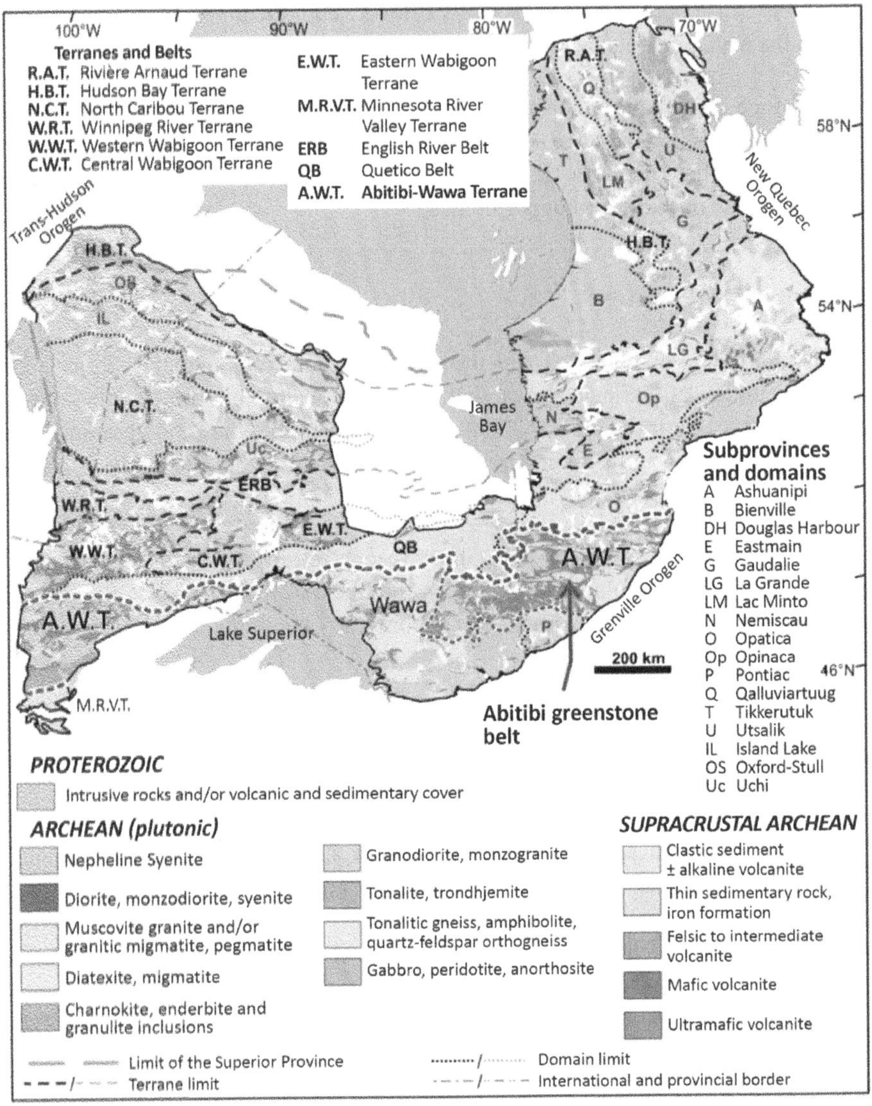

Figure 2.1. *Geological map of the Superior Province in Canada, taken from Percival et al. (2012). The map shows the location of the Abitibi-Wawa terrane in the southern part of the Superior, and also the limits of the Abitibi greenstone belt that forms the western segment of the Abitibi-Wawa terrane*

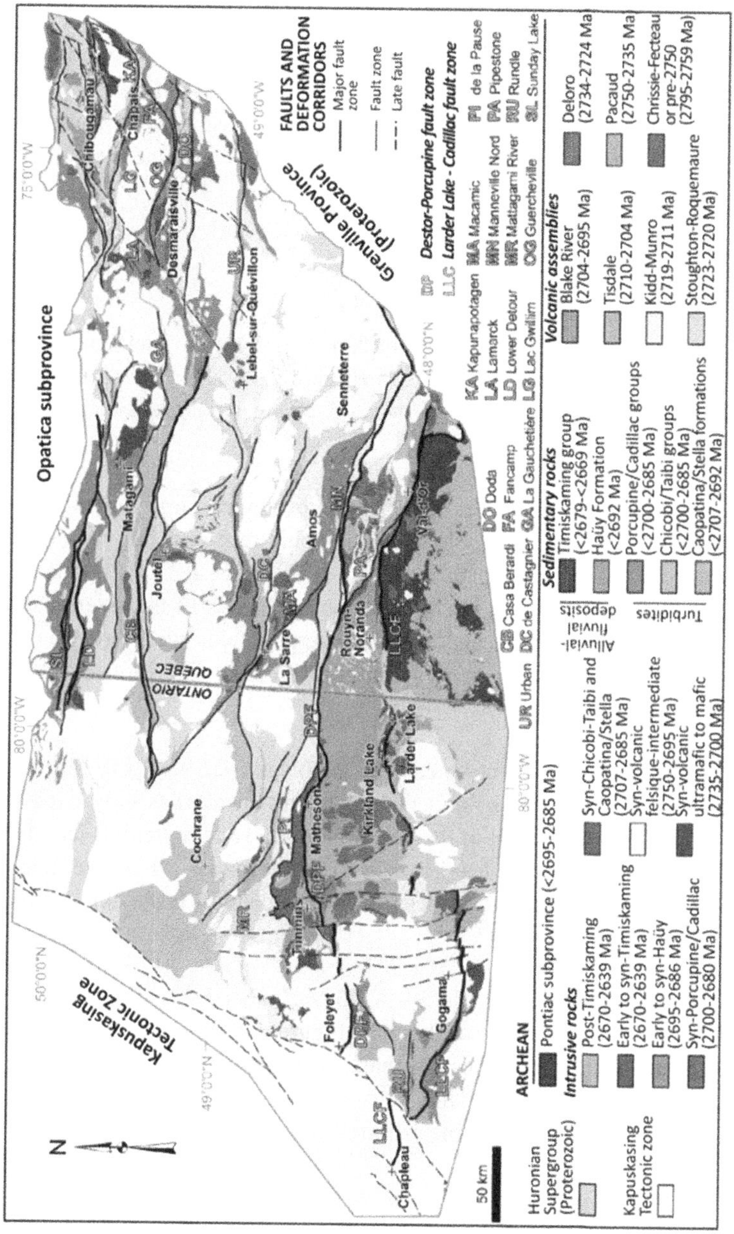

Figure 2.2. *Geological map of the Abitibi greenstone belt. Taken from Dubé and Mercier-Langevin (2020b). The names of some units have been simplified to lighten the legend*

The Abitibi greenstone belt comprises several sequences of volcanic and sedimentary rocks generally oriented east-west, folded and intersected by several faults and intrusive bodies (Figure 2.3). The primary characteristics of rocks are generally well preserved despite the deformation and the low to moderate metamorphism affecting the entire belt (Monecke et al. 2017). Significant corridors of ductile to brittle faults intersect the rocks of the belt from east to west. These key elements of the Abitibi greenstone belt are briefly described below, and their role(s) in the metallogenic evolution are presented in detail in section 2.4.

2.3.1. *Volcanic rocks*

Predominantly emplaced in a submarine context, volcanic rocks represent approximately one-third of all the Abitibi rocks (Figure 2.3). Their age varies from ~2,795 to 2,695 Ma, with the exception of several younger units associated with sedimentary basins.

These volcanic rocks have been grouped into seven large assemblages (Figure 2.2) depending on their distribution, stratigraphic relations, but most of all based on U-Pb ages on zircons (Ayer et al. 2002; Thurston et al. 2008; Dubé and Mercier-Langevin 2020b). The nature of contacts between each of the assemblages varies from concordant to tectonic, and various interpretations were proposed to explain the stratigraphy and the volcanic evolution of the belt (see Monecke et al. (2017) and the references therein).

Chrissie-Fecteau, the most ancient volcanic assemblage (~2,795–2,759 Ma) (see Dubé and Mercier-Langevin (2020b) and the references therein) is only exposed very locally in the north-western part of the belt (Figure 2.2) and represents less than 1% of the total surface area of volcanic rocks (Figure 2.4(a)). This assemblage predominantly comprises volcaniclastic rocks of intermediate to felsic composition and calc-alkaline affinity.

The second assemblage, Pacaud (2,750–2,735 Ma), also has a limited extent (about 1.3% of all the volcanic rocks of the belt (Figure 2.4(a)) and is present only locally in the central and south-western parts of Abitibi (Figure 2.2). This assemblage comprises ultramafic, mafic and massive felsic to volcaniclastic volcanic rocks of tholeiitic to calc-alkaline affinity.

The third assemblage, Deloro (2,734–2,724 Ma), has a far more significant extent (Figure 2.2), representing about a quarter of all the volcanic rocks of the belt (Figure 2.4(a)).

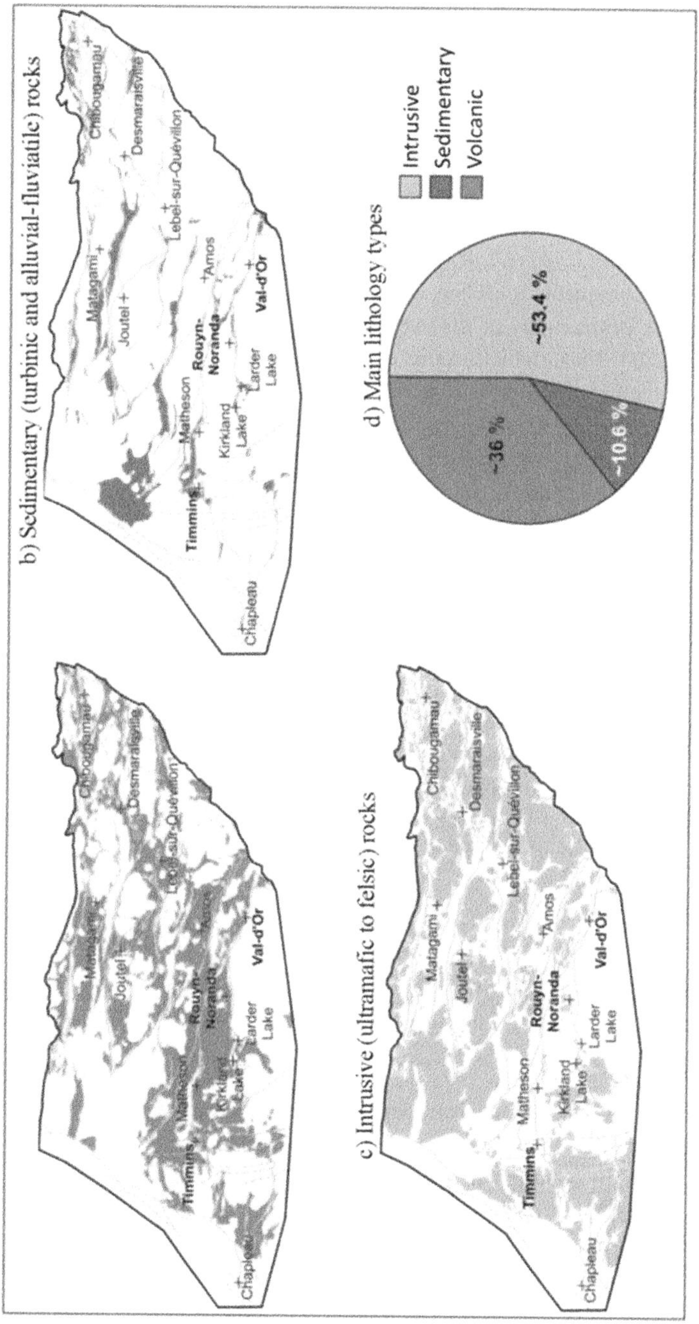

Figure 2.3. *Spatial distribution of the three main types of rocks of the Abitibi greenstone belt, including (a) volcanic rocks; (b) sedimentary rocks; (c) intrusive rocks; and (d) relative surface of each of the three significant rock types of the Abitibi greenstone belt. Modified from Dubé and Mercier-Langevin (2020b)*

The Deloro assemblage comprises mainly mafic to felsic calc-alkaline rocks with a lesser part of tholeiitic mafic rocks. Though emplaced during a relatively short period (2,723–2,720 Ma), the Stoughton-Roquemaure assemblage is also common, particularly in the Northern half of the belt (Figure 2.2), representing about one-third of all the volcanic rocks of the Abitibi (Figure 2.4(a)). Tholeiitic basalts, and to a lesser extent komatiite and felsic volcanic rocks, represent the main components of the Stoughton-Roquemaure assemblage. The Kidd-Munro assemblage (2,719–2,711 Ma) (Figure 2.2), following the Deloro assemblage, composes about 15% of all the volcanic rocks of Abitibi (Figure 2.4(a)). The Kidd-Munro assemblage rocks vary in composition from komatiitic to felsic, have a tholeiitic affinity, and are mainly concentrated in the central part of the Abitibi (Figure 2.2). The sixth assemblage, Tisdale (2,710–2,704 Ma), which represents slightly less than 15% of all the volcanic rocks of the belt (Figure 2.4(a)), is mainly located in the south-western part of the Abitibi (Figure 2.2). This assemblage locally contains komatiitic ultramafic and tholeiitic mafic volcanic rocks in the lower part and intermediate to calc-alkaline felsic rocks in the upper part. The most recent of the seven large volcanic assemblages (Blake River: 2,704–2,695 Ma) (Figure 2.2) represents about one-tenth of the surface covered by volcanic rocks (Figure 2.4(a)) and is concentrated in the southern part of the belt. The composition of this assemblage is highly variable, spanning from tholeiitic basalts to calc-alkaline rhyolites.

More recent periods of volcanic activity, limited in time (Figure 2.5) and highly localized, are associated with the development of turbiditic and alluvial-fluviatile sedimentary basins (see below), including the Krist Formation (massive and volcaniclastic rocks of felsic and calc-alkaline composition) in the Timmins area and several alkaline volcanic units intercalated with alluvial-fluviatile deposits of the Timiskaming Group (south of Abitibi; Figure 2.2) and Haüy Formation (north of Abitibi; Figure 2.2).

The seven large assemblages define a period of volcanic activity extending quasi-continuously over about 100 million years, but maybe more intensively after 2,730 Ma (Figure 2.5). The production rate of volcanic rocks has however widely varied during this period (Figure 2.5).

The various assemblages result from partly similar volcanic cycles with tholeiitic mafic rocks at the base, including, in some cases, ultramafic rocks (komatiites; Figure 2.6(a)), overlain by units becoming increasingly felsic. This evolution and stratigraphy is characteristic of the evolution of most Archean greenstone belts.

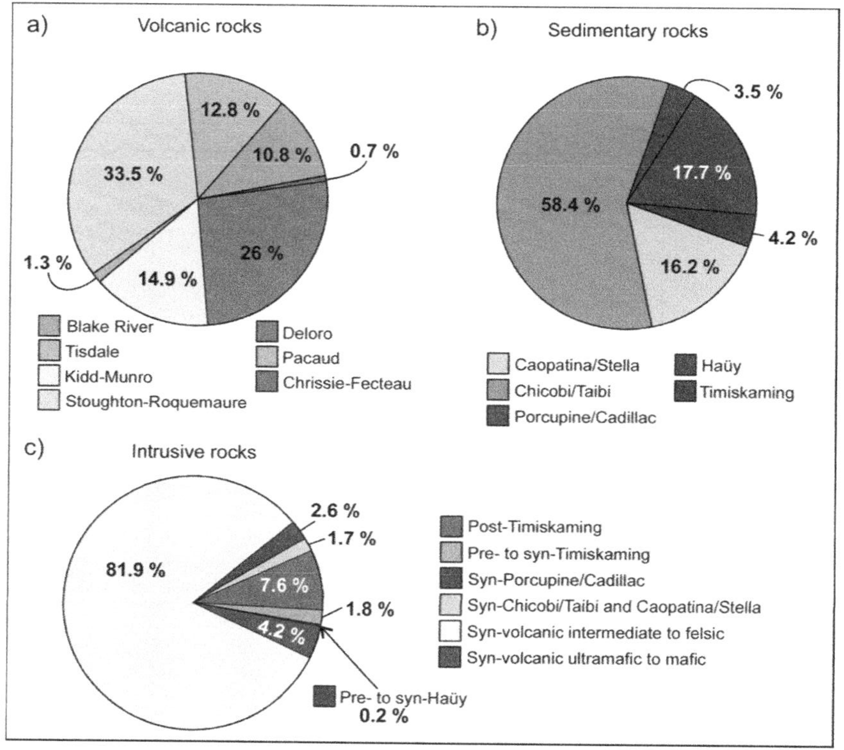

Figure 2.4. *Relative surface area for the seven volcanic assemblages (a), for each group of sedimentary rocks (b), and for each group of intrusive rocks (c) of the Abitibi greenstone belt. Sedimentary and intrusive rocks groupings from Dubé and Mercier-Langevin (2020b)*

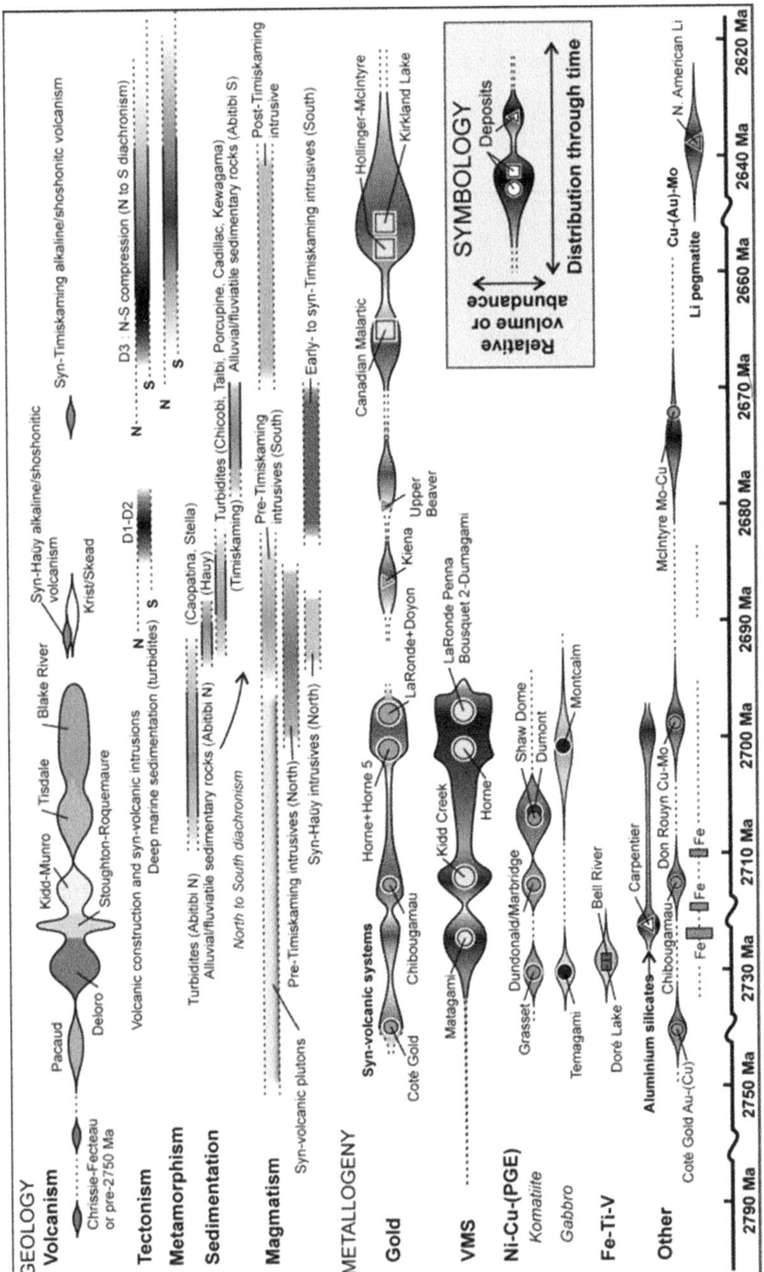

Figure 2.5. *Schematic representation of key elements of the geology and metallogeny of the Abitibi greenstone belt and their evolution though time. See text for details*

2.3.2. *Sedimentary rocks*

Sedimentary rocks represent about 10% of all Abitibi rocks and form bands oriented from east to west that consist of relics of sedimentary basins now lined by large faults (Figures 2.2 and 2.3). The thick successions of fine-grained sediments (turbidites) and also the fluvial-alluvial deposits that lie unconformably on the turbidites postdate volcanism, with the exception of certain minor volcanic units intercalated with sedimentary rocks (Figure 2.6(b)). Turbidite sequences, including graywacke, mudstone and local conglomerates and thin iron formations (Thurston et al. 2008; Monecke et al. 2017), are represented by Porcupine, Cadillac and Kewagama groups in the southern part of the Abitibi, by the Chicobi and Taibi groups in the northern-central part and by the Caopatina and Stella formations in the northern part of the belt (Figure 2.2). Turbidites have the characteristics of typical Bouma sequences, with centimetric to metric beds and locally well-preserved graded beds. The fluvial-alluvial deposits consist mainly of polymictic conglomerates (Figure 2.6(c)) and sandstones locally associated with levels of thin sedimentary rocks and alkaline/shoshonitic volcanic rocks (Dubé and Mercier-Langevin 2020b).

The Timiskaming Group in the south of the Abitibi and the Haüy Formation in the north represent this sedimentary phase in an alluvial to fluviatile context (Figure 2.2). Turbidite sedimentary rocks are largely dominant compared to the bands of fluvial-alluvial sediments (Figure 2.4(b)). The fluvial-alluvial sedimentary bands, known as *Timiskaming type* (Timiskaming Group and Haüy Formation), represent a very important marker for gold deposits exploration in the Abitibi, given their association with major faults controlling the distribution of the main auriferous deposits of the belt and the fact that they mark the period of transition between compressive deformation phases D_1-D_2 and D_3 (Dubé and Mercier-Langevin 2020b) (see section 2.4.1).

Turbidite sequences, as well as the layers of fluvial-alluvial deposits are, depending on the available geochronological constraints, slightly older in the northern part of the Abitibi greenstone belt than in the southern part (Figure 2.5) (Goutier et al. 2009, 2010; Goutier and Melançon 2010; Monecke et al. 2017; Dubé and Mercier-Langevin 2020b). Turbidite sequences mark the transition between the volcanic buildup of the belt and the first phases of regional deformation (D_1 and D_2: imbrication, folding and thrusting) (Robert et al. 2005; Bleeker 2015). This early deformation was followed by a period of extension and uplift (emergence) that led to the erosion of the previously folded volcanic and turbidite sequences, to the development of subaerial or shallow sedimentary basins, and to the deposit of fluvial-alluvial conglomerates of Timiskaming type unconformably (Figures 2.5 and 2.6(b)).

Figure 2.6. *Several examples of sedimentary rocks*

COMMENT ON FIGURE 2.6.– *(a) Komatiitic flow with spinifex (Sx) texture, Pyke Hill area, near Matheson in Ontario; (b) angular unconformity between the turbidites (Tur) of the Porcupine Group and the Timiskaming conglomerates (Cgl) in Timmins; (c) Timiskaming polygenic conglomerates in Kirkland Lake; (d) corridor of strong deformation and iron carbonate alteration affecting Timiskaming conglomerates near the Destor-Procupine fault in Timmins, Buffalo Ankerite mine.*

2.3.3. *Intrusive rocks*

Slightly over half of the surface of the Abitibi greenstone belt consists of intrusive rocks (Figure 2.3(d)). These sometimes polyphased intrusive bodies are present all over the belt (Figure 2.2) and their age varies from about 2,750 to 2,640 Ma (Goutier and Melançon 2010; Dubé and Mercier-Langevin 2020(b)) (Figure 2.5). Their composition varies from ultramafic to felsic, and from subalkaline to alkaline, including peridotite, pyroxenite, gabbro, anorthosite, diorite, granodiorite, tonalite, trondhjemite, granite, monzodiorite, monzonite, quartz monzonite, syenite, carbonatite and lamprophyre (Dubé and Mercier-Langevin 2020b). Due to the link between intrusive rocks and the tectonic evolution of the belt, these are commonly grouped depending on the formation setting. Hence, a significant part of intrusive rocks is "syn-volcanic" and grouped according to their main composition, either as ultramafic to mafic, or intermediate to felsic (Figure 2.2). Post-volcanic intrusive rocks are grouped according to their age with respect to large sedimentary assemblages: syn-turbidites (syn-Chicobi/Taibi and syn-Caopatina/Stella) and syn-fluvial-alluvial sedimentation (early to syn-Haüy in the north of Abitibi and early to syn-Timiskaming in the south of Abitibi), while the more recent intrusive rocks are "post-Timiskaming", according to the classification in Dubé and Mercier-Langevin (2020b; Figures 2.2 and 2.5). Syn-volcanic intrusive bodies represent about 80% of all intrusive rocks of Abitibi, while the post-Timiskaming ones represent about 8% (Figure 2.4(c)). Intrusive syn-turbidites and especially syn-fluvial-alluvial deposits are particularly important in the context of auriferous mineralization (see section 2.4.1). Syn-volcanic intrusive rocks are sometimes associated with volcanogenic massive sulfide mineralization (see section 2.4.2), Cu-(Au)-Mo magmatic-hydrothermal mineralization (see section 2.4.5.1) and Ni-Cu-(PGE)-Cr-V mineralization (see section 2.4.3), while post-Timiskaming rocks are associated with Mo-Bi mineralization and some rare-metal deposits (see sections 2.4.5.2 and 2.4.5.4).

2.3.4. *Deformation and metamorphism*

The Abitibi greenstone belt underwent several stages of major deformation associated with structures that control the current geometry of the belt. A significant part of mineral resources resulted from deformation and metamorphism, while those previously formed underwent deformation and metamorphism.

The formation of thick turbidite sequences at about ≤ 2,710–2,685 Ma in the northern part of the Abitibi is ≤2,700–2,685 Ma in the southern part, after a long period of submarine volcanism, and indicates the emergence of a hinterland whose erosion generated large amounts of fine-grained sediments that were carried by turbiditic currents toward deep basins. This suggests a first episode of major deformation (D_1) linked to the formation of turbidites (Robert et al. 2005; Monecke et al. 2017). The age decrease in these sequences toward south suggests the gradual propagation of the deformation front in this direction, in agreement with the age of the deformation episodes that appear to be slightly older in the north than in the south of the Abitibi (Daigneault et al. 2004). This first deformation episode is responsible for the formation of thrusts and folds, particularly in the Timmins sector (Robert et al. 2005; Bateman et al. 2008; Bleeker 2015). Other early thrusts may be present elsewhere in the belt, particularly in the Val-d'Or sector (Figure 2.2) (Dubé and Mercier-Langevin 2020b). These first-generation faults and folds are overprinted by a second generation of folds and deformation (D_2), well represented in the Timmins area in Ontario (Dubé et al. 2020). These two generations of deformation are cut by the (angular) unconformity at the base of fluvial-alluvial sequences of the Timiskaming Group (Poulsen 2017; Dubé et al. 2020) (Figure 2.6(c)), indicating that D_1 and D_2 predate the extension and exhumation related to the development of these sedimentary sequences (Figure 2.5).

The main episode of regional deformation (D_3) affects the fluvial-alluvial sedimentary rocks of the Timiskaming Group and it is therefore younger than 2,669 Ma, the youngest age obtained for these sedimentary rocks (Figure 2.5). However, at the Abitibi scale, this deformation episode is commonly referred to as D_2 outside of the Timmins mining camp, as to this day this mining camp is the only place where these three deformation phases are clearly recognized. The episode of main deformation is denoted by D_3 in the rest of this text. The D_3 episode represents a shortening event that progressed from north to south (Daigneault et al. 2002). The intensity of D_3 deformation varies at the belt scale, alternating between slightly deformed corridors of east-west dominant orientation separated by zones of very strong deformation (Figure 2.2). Strongly plunging folds are common and associated with a main east-west schistosity, particularly near large areas of faults or in the sectors previously affected by a strong hydrothermal alteration, for example, near syn-volcanic deposits.

Major east-west oriented and generally steeply dipping fault zones intersect the rocks of the belt. These faults, associated with the D_3 deformation episode, are known as "first-order" faults, as they dominate the structural pattern of the belt and reach a length from several dozen to several hundred kilometers. The Larder Lake-Cadillac and Destor-Porcupine fault zones are the main examples (Figure 2.2). The intensity of D_3 deformation (schistosity and elongation lineation) increases progressively when approaching these fault zones (Figure 2.6(d)). These fault zones are associated with high-strain corridors or "second-order" faults that trend east-west to east-north-east and east-south-east with variable dips and are commonly affected by iron carbonate alteration (Figure 2.6(d)). These fault zones locally exhibit slight variations of orientation and evidence of thrust and strike-slip motions, the latter being considered late with respect to the main reverse vertical component. The large fault zones, whose position and geometry are probably partly inherited from the volcanic architecture and the faults associated with D_1 and D_2 (Dubé and Mercier-Langevin 2020b), have a polyphased history and play a fundamental role in the evolution of the Abitibi. They control the distribution of Timiskaming-type fluvial-alluvial sedimentary rocks, as well as early to syn-Timiskaming intrusive rocks, and orogenic gold deposits (see section 2.4.1).

The rocks of the Abitibi greenstone belt were subjected to low to moderate metamorphism that varies from prehnite-pumpellyite facies to lower amphibolite facies, greenschist facies being dominant (Jolly 1978). The minerals associated with metamorphism define the main schistosity and the peak of metamorphism is generally considered as being largely synchronous with the main deformation (D_3; Powell et al. 1995; Robert et al. 2005; Dubé and Mercier-Langevin 2020b) (Figure 2.5).

2.4. Mineral resources and metallogeny

The Abitibi greenstone belt is a region with an exceptional wealth of mineral resources. It contains a variety of metals, which are concentrated in various types and styles of deposits. Gold is the main metal of the belt, though Cu, Zn, Ni and Ag are also very important from an economic standpoint. Other significant metals and substances have been or are extracted (e.g. elements of the platinum group, Li, Fe, Mo and Bi), while there is significant potential for other metals and materials (e.g.V, Ti, rare earth elements and alumina silicates). The main characteristics of the large types of deposits and ore deposits of the Abitibi greenstone belt are summarized below. It is important to note that the term “ore deposit” designates the concentration of metals whose extraction is economically profitable, which therefore includes former and current mines and those that are currently under development, while the term “deposit” applies here to any anomalous concentration of metals or mineral substances, either economically minable or not. The term deposit is therefore used in the broad sense in the text below, while the term ore deposit applies to the deposits that have been or are currently being mined.

2.4.1. *Auriferous deposits*

The typology of gold deposits of the Abitibi greenstone belts is closely related to its volcanic, plutonic and tectonic evolution. There are thus various types and styles of auriferous mineralization formed at various times and at various crustal levels

(Dubé and Gosselin 2007; Dubé and Mercier-Langevin 2020b). The classification of these deposits must therefore take into consideration the nature of mineralization, but also the geological context in which they formed, since two mineralized zones of similar *styles* may result from different processes, and two mineralized zones simultaneously formed and resulting from the same process (of the same type) may have very different characteristics (*styles*). The reader is referred to Dubé and Mercier-Langevin (2020b) for an exhaustive review of the gold metallogeny of the Abitibi. Here, the deposits are grouped depending on the mineralization style and on the relative timing of emplacement. Four principal styles of mineralization are predominant: (1) auriferous sulfide lenses and auriferous ± polymetallic veins hosted in volcanic rocks, (2) sulfide-rich Cu-Au-Ag veins, (3) *stockwork*, veinlets and replacement ± intrusive (pre- and syn-Timiskaming) rocks associated veins and (4) orogenic quartz and carbonate veins (Figure 2.7).

Several other mineralization styles or sub-styles exist besides those listed here, and more than one can be found within the same gold deposit. Each style is associated with a lithological and structural context and specific genetic processes, and also to a distinct hydrothermal footprint. The fact that certain deposits have hybrid and/or more than one mineralization style may increase the complexity of their classification and induce an uncertainty about their genesis.

The orogenic ore deposits of quartz and carbonate vein style represent the main source of gold in the Abitibi (Dubé and Mercier-Langevin 2020b). The *stockwork*-dissemination, veinlets and replacement zones associated with syn-volcanic intrusive and pre-Timiskaming rocks, as well as the auriferous sulfide lenses and the auriferous ± polymetallic veins hosted in volcanic rocks also represent a significant part of the total gold content and comprise some of the largest deposits of the belt (Figure 2.8; section A2.2). Though they are not as significant in terms of production, the sulfide-rich Cu-Au-Ag veins deposits of the Chibougamau region (Figure 2.8), often considered as a distinct type of mineralization (Cu ore deposits of "Chibougamau type"), are included here due to their gold content and because they are an interesting auriferous exploration target.

Certain gold ore deposits of the Abitibi greenstone belt are among the most significant in the world (Figure 2.9; section A2.2). Out of the 127 deposits of over 100,000 ounces (3.1 t) of gold of Abitibi, 40 contain over 1 million ounces (31.1 t), among which 14 have more than 5 million ounces (155.5 t) of gold (Figure 2.9(a)). These 14 ore deposits contain about 65% of the total auriferous budget of the belt (Figure 2.9(b)), which is a good illustration of the importance of large ore deposits on the metallic budget of mining districts.

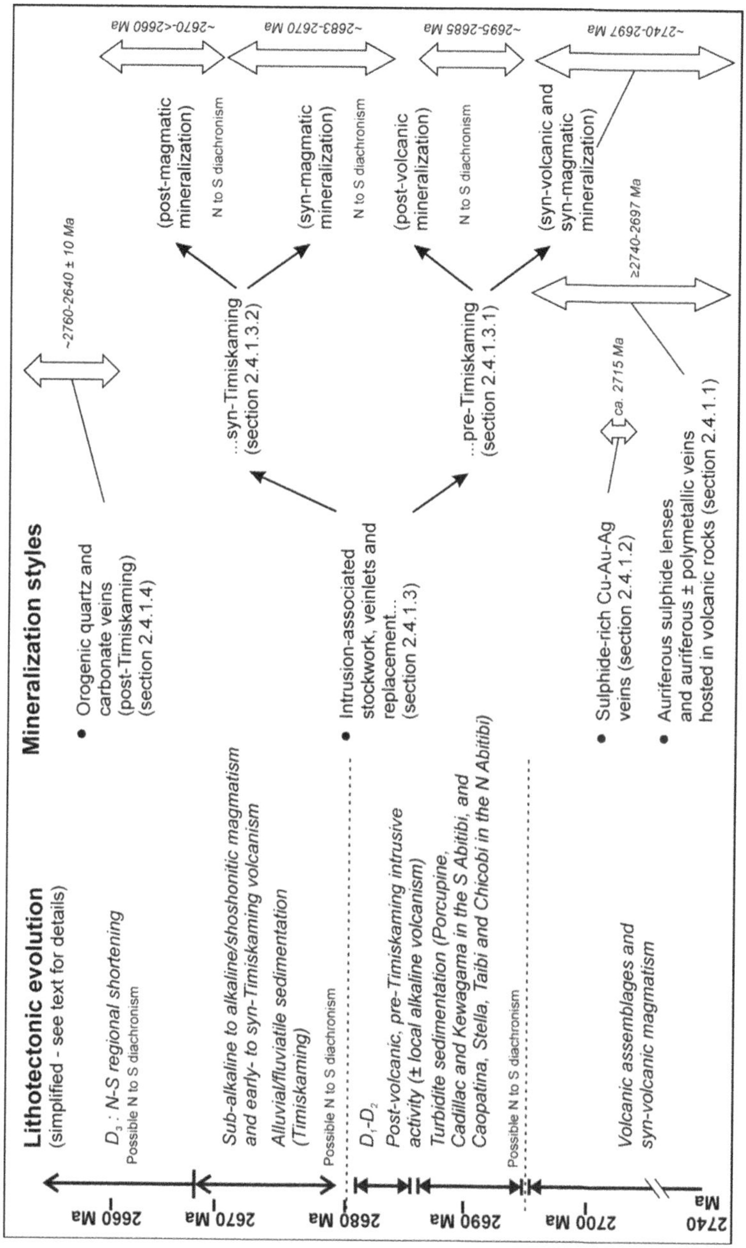

Figure 2.7. *Simplified summary of the different auriferous mineralization styles of the Abitibi greenstone belt in relation with the lithotectonic evolution. Indicated age ranges are representative of the southern part of the belt*

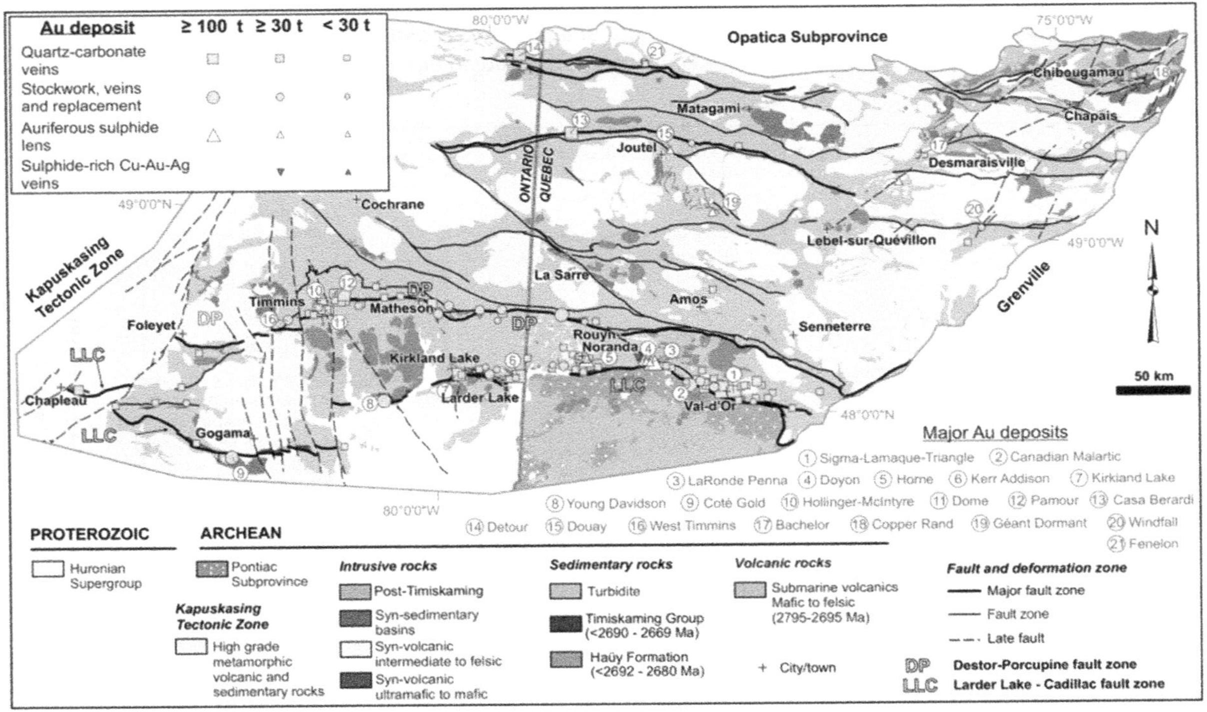

Figure 2.8. *Geological map of the Abitibi greenstone belt showing the distribution and the location of deposits with more than 3 Mt of gold (production, reserves and resources). Modified from Dubé and Mercier-Langevin (2020b) and deposit data from Bécu et al. (forthcoming)*

2.4.1.1. *Sulfide lenses and auriferous ± polymetallic vein deposits hosted in volcanic rocks*

Sulfide lenses and auriferous ± polymetallic vein deposits hosted in volcanic rocks span over a wide range of size and content (section A2.2). They can be found in several volcanic assemblages of the Abitibi greenstone belt (Figures 2.2 and 2.8). Despite this wide distribution in space and time, these deposits are more significant and numerous in the southern part of the belt (Figure 2.8).

The primary characteristics of auriferous sulfide lenses and auriferous ± polymetallic vein deposits hosted in volcanic rocks, which in most cases consist of volcanogenic massive sulfide (VMS) systems, are commonly obscured by the deformation and metamorphism affecting the entire belt. This represents an additional complexity that explains why the origin and the classification of some of these deposits remain disputed. Recent research works have shown the syn-genetic nature of most of these zones of sulfides and auriferous ± polymetallic veins, and the study of active systems confirmed that gold can be primarily enriched in submarine volcanogenic hydrothermal systems (Hannington et al. 2005). As for the reasons explaining the anomalous gold content in these systems, several processes have been identified, which can be associated with two main phenomena: (1) a source and/or gold-enriched fluids either due to a specific geodynamic context or to a direct magmatic contribution, and (2) particularly efficient processes of gold transportation and concentration (e.g. physicochemical conditions, presence of ligand agents, boiling and phase separation, and colloidal preconcentration). These processes are discussed in detail in a number of publications, including (Poulsen and Hannington 1996; Hannington et al. 1999; Huston 2000; Dubé et al. 2007a; Mercier-Langevin et al. 2011, 2015). Several VMS deposits of the Abitibi are included in the classification of auriferous ore deposits as they were or are mainly mined for their gold content.

The gold-rich VMS sometimes form complex groups of lenses along specific volcanic levels (e.g. lens 20 North of LaRonde Penna, main lens at Bousquet 2-Dumagami and Warrenmac lens at Westwood mine in the Doyon-Bousquet-LaRonde mining camp) (Mercier-Langevin et al. 2007a) (Figure 2.14). These lenses can also be stacked in the host stratigraphic sequence (e.g. lenses 6, 7, 20 North and 20 South of LaRonde Penna) (Mercier-Langevin et al. 2007b). Several mineralized lenses can be part of the same hydrothermal system (e.g. LaRonde Penna and Bousquet 2-Dumagami ore deposits) (Dubé et al. 2007b), though certain deposits that are close to one another can be part of distinct hydrothermal systems (e.g. Horne and Quémont deposits) (Krushnisky et al. 2022). Mineralized lenses, mainly composed of pyrite with a variable abundance of sphalerite, chalcopyrite, pyrrhotite, galena and accessory minerals (e.g. sulfides, sulfosalts and tellurides) vary individually from several dozen to hundreds of thousands of tons (for example, the Lemoine mine) (Mercier-Langevin et al. 2014b) up to several dozen million tons (for example, the Upper and Lower H lenses at the Horne mine) (Barrett et al. 1991). The auriferous VMS are composed of stratiform to stratabound, semi-massive (40–70% vol.) to massive (>70% vol.) concentrations of sulfides overlying zones of veins (*stringers*) of discordant sulfides (generally deformed; Figure 2.10(a)) of variable lateral and vertical extents. Other deposits are rather characterized by stratabound to discordant zones of auriferous ± polymetallic sulfide veinlets (pyrite and envelopes of disseminated pyrite containing between 10 and 50% vol. sulfides) (Valliant et al. 1983; Tourigny et al. 1989; Boily-Auclair et al. 2020) (Figure 2.10(b)). Volcaniclastic levels containing fragments of auriferous massive and semi-massive sulfides are commonly present near mineralized lenses (Dubé et al. 2007b; Monecke et al. 2008; Yergeau et al. 2015; Krushnisky et al. 2020).

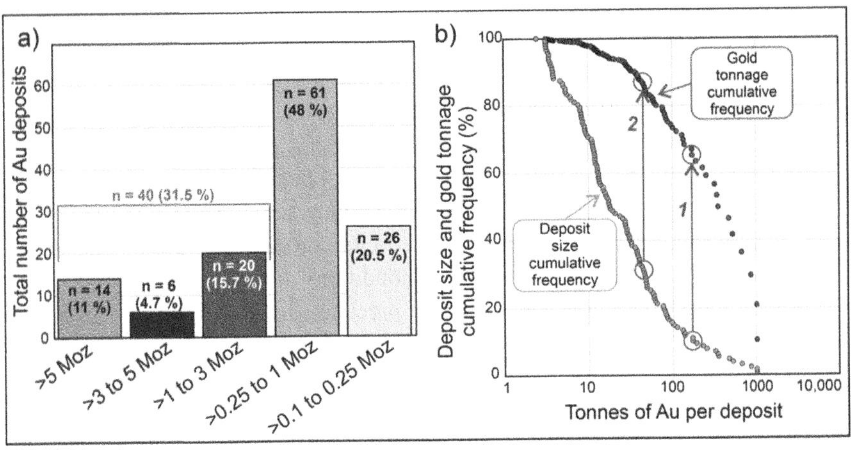

Figure 2.9. *Ore deposits of over 100,000 ounces (3.1 Mt) of gold of the Abitibi greenstone belt, including production, reserves, and measured and indicated resources*

COMMENT ON FIGURE 2.9.– *Abitibi greenstone belt deposits with over 100,000 ounces (3.1 t) of gold, including production, reserves, and measured and indicated resources. (a) Total number of depsits as a function of total gold content (n = 127). Twenty deposits are considered world-class with more than 3 million ounces of gold. (b) Cumulative frequency graph of deposits and their gold content. Line 1 shows that the 14 deposits with over 5 million ounces each (11% of the deposits) collectively contain approximately 65% of the total Abitibi belt gold, whereas line 2 shows that the 40 deposits containing over 1 million ounces each (31.5% of the deposits) collectively contain more than 85% of the total Abitibi belt gold. Data from Bécu et al. (forthcoming).*

Figure 2.10. *Metamorphosed sulfide-bearing mineralization and alteration assemblages*

COMMENT ON FIGURE 2.10.– *(a) Zone of strongly deformed pyrite ± chalcopyrite veinlets in mafic to chloritized intermediate rocks, LaRonde Zone 5 mine, Doyon-Bousquet-LaRonde camp; (b) layer of auriferous granoblastic pyrite and quartz, Bousquet 1 mine, Doyon-Bousquet-LaRonde camp; (c) garnet, biotite, muscovite, quartz and chlorite metamorphosed alteration assemblage, Westwood mine, Doyon-Bousquet-LaRonde ; (d) advanced argillic-style alteration assemblage metamorphosed to andalusite ± kyanite porphyroblasts, quartz, pyrite and sphalerite, LaRonde Penna mine, Doyon-Bousquet-LaRonde camp. And = andalusite, Bo = biotite, Gt = garnet, Mv = muscovite, Py = pyrite, Qz = quartz, Sph = sphalerite.*

Auriferous lenses are hosted in mafic (basalt) to felsic (rhyolite) volcanic sequences, commonly in thick (up to several dozen meters) intermediate to felsic units. These sequences are composed of massive lava flows (flows, domes and lobes) with volcaniclastics intersected by dykes and sills of similar composition. The thick (several dozen meters) volcaniclastic sequences are generally hosting larger lenses. Some deposits are associated with sequences that are bimodal in composition, whereas others belong to continuous sequences (basalt-andesite-dacite-rhyodacite-rhyolite). The magmatic affinity of volcanic rocks varies from tholeiitic to calc-alkaline, though the most significant deposits (e.g. LaRonde Penna, Bousquet 2-Dumagami, Horne and Horne 5) seem preferentially associated with rocks of transitional to calc-alkaline affinity (Mercier-Langevin et al. 2015; Krushnisky et al. 2020).

The most common type of alteration is dominated by muscovite. Muscovite, most commonly referred to as sericite, is associated with variable abundances of chlorite, carbonates and disseminated sulfides, and generally form large stratabound to discordant halos intersected by more limited chlorite-altered discordant corridors. Overprinting metamorphism induced the crystallization of biotite, manganese-bearing garnet and chloritoid in certain cases (Figure 2.10(c)). An acid clay style (or advanced argillic-style) alteration is associated with several ore deposits of Doyon-Bousquet-LaRonde mining camp (Dubé et al. 2014; Yergeau et al. 2015) (Figure 2.8). Metamorphism overprinting this alteration is responsible for the formation of a granoblastic assemblage containing quartz, muscovite and aluminosilicates (andalusite, kyanite, staurolite and pyrophyllite with, very locally, diaspore and topaz) (Figure 2.10(d)).

The ore deposit of the Géant Dormant mine (Figure 2.8) and the mineralized zones of the Northern Corridor at the Westwood mine are also syn-volcanic, but their characteristics differ from those previously described. The ore has the form of veins of maximal thickness 2 m (Géant Dormant). These veins are mainly composed of quartz and sulfides (pyrite ± pyrrhotite, sphalerite, galena, chalcopyrite and visible gold) in variable relative abundance and are hosted in mafic to felsic volcanic and volcaniclastic rocks emplaced in a submarine context, sometimes in association with felsic porphyritic dykes (Gaboury and Daigneault 1999; Yergeau et al. 2015). The dominant alteration is composed of sericite (Géant Dormant), or of sericite, biotite and manganese-bearing garnet (Westwood).

2.4.1.2. *Sulfide-rich Cu-Au-Ag vein deposits*

The sulfide-rich Cu-Au-Ag veins of the Chibougamau mining camp (section A2.2; (Allard 1976)), commonly known as "Chibougamau-type" veins, have a relatively low total gold content (~120 t Au, average content ~1.76% Cu and 2.08 g/t Au) (Pilote and Guha 1995). Sulfide veins are hosted by anorthosite and gabbro of a large intrusive complex dated at 2,728 ± 1 Ma (Mortensen 1993). The mineralization consists of lenses and semi-massive to massive chalcopyrite-pyrite-pyrrhotite veins hosted by north-west and north-east oriented shear zones consisting of quartz-carbonate-sericite and/or chlorite (± chloritoid and paragonite; Figure 2.11(a)) schists (Pilote and Guha 1995). Dykes of variable composition are common in the vicinity of mineralized zones (Allard 1976). Some of these dykes are cut by sulfide veins, while others cut the mineralization and contain the same schistosity as the host shear zones, which indicates that the Cu-Au-Ag sulfide veins and the dykes predate the development of shear zones (Robert 1994a; Pilote and Guha 1995). Pilote et al. (1994, 1995) and Pilote and Guha (1995) link the Cu-Au-Ag sulfide veins to the magmatic-hydrothermal activity associated with the late porphyric tonalite phase of the Chibougamau syn-volcanic pluton.

2.4.1.3. *Stockwork, veinlets and replacement-style auriferous deposits associated with intrusive rocks*

Many auriferous deposits characterized by *stockwork*, veinlets, breccia, dissemination and/or replacement zones are spatially associated with intermediate to felsic intrusive bodies. These deposits are present throughout the Abitibi greenstone belt, but are mainly located several kilometers (≤5 km) from the east-west oriented major faults, or in certain cases, along the east-north-east and east-south-east oriented fault zones (Figure 2.8). Intrusive rocks associated with this mineralization style normally host at least a part of the mineralization. This auriferous mineralization style can be divided into two sub-groups: (1) deposits associated with pre-Timiskaming intrusive bodies; (2) deposits associated with syn-Timiskaming intrusive bodies (sensu lato).

2.4.1.3.1. Deposits in pre-Timiskaming intrusive rocks

Auriferous deposits associated with pre-Timiskaming intrusive rocks include significant ore deposits such as Côté Gold and Doyon (Figure 2.8). In most cases, the associated intrusions have a subalkaline affinity. Two subgroups can be defined: (1) deposits associated with syn-volcanic (~2,740–2,695 Ma) pre-Timiskaming intrusive rocks, and (2) deposits associated with post-volcanic (2,695 to ~2,685 Ma) pre-Timiskaming intrusive rocks (Figure 2.7).

The Côté Gold ore deposit is a high tonnage and low grade (310 t Au, ~0.9 g/t Au) deposit characterized by Au ± Cu zones of disseminated pyrite and chalcopyrite, veins, *stockworks* and breccias with pyrite-chalcopyrite-sphalerite (± galena-arsenopyrite-fluorite-molybdenite), and auriferous veinlets with quartz ± carbonate ± sulfides centered on a system of magmatic and hydrothermal breccias developed in the tonalitic and dioritic phases of a syn-volcanic intrusive complex (Figures 2.8 and 2.11(b)). The main types of alteration are potassic (biotite), phyllic (muscovite-quartz-carbonate) and sodic ± siliceous (quartz-albite-carbonate) (Katz et al. 2021). The mineralization is dated at 2,740 ± 6 Ma, therefore contemporary with the phases hosting the intrusive complex that were dated between 2,741 and 2,739 ± 1 Ma (Katz et al. 2021).

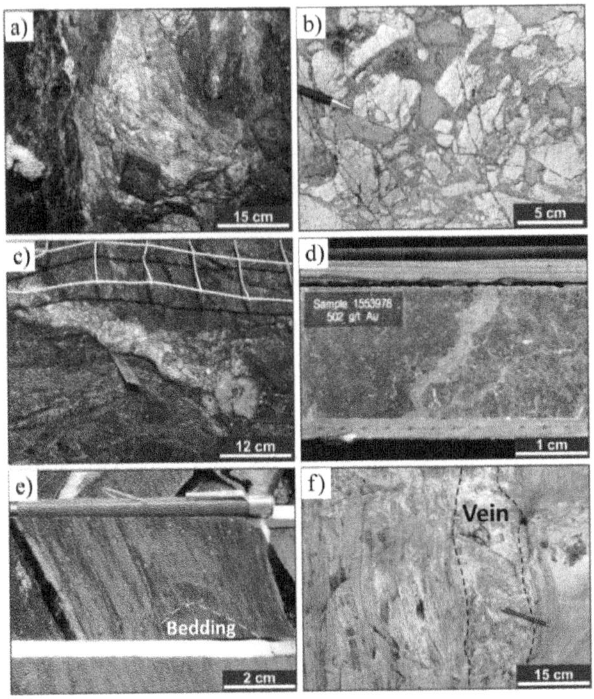

Figure 2.11. *Examples of pre-Timiskaming deposits*

COMMENT ON FIGURE 2.11.– *(a) Massive pyrite-pyrrhotite-chalcopyrite vein, Portage mine, Chibougamau (source: A. Tessier); (b) magmatic-hydrothermal breccia with biotite-chlorite matrix and traces of chalcopyrite, Côté Gold mine, Swayze; (c) quartz-chalcopyrite vein with high gold content, Doyon mine, Doyon-Bousquet-LaRonde camp; (d) strong silicification zone with pyrite, 500 g/t Au, Windfall project, Lebel-sur-Quévillon/Desmaraisville; (e) graywacke (≥1 g/t Au) with quartz-calcite-pyrite veinlets perpendicular to bedding, East Gouldie zone, Canadian Malartic mine, Malartic; (f) quartz vein with traces of auriferous pyrite in a shear zone, Gilbert zone, Canadian Malartic mine, Malartic.*

The Doyon mine located in the Doyon-Bousquet-LaRonde mining camp (Figure 2.8) produced 168 t Au. The Doyon ore deposit is partly hosted by calc-alkaline tonalite-trondhjemite phases of a synvolcanic intrusive complex dated at about 2,697 Ma (Galley and Lafrance 2014; McNicoll et al. 2014) and by adjacent volcanic rocks that are part of the Blake River Group (Bousquet Formation; 2,699–2,697 Ma) (Lafrance et al. 2005). The deposit is mainly composed of quartz, pyrite and chalcopyrite veins and veinlets with high gold content (Figure 2.11(c)) (Gosselin 1998). When hosted in volcanic rocks, the veins and veinlets are partially to fully transposed in zones of felsic and mafic schists, altered in quartz, muscovite, aluminosilicates and chlorite (Savoie et al. 1990). Pyrite-rich auriferous veins with variable orientation, generally dipping steeply to the east, are hosted by the calc-alkaline tonalite phase of the intrusive complex (Gosselin 1998). A hydrothermal alteration assemblage composed of chlorite, biotite, epidote, muscovite and a small quantity of carbonates is present along the vein selvages. Anhydrite-quartz-pyrite-chalcopyrite veins are present outside the ore deposit (Galley and Lafrance 2014; Yergeau et al. 2015). The Doyon ore deposit vein system extends toward the east, up to the Westwood mine, where a deformed mafic dyke dated at 2,699 Ma cuts the veins and their alteration envelope (Yergeau et al. 2015). Doyon is interpreted as an ore deposit of magmatic-hydrothermal origin predating the main deformation (Gosselin 1998; Galley and Lafrance 2014; Yergeau et al. 2022a, 2022b).

Quartz (± carbonate) veins and veinlets and auriferous replacement zones associated with postvolcanic but pre-Timiskaming dykes of intermediate (dioritic) composition characterize some important deposits (Dubé and Mercier-Langevin 2020b). The mineralization is present in fault and/or deformation zones that predate the main shortening phase (D_3). Examples include the Kiena, Norlartic-Kierens and Siscoe mines in the mining camps of Val-d'Or and Malartic (Figure 2.8), where auriferous zones are cut by dykes dated between 2,690 and 2,686 Ma (Couture et al. 1994; Robert 1994b; Morasse et al. 1995; De Souza et al. 2020). Similar replacement, *stockwork* and quartz-sulfides systems include Barry and Windfall deposits (Figure 2.11(d)) in the north-eastern part of the Abitibi, formed at about 2,698–2,697 Ma (Kitney et al. 2011; Padilla and Savard 2018). These auriferous systems have recorded the main regional shortening (D_3), and the original characteristics were partly obliterated by the deformation, complicating the interpretation of the chronology of the mineralizing events. However, the geological context, the chronology and nature of host structures, and the mineralization style suggest they may have been formed during the pre-Timiskaming deformation (D_1-D_2) (Kitney et al. 2011; Dubé and Mercier-Langevin 2020b).

2.4.1.3.2. Deposits in syn-Timiskaming intrusive rocks

Auriferous deposits associated with syn-Timiskaming intrusive rocks (sensu lato) add about 1,530 t of gold to the auriferous budget of the Abitibi (Dubé and Mercier-Langevin 2020b). Many examples of this style are relatively small, but there are notable exceptions, such as Canadian Malartic (>520 t Au) and Young-Davidson (>195 t Au; Figure 2.8). This mineralization style is generally centered on mafic to felsic, multiphased intrusive complexes forming clusters (≤25–40 km^2) of small intrusions of subalkaline to alkaline/shoshonitic affinity and dyke swarms (Robert 2001; Dubé and Mercier-Langevin 2020b). Two subgroups of *stockwork*, veinlets and replacement style deposits associated with syn-Timiskaming intrusive rocks can be distinguished: (1) those potentially syn-magmatic and (2) post-magmatic (Figure 2.7). Most of these deposits are located along the Larder Lake-Cadillac and Destor-Porcupine fault zones, or along second or third order, subparallel structures (e.g. Garrcon and Holt-McDermott deposits in Ontario) (Figure 2.8). In the northern part of the Abitibi, most of the intrusions associated with this mineralization style are 10–15 million years older than the similar intrusions located on the southern part of the Abitibi (which is ~2,695–2,688 Ma in the northern part versus ~2,683–2,670 Ma in the southern part) (Dubé and Mercier-Langevin 2020b) and are defined as being syn-Haüy (e.g. Shortt lake and Bachelor Lake) (Figure 2.8). One exception is the Douay syenite, hosting the Douay ore deposit, dated at 2,676 +6/–5 Ma (Davis et al. 2000), though located in the northern part of the Abitibi. For the sake of simplicity, all the deposits characterized by this style of mineralization are located in the northern part of the Abitibi belt, and are included in this subgroup, even though the associated intrusions are not syn-Timiskaming (sensu stricto) but mostly syn-Haüy. Due to similar characteristics and geological context, they are considered syn-Timiskaming (*sensu lato*) (Dubé and Mercier-Langevin 2020b).

The intrusive rocks are equigranular to porphyritic and include syenite, quartz monzonite, quartz monzodiorite, granodiorite, carbonatite and lamprophyre dykes. Though they predate the main deformation, syn-Timiskaming intrusions are often not, or only weakly foliated due to their high competence compared to that of the hosting rocks and deformation partitioning. The main deformation (D_3) in these competent intrusive bodies commonly forms brittle faults and/or more locally various networks of fractures, and quartz ± pyrite veins and veinlets and associated replacement zones.

Mineralization is in the intrusions and/or their extrusive equivalent (e.g. the Upper Canada deposit), but also in the proximity of intrusions, in deformed mafic volcanic and volcaniclastic rocks and in fluvial-alluvial and turbidite sedimentary rocks of the Porcupine, Cadillac and Timiskaming groups (Dubé and Mercier-Langevin 2020b). In certain cases, the major part of the ore is located along east-north-east to east-west deformation zones and more locally, east-south-east, often in corridors subparallel to the main foliation (e.g. the Canadian Malartic ore deposit) (Figures 2.11(e) and (f)). In some deposits, the auriferous replacement zones are in the shears and early to syn-D_3 mylonite zones. Some of these auriferous replacement zones are transposed and boudinaged and have registered the main deformation D_3. Barren extension quartz veins, oriented at a sharp angle with respect to the mylonitic foliation and to the boudinaged dykes, cut the mineralized zones locally.

Mineralization forms replacement zones, with quartz ± sulfides and/or quartz ± carbonate veins and veinlets of millimetric to centimetric thickness, *stockwork* zones and breccia (Figures 2.12(a) and (b)). The associated alteration zones are generally light brown, gray or brick-red and composed of an assemblage containing variable proportions of quartz, carbonates (calcite, dolomite and/or ankerite), feldspars (albite and/or potassic feldspar), sericite, and fine pyrite (≤2 mm, 1–5% vol.). Traces of hematite in the form of fine particles are present within or in the vicinity of mineralized zones and are responsible for the often characteristic brick-red color (e.g. Young-Davidson and West Timmins) (Figure 2.12(a)). This red coloration is mainly developed in the alkaline/shoshonitic intrusions, but sometimes also in the mafic volcanic and sedimentary rocks (e.g. Canadian Malartic and Bachelor ore deposits) (Figure 2.12(b)). Anhydrite, scheelite and/or fluorite are present in minor amounts. Trace amounts of tellurides, molybdenite, galena, chalcopyrite and arsenopyrite are also commonly present. At the Upper Beaver deposit, which is centered on a complex of multiphased quartz-bearing monzodiorite (Figure 2.8) (Feick 2016), distinct alteration assemblages containing actinolite, epidote, magnetite and chalcopyrite are associated with several Cu-Au ore zones and share certain similarities with certain gold skarns, with the magnetite subgroup of iron oxide copper-gold ("IOCG") deposits (Kontak et al. 2013), and also with Cu-Au alkaline ore deposits (Sillitoe 2002; Pacey et al. 2019).

The genesis of auriferous systems associated with syn-Timiskaming intrusions is controversial, mainly due to complex temporal relations between magmatism, deformation and hydrothermal activity, and the scarcely available precise geochronological data (Robert 2001; Dubé and Mercier-Langevin 2020b). In certain cases, mineralization is related to syn-Timiskaming intrusive activity and is probably of magmatic-hydrothermal origin (Lauzière 1989; Robert 2001). At Upper Beaver, the overlapping of the age of the syn-Timiskaming multiphased quartz-bearing monzodiorite complex (2,678 ± 1 Ma) and Re-Os age of 2,680 ± 6 Ma of the molybdenite associated with auriferous mineralization is compatible with a pre-D_3 magmatic-hydrothermal auriferous system (Mercier-Langevin et al. 2021). There are however examples where mineralization is syn-main deformation (syn-D_3), with syn-Timiskaming intrusions having then played a passive role. The geometry of auriferous quartz ± sulfides or quartz-carbonate ± pyrite veinlets and veins, with which the zones of auriferous replacement with fine pyrite are associated, is at a moderate to sharp angle with respect to the elongation lineation and compatible with the orientation of syn-main deformation structures (e.g. Young-Davidson, Matachewan and West Timmins). The formation of auriferous zones is then interpreted as contemporary with the main deformation (Campbell 2014; Zhang et al. 2014; Dubé and Mercier-Langevin 2020b). At Canadian Malartic, where syn-Timiskaming intrusions host a part of the mineralization, auriferous zones are controlled by a foliation and folds associated with the main deformation (Figures 2.11(e) and (f)), and by the east-trending Sladen fault (De Souza et al. 2017, 2019; Perrouty et al. 2017). The molybdenite directly associated with auriferous mineralization at Canadian Malartic has given a Re-Os age of 2,664 ± 11 Ma, compatible with a syn-main deformation emplacement (De Souza et al. 2020).

2.4.1.4. *Orogenic (post-Timiskaming) quartz and carbonate vein-style deposits*

Orogenic post-Timiskaming quartz and carbonate vein-style deposits represent the most common mineralization style and the main source of gold in the Abitibi. The deposits are grouped in mining camps, the largest being concentrated in the Southern Abitibi and distributed along the Larder Lake-Cadillac and Destor-Porcupine fault zones. This distribution is particularly well illustrated by world-class auriferous systems (containing over 100 t Au), such as Hollinger-McIntyre in Ontario and Sigma-Lamaque-Triangle in Quebec (Figure 2.8). Though several deposits are located in first-order fault zones (e.g. Kerr Addison-Chesterville and Lapa ore deposits) (Figure 2.8), most of the quartz and carbonate veins systems are associated with the second- and third-order oblique-reverse motion faults and shear zones located at less than 5 km from first-order structures (Robert 1990; Robert et al. 2005; Dubé and Gosselin 2007). Most major deposits are located in greenschist facies metamorphosed rocks, and less frequently in lower amphibolite facies rocks. At regional scale, quartz and carbonate veins are hosted in a wide variety of host rocks (Hodgson and MacGeehan 1982) and generally exhibit lithologic associations specific to mining camps. The deposits that combine the highest tonnage with the highest average grade are hosted in mafic and ultramafic volcanic rocks (e.g. Hollinger-McIntyre and Kerr Addison-Chesterville ore deposits) (Figure 2.8) and alkaline/shoshonitic to syn-Timiskaming subalkaline intrusions (Kirkland Lake) (Dubé and Mercier-Langevin 2020b) (Figure 2.8). Several others are partly hosted or adjacent to the conglomerate of the Timiskaming Group (e.g. Dome, Kirkland Lake and Pamour) (Figure 2.8), suggesting an empirical and spatial relationship between the large systems of quartz and carbonate veins and the regional unconformities (Robert et al. 2005; Dubé and Gosselin 2007; Bleeker 2015; Dubé et al. 2020). However, Hollinger-McIntyre, the largest Canadian gold deposit, is an exception, as it is located at about 5 km North of the Destor-Porcupine fault zone and the conglomerates of the Timiskaming Group (Dubé et al. 2020).

At the deposit scale, mineralization consists of networks of quartz and carbonate (± tourmaline) veins that vary in thickness from a few centimeters to a couple of meters and have relatively simple to very complex geometry. Anisotropy, the contrast of competence and the superposition of generations of structures may influence and complexify the geometry of the network of veins. The networks of veins extend vertically and laterally up to 2,000 m. The veins are banded/laminated and generally present in the central part of the faults and the shear zones, and dip moderately to steeply (Figure 2.12(c)). Extension veins and hydrothermal breccia are also commonly present. The associated extension veins have a low to moderate dip and form sigmoid ± transposed networks when they are in the deformation zone, and rather planar when they extend outside of it (Robert and Poulsen 2001). The geometry and the kinematics of these networks of veins indicate that most of them were formed in active deformation zones (Robert and Poulsen 2001; Dubé and Gosselin 2007) during the main deformation and shortening (D_3). The quartz and carbonate vein-style deposits are therefore of orogenic origin (Groves et al. 2003; Goldfarb et al. 2005). In some cases, massive colloform-crustiform ankerite veins, barren or at low grade, are also present (Stromberg et al. 2019; Dubé et al. 2020).

Quartz and carbonate veins of the Abitibi generally contain 1–5% vol. of pyrite, with varied but generally low proportions of chalcopyrite, arsenopyrite and tellurides, and more locally, trace amounts of scheelite. At Kirkland Lake (Figure 2.8), centimeter to meter-thick laminated quartz veins and brecciated quartz veins with high Au-Te content (average >14 g/t Au) are present in the brittle faults that intersect an alkaline multiphased intrusive complex (mafic syenite and syenite) to subalkaline (quartz monzonite and/or quartz monzodiorite) (Figure 2.12(d)). The zones with higher gold content are characterized by brecciated quartz veins and contain fragments of host rocks and veins in a dark gray matrix composed of fine-grained quartz, sericite, graphite and/or molybdenite, with traces of pyrite, gold and tellurides (Ispolatov et al. 2008). The metallic signature of quartz and carbonate vein systems is influenced by the nature of host rocks. However, various elements (Ag, As, Te, W, B, Sb and Mo) are generally associated with gold, with anomalous concentrations of base metals in some deposits. In general, gold is mainly present in veins, but can also be present in pyritized and carbonatized host rocks. The average grade of deposits varies between about 1 and 15 g/t Au (section A2.2), while the tonnage varies from several thousand tons to >100 million tons of ore.

The nature, distribution and intensity of hydrothermal alteration are influenced by the composition and rheology of the host rocks and metamorphic facies (Goldfarb et al. 2005; Robert et al. 2005; Dubé and Gosselin 2007). Typically, greenschist facies alteration postdates regional metamorphism and is characterized by a proximal assemblage dominated by Fe carbonate, sericite ± albite and sulfides (pyrite ± arsenopyrite) (Figure 2.12(e)), with corresponding enrichments in CO_2, K_2O and S, and Na_2O leaching. Distal hydrothermal alteration consists of variable abundances of chlorite, calcite/dolomite and locally, magnetite (Hodgson 1993). The ultramafic rocks proximal to veins are strongly carbonatized, generally contain fuchsite or roscoelite (green micas rich in Cr or V, respectively) and have low gold contents (≤1–2 g/t). However, locally, these ultramafic rocks replaced by such a hydrothermal assemblage represent a large part of the ore, as illustrated by the Kerr Addison-Chesterville ore deposit (Figure 2.12(f)) (Smith et al. 1993). At upper greenschist and lower amphibolite facies, the proximal alteration is characterized by an assemblage of biotite-quartz-sericite-carbonate-actinote-K-feldspar-albite and pyrrhotite ± chalcopyrite, and generally overprints the regional metamorphic assemblage (e.g. Detour ore deposit) (Oliver et al. 2012; Castonguay et al. 2020).

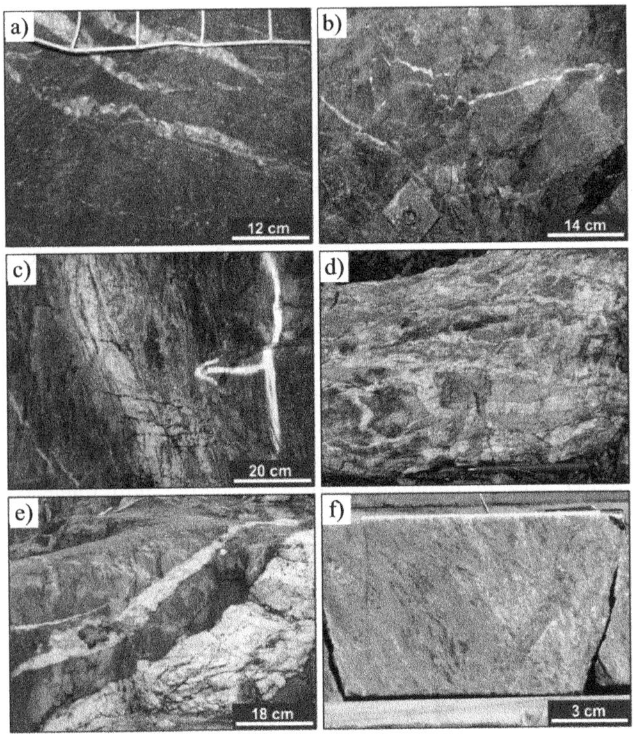

Figure 2.12. *Examples of auriferous orogenic quartz and carbonate veins*

COMMENT ON FIGURE 2.12.– *(a) Quartz veinlets with pyrite traces in brick-red syenite with disseminated pyrite, vertical section, Young Davidson mine, Kirkland Lake; (b) auriferous stockwork and replacement and quartz veinlets zone, vertical section, Bachelor mine, Desmaraisville; (c) layered quartz-carbonate vein in shear zone, Triangle mine, Val-d'Or; (d) auriferous quartz vein with many fragments of carbonatized host, plan view, vein #1, Teck-Hughes mine, Kirkland Lake; (e) quartz veins in strongly carbonatized gabbro, Lucien-Béliveau mine, Val-d'Or; (f) strongly carbonatized ultramafic rock of Piché Group, with green micas and fine disseminated pyrite, 7 g/t Au zone, Kerr Addison mine, Larder Lake.*

2.4.1.5. *Abitibi gold: evolution in time and space*

The gold deposits of the Abitibi show a wide variety of styles. This diversity, partly due to local parameters, is most of all due to the evolution of the geological context in which the deposits were formed: (1) during the main volcanic construction phase of the belt (auriferous sulfide lenses and auriferous ± polymetallic vein-style deposits hosted in volcanic rocks, sulfide-rich Cu-Au-Ag veins, and *stockwork*, veinlets and replacement associated with syn-volcanic pre-Timiskaming intrusive rocks, between ~2,740 and 2,695 Ma), (2) during the first regional deformation phases D_1 and D_2 (*stockwork*, veinlets and replacement-style deposits associated with post-volcanic pre-Timiskaming intrusive rocks and some syn-magmatic *stockwork*, veinlets and replacement deposits associated with syn-Timiskaming intrusive rocks, between ~2,700 and ~2,670 Ma) and (3) during the main phase (D_3) of N-S compression (post-magmatic *stockwork*, veinlets and replacement deposits associated with syn-Timiskaming intrusive rocks and orogenic quartz and carbonate vein-style deposits, syn- to late-D_3 ~2,660 to ~2,640 ± 10 Ma) (Dubé and Mercier-Langevin 2020b) (Figures 2.5 and 2.7).

The major part of gold in the Abitibi greenstone belt is distributed along the Larder Lake-Cadillac and Destor-Porcupine fault zones (Figure 2.8) and was emplaced during the main phase of regional shortening D_3 (Figures 2.5 and 2.7). At the belt scale, a relatively gradual transition can however be noted in the styles of early to syn-orogenic ore distributed along the Larder Lake-Cadillac and Destor-Porcupine fault zones as the shortening associated with the main regional deformation D_3 intensifies.

Though similar throughout the camps, the sequence of tectonometamorphic and hydrothermal events was probably not perfectly synchronous at the belt scale (Figure 2.5), but has migrated with time from north to south, laterally along the main deformation corridors, and also possibly as a function of the crustal depth, as discussed in Dubé and Mercier-Langevin (2020b). The fact that a significant part of the auriferous mineralization is of syn-volcanic and/or syn-magmatic nature and was emplaced prior to the episodes of regional deformation (Figures 2.5 and 2.7) represents another specificity of the Abitibi greenstone belt. This diversity of auriferous mineralization styles reflects a complex magmatic, tectonic, metamorphic and hydrothermal evolution.

2.4.2. *Cu-Zn-(Au-Ag-Pb) volcanogenic massive sulfide deposits*

Volcanogenic massive sulfide deposits, including those of the Abitibi greenstone belt, form lenses or mounds of polymetallic sulfides on or near the seafloor in a dominantly volcanic environment (Galley et al. 2007). These metal accumulations result from the convective circulation of hydrothermal, metal-enriched fluids in the oceanic crust and their focused discharge in one or several specific places in the presence of favorable conduits (e.g. syn-volcanic faults and fractures, permeable units, dykes, intrusive bodies and so on).

Most VMS deposits are generally defined by two main components: a mound, stratiform or strata-bound tabular body comprising over 40% vol. sulfides with variable amounts of quartz, oxides, phyllosilicates and relic fragments of altered host units (Figure 2.13(a)) overlying a discordant to semiconcordant zone of sulfides and quartz veins and veinlets (*stockwork*) (Figures 2.13(a) and (b)) and disseminated sulfides. VMS are typically associated with large areas of hydrothermal alteration defining the extent of the zone of circulation of hydrothermal fluids. Sulfide lenses are generally formed along specific stratigraphic levels corresponding to a volcanic hiatus sometimes occupied by very thin layered tuffs or sediments that may include a distal hydrothermal component ("exhalite"; Figures 2.13(c) and (d)), defining significant stratigraphic markers. The VMS-bearing volcanic sequences are also commonly associated with volcaniclastic units containing fragments of massive sulfides (Figure 2.13(e)), also forming marker levels that may serve as exploration vectors. The metals in VMS result from the leaching of rocks of the host volcanic sequence whose composition strongly influences the metallic budget of deposits (Franklin et al. 1981). However, a magmatic contribution in volatile elements and metals is suggested in many cases, particularly in the case of gold-enriched deposits. A magmatic contribution is also invoked in the deposits enriched in elements associated with felsic magmas (e.g. Bi, In and Sn) or in elements of the "epithermal suite" (As, Sb, Hg et Ag), as well as in the case of deposits formed in the context of volcanic arc (see Mercier-Langevin et al. (2015) and the references therein).

The Archean, which represents about one-third of the history of the Earth, was a prolific period for the formation of VMS deposits, notably at the Neoarchean, and particularly in the Abitibi greenstone belt. The Archean VMS represent a major source of base, precious and trace metals, which present an average concentration in metals that is higher than the VMS deposits of other geological periods (Mercier-Langevin et al. 2014a).

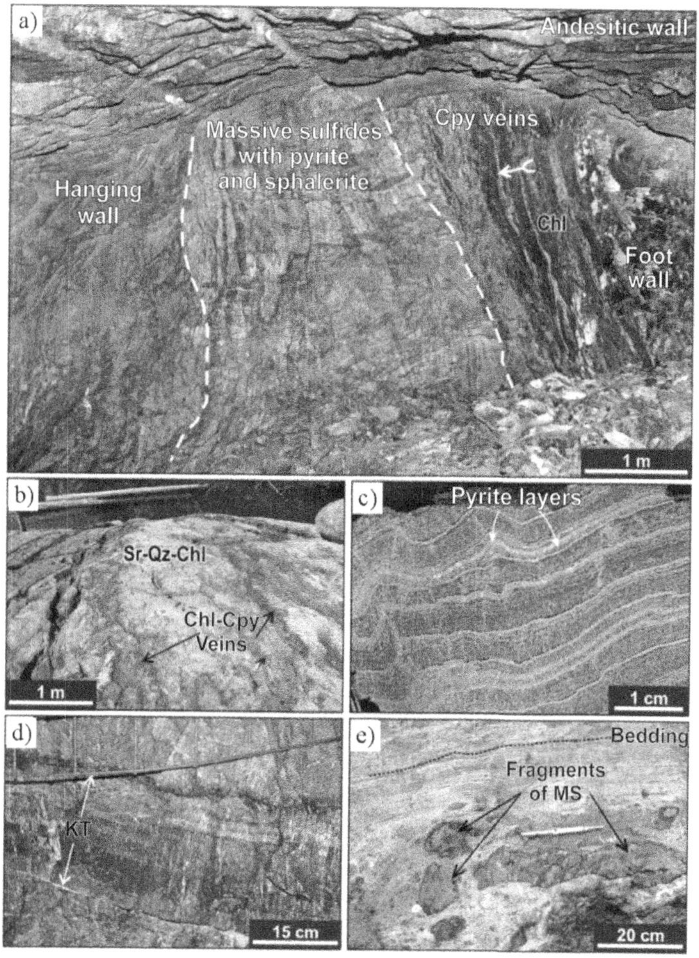

Figure 2.13. *Examples of VMS-type Cu-Zn deposits and key features*

COMMENT ON FIGURE 2.13.– *(a) Massive sulfides lens with Zn (pyrite and sphalerite) topping a* stockwork *of chalcopyrite veins (Cu zone) in the chloritized foot wall. Sulfides layering represents the regional foliation. Chalcopyrite veins are transposed in this intense foliation, and the lensis tipped over and even slightly reversed, Bracemac-McLeod mine, Matagami mine; (b) chlorite and chalcopyrite veins intersecting a typical sericitized rhyolite of the foot wall of VMS, Inmont mine, Noranda camp; (c) "exhalite", Noranda camp; (d) layered level of thin siliceous tuff known as "key tuffite" (KT) in the Matagami mining camp. This "exhalite" is a stratigraphic marker at the top of Watson rhyolite above which there are several VMS of the camp, Perseverance mine; (e) layered volcaniclastic deposits with block tuff level with fragments of massive sulfides (MS), West Horne, Horne mine, Noranda camp. Chl = chlorite, Cpy = chalcopyrite, Qz = quartz, Sr = sericite.*

2.4.2.1. *VMS deposits of the Abitibi greenstone belt*

Over one hundred major VMS deposits are found in the Abitibi greenstone belt (Figure 2.14; section A2.3). These add up to nearly 760 million tons of ore (Figures 2.15 and 2.16). Two of the Abitibi VMS deposits can be considered as super-giants (>150 Mt: Horne and Horne 5 with ~146 Mt in total, and Kidd Creek with ~170 Mt), two can be seen as very large (50–100 Mt: LaRonde Penna with ~83 Mt and Selbaie with ~57 Mt, though the latter has hybrid characteristics as discussed below), while two can be considered as large (25–50 Mt: Mattagami Lake and Bousquet 1-LaRonde Zone 5, both with ~26 Mt) (section A2.3). Moreover, nine of the VMS of the Abitibi greenstone belt are rich in gold. Out of these nine deposits, seven are in the Blake River Group, of which four represent world-class gold deposits (>100 t or 3.1 Moz Au; see section 2.4.1). The Horne complex (Horne and Horne 5, ~467 t Au) and LaRonde Penna (~249 t Au) are the two most important gold-rich VMS in the world (Mercier-Langevin et al. 2020). Figure 2.16(a) shows that the majority of VMS have a tonnage varying between 0.1 and 10 Mt, and that the 10 largest ore deposits contain nearly 75% cumulative tonnage, illustrating the enormous impact of the very large ore deposits on the metallic budget.

The VMS deposits of the Abitibi greenstone belt form, in several cases, clusters of deposits associated with a specific magmatic center, or with a particular area, defining a certain number of mining camps (Figures 2.16(b) and (c)). Noranda, Matagami, Doyon-Bousquet-LaRonde, Amos-Barraute, Timmins and Val-d'Or camps (Figure 2.14) are the most important in terms of the number of deposits and total tonnage (Figures 2.16(b) and (c)). Except for a few deposits, the VMS of the Abitibi have been strongly deformed and metamorphosed, complicating their geometry (Figure 2.13(a)), though the detailed mapping done at most of these deposits has allowed the reconstruction of their original geometry.

Since the VMS deposits result in a large part from similar genetic processes, the deposits are reviewed here mainly based on their age or the volcanic assemblage to which they belong. VMS deposits are found in the seven volcanic assemblages of the Abitibi belt (Goutier et al. 2009) (Figure 2.14), though the abundance and the total tonnage vary widely from one assemblage to the other (Figures 2.15(a) and (b)), independently from the extent of these assemblages (Figure 2.15(c)). Nearly half of the total VMS tonnage can be found in the most recent volcanic assemblage, that is, in the 2,704–2,695 Ma Blake River assemblage. Some of the main characteristics of the VMS-type deposits formed at certain specific eras during the volcanic construction of the Abitibi greenstone belt are presented below.

2.4.2.2. *Pre-Blake River VMS deposits (>2,704 Ma)*

The Chrissie-Fecteau assemblage (~2,795–2,750 Ma) contains only very few VMS-type mineralized zones. The main showing (Lac-des-Vents), located at about 50 km south-east of Chibougamau, has only several dozen thousand tons and is rather cupriferous, associated with mafic to felsic volcanic rocks and pyroclastic felsic units (Leclerc et al. 2012). However, numerous barren sulfide zones mineralized with pyrite and pyrrhotite, which can, in certain cases, reach several million tons, are also present (Leclerc et al. 2012).

The Gemini-Turgeon area, included here in the Casa Berardi camp, located in the north of the Abitibi (Figure 2.14), contains several showings and Zn-rich polymetallic VMS deposits hosted in the Pacaud volcanic assemblage (2,750–2,735 Ma). An extensive Fe carbonate alteration halo is associated with Gemini showing, as well as a discordant proximal chlorite and sericite alteration. These alteration assemblages were developed in volcaniclastic felsic units of calc-alkaline affinity dated at 2,736 ± 1 Ma (Davis et al. 2005).

The third volcanic period associated with VMS (Deloro: 2,734–2,724 Ma) is more fertile with nearly 40 major deposits, totaling ~148 Mt of ore, or about one-fifth of the total VMS tonnage of the Abitibi (Figure 2.15). The mining camps of Matagami, Normétal, Joutel, Selbaie and Chibougamau-Chapais are part of this period, in addition to several deposits of the Swayze and Desmaraisville areas (Figures 2.14 and 2.16). They are characterized by a wide variety of mineralization: Zn-Cu (Matagami and Normétal), Cu-Zn (Joutel and Chibougamau-Chapais), Zn-Cu-Au-Ag (Chibougamau-Chapais) and possibly a hybrid polymetallic type (Selbaie). The deposits in the Matagami camp are particularly rich in Zn and poor in gold, whereas those of Chibougamau are relatively enriched in gold. Several of these deposits are associated with rocks dated between 2,729 and 2,724 Ma (Mortensen 1993; Ross et al. 2014, 2020).

The deposits of the Matagami camp are characterized by proximal and discordant chlorite and talc alteration zones associated with discordant to stratiform massive sulfide lenses (Figure 2.13(a)) containing magnetite-rich portions (Piché et al. 1993). The VMS lenses at Matagami were largely formed along a sedimentary and/or exhalative marker level known as "key tuffite", located between a tholeiitic rhyolite (footwall; Figure 2.13(d)) and an andesite with dacite and rhyolite locally (hanging wall) (Debreil et al. 2018). There are also, though to a lesser extent, mineralized levels higher up in the stratigraphy (e.g. Bracemac-MacLeod deposit) (Debreil et al. 2018). The Normétal deposit is associated with a regional carbonate-chloritoid alteration, a proximal sericite alteration and a discordant black chlorite alteration. The host sequence is composed of tuffs and mafic to felsic flows that overlie intermediate volcanic rocks of tholeiitic to transitional affinity (Lafrance et al. 2000).

The Joutel camp deposits are associated with distal discordant sericite alteration and proximal chlorite and talc alteration zones, mainly developed in volcanic and volcaniclastic units of felsic composition and transitional affinity (Legault et al. 2002). The Selbaie polymetallic deposit is characterized by Ag-rich sphalerite, chalcopyrite and galena veins associated with large lenses of massive pyrite surrounded by an external halo of sericite alteration and zones of proximal chlorite-carbonate and quartz alteration. These zones are mainly developed in felsic volcaniclastic rocks of transitional calc-alkaline affinity (Larson and Hutchinson 1993).

The Selbaie deposit is considered by some as being of a hybrid nature between VMS and epithermal deposit (Larson and Hutchinson 1993). The VMS deposits of the Selbaie camp are particularly rich in Ag (e.g. the B26 deposit) (Fayard et al. 2020a, 2020b). The several VMS deposits of the Chibougamau mining camp are hosted in rhyolites of tholeiitic affinity and span over a wide range of mineralization styles between unique and tabular lenses (Lemoine) and mineralized zones with complex geometry (Lac Scott). The Lemoine auriferous VMS is linked to significant halos of sericite-carbonate-quartz and sericite-chlorite-chloritoid alteration, and to proximal discordant chlorite ± carbonates alteration (Mercier-Langevin et al. 2014b).

The Stoughton-Roquemaure assemblage (2,723–2,720 Ma), one of the most important in terms of volume or geographical extent (Figures 2.2 and 2.14), is underrepresented in terms of number of VMS (Figure 2.15), with the presence of only five modest deposits distributed in various sectors. The Estrades deposit, and several deposits along the same level (e.g. the Caribou deposit), form thin mylonitized Zn-Au-Cu-Ag lenses associated with a strong proximal sericite-quartz alteration and developed in a bimodal assembly of tholeiitic affinity (Welch 1995) dated at 2,720 ± 3 Ma (David et al. 2007). The Coniagas mine, in the Desmaraisville area, exploited a small ore deposit associated with a thin sequence of mainly tholeiitic mafic composition (flows and volcaniclastic units), with dominant chlorite alteration (Doucet et al. 1998).

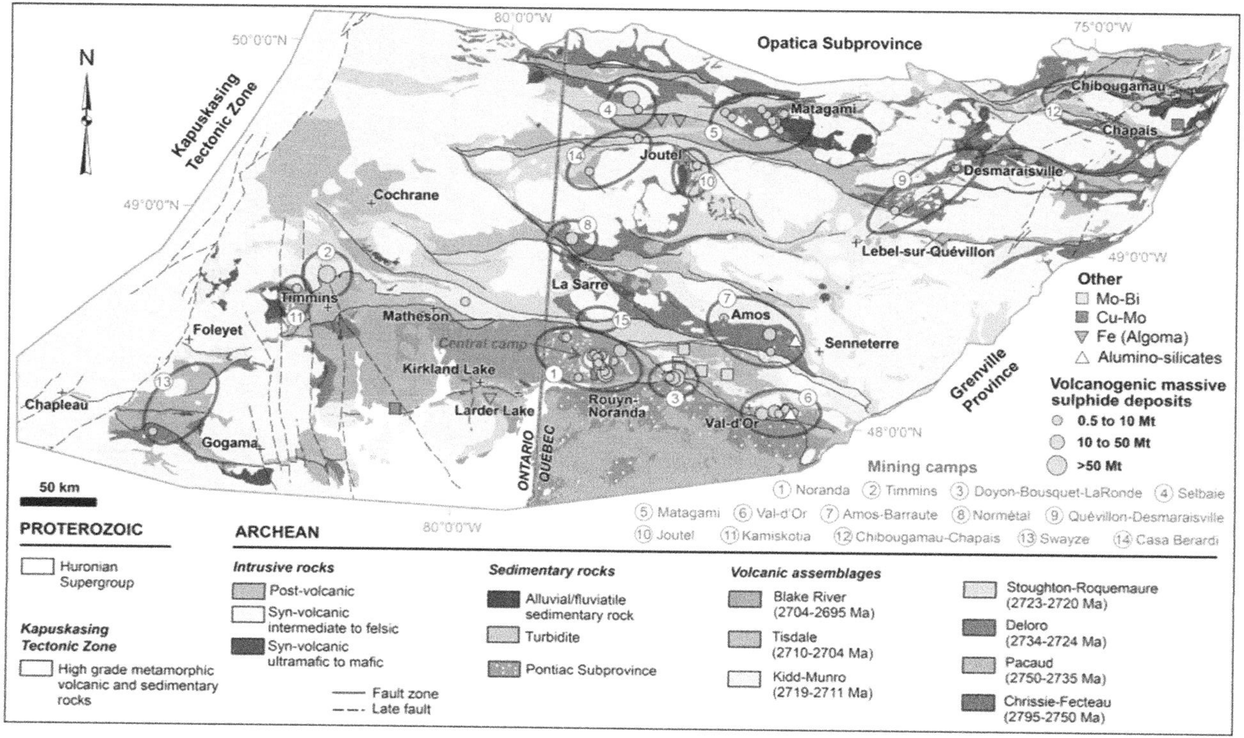

Figure 2.14. *Geological map of the Abitibi greenstone belt showing the distribution and the location of the volcanogenic massive sulfide camps and deposits with a tonnage of more than 0.5 million tons. Modified from Dubé and Mercier-Langevin (2020b) and deposit data from Krushnisky et al. (forthcoming)*

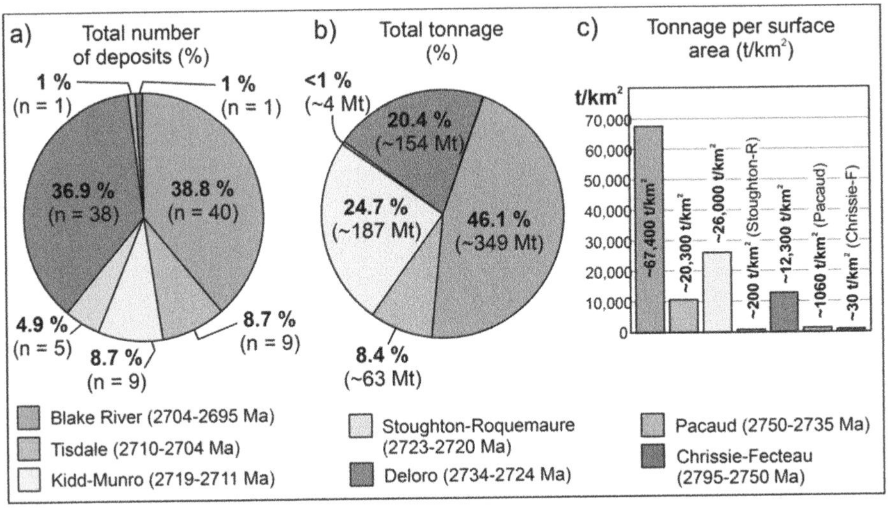

Figure 2.15. *Distribution of VMS deposits associated with the Abitibi volcanic assemblages*

COMMENT ON FIGURE 2.15.– *(a) Total number of deposits and their relative abundance for each of the seven main volcanic assemblages of the Abitibi greenstone belt; (b) relative total tonnage for each volcanic assemblage; (c) total ore tonnage per km² for each volcanic assemblage. Data from Krushnisky et al. (forthcoming).*

The Kidd-Munro assemblage (2,719–2,711 Ma) hosts the super-giant Kidd Creek ore deposit in the Timmins camp (Figures 2.15 and 2.16). This ore deposit is hosted in a folded sequence of ultramafic volcanic rocks overlain by massive to brecciated tholeiitic rhyolites (mineralized level) and porphyritic massive rhyolites and basalts (Bleeker 1999). The lenses are associated with a stratiform distal silicification, to a stratiform proximal sericitisation overprinted by a strong discordant chloritization and by a local carbonate alteration in the immediate footwall (Koopman et al. 1999). This volcanic episode also hosts the Potter deposit east of Timmins in Ontario and the Langlois deposit near Lebel-sur-Quévillon in Quebec. The Potter ore deposit, particularly rich in Co, is hosted in a sequence of tholeiitic mafic volcaniclastic rocks and komatiitic volcanic rocks intercalated with argilite units (Préfontaine et al. 2008). Chlorite is the main alteration mineral. The Langlois ore deposit is composed of several strongly deformed and transposed lenses of Zn-rich sulfides hosted in mafic to felsic volcaniclastic units of tholeiitic to transitional affinity, silicified and associated with transposed halos of black chlorite and sericite alteration (Théberge et al. 1999). The Kidd-Munro assemblage also comprises some Zn-Cu VMS deposits of the Amos-Barraute mining camp (Figures 2.14 and 2.16), which is characterized by a complex geology and the intercalation of units of different ages.

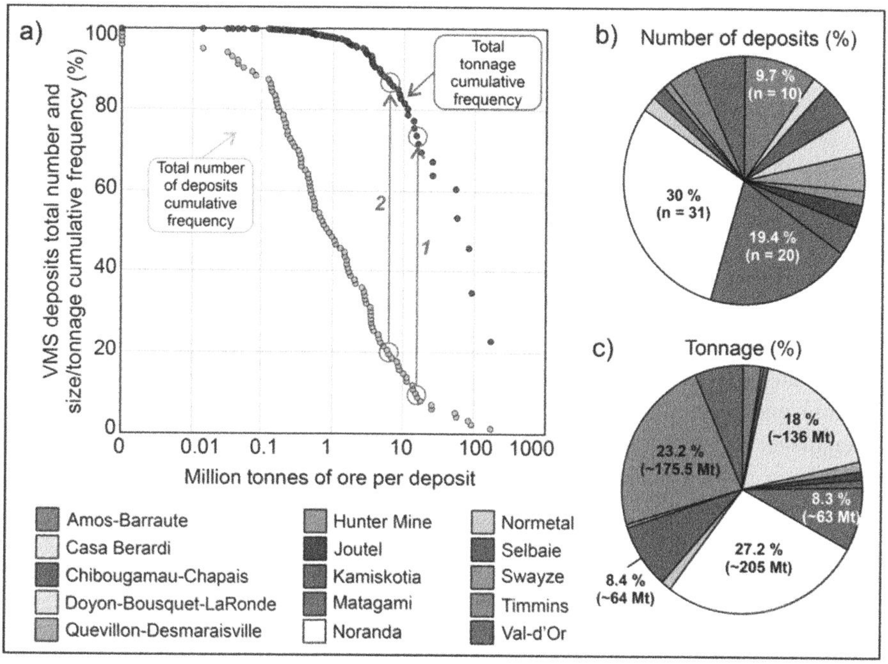

Figure 2.16. *Data on the number of VMS deposits and total tonnage per district or camp*

COMMENT ON FIGURE 2.16.– *(a) Graphic representation of the cumulative frequency of the ore deposits and their total tonnage. Line 1 shows that the 10 largest ore deposits (>15 Mt) contain about 74% of the total VMS tonnage of the Abitibi, while line 2 shows that the 20 largest ore deposits contain nearly 87% of the VMS ore tonnage of the Abitibi; (b) number of VMS ore deposits and their relative abundance per mining camp; (c) relative total VMS tonnage per mining camp. Data from Krushnisky et al. (forthcoming).*

The Tisdale assemblage (2,710–2,704 Ma) contains some of the deposits of the Amos-Barraute camp and those of Val-d'Or mining camp (Figures 2.14 and 2.15).

The Amos-Barraute district contains the Abcourt-Barvue mine, an Ag-rich Zn ore deposit, of Tisdale age that was one of the first open-pit mines in Abitibi. The Abcourt-Barvue ore deposit has a tabular form with disseminated to massive replacement-style mineralization of intermediate polymictic tuffs, of tholeiitic to transitional affinity, dated at 2,706 ± 3 Ma (David et al. 2007). These tuffs were affected by a carbonate and chlorite ± sericite alteration (Labbé and Couture 1999). The Zn-Cu deposits of the Val-d'Or district, one of the main VMS mining camps of the Abitibi (Figures 2.14 and 2.16), form mounds associated with a distal sericite alteration and proximal discordant black chlorite feeder zones. These zones grade into green chlorite and carbonate local semi-concordant zones (e.g. the Louvicourt Cu-rich deposit; Moorhead et al. 2000) developed in intermediate to felsic volcaniclastic units of transitional to calc-alkaline affinity.

2.4.2.3. *2,704–2,695 Ma VMS deposits (Blake River assemblage)*

The Blake River assemblage (2,704–2,695 Ma) contains about 40 VMS ore deposits that represent nearly half of the total VMS tonnage of the Abitibi (Figures 2.15(a) and (b)). They are also overall richer in metals than those of the other assemblages (Figure 2.15(c)). The VMS ore deposits associated with this volcanic assemblage are concentrated in the Noranda, Doyon-Bousquet-LaRonde and Kamiskotia mining camps (Figure 2.14), where VMS lenses are present at several stratigraphic levels.

The VMS ore deposits of the Noranda mining camp, especially those of the "central camp" (Figure 2.14), define the classical Noranda VMS model that is characterized by lenses or piles of sulfides formed in the proximity of tholeiitic to transitional effusive centers, above a discordant and limited zone of sulfide veinlets (Franklin et al. 1981; Gibson and Galley 2007). These veinlets are associated with thin discordant zones of iron-bearing chlorite surrounded by magnesium chlorite and sericite halos (Figure 2.13(b)). The alteration is developed in the proximity of felsic domes and along synvolcanic faults or fractures in mafic to intermediate horizons dominated by flow facies (Gibson 1990). The ore deposits located outside the central camp (e.g. the Horne and Bouchard-Hébert ore deposit) (Riopel et al. 1995; Gibson et al. 2000) have a tabular form and result, at least partly, from the replacement of felsic volcaniclastic units (Monecke et al. 2008), of transitional to slightly calc-alkaline affinity (Krushnisky et al. 2020). These deposits are associated with very extensive, concordant to discordant zones of mainly sericite and quartz alteration (distal envelope) and to proximal chlorite ± carbonate zones of variable intensity.

The Doyon-Bousquet-LaRonde district is characterized by gold-rich large tonnage VMS deposits (see section 2.4.1.1) comprising strongly deformed lenses of disseminated, semi-massive and massive sulfides, and networks of veins and veinlets, locally rich in base metals (Mercier-Langevin et al. 2007a). These lenses are associated with important semi-concordant to discordant alteration zones that were metamorphosed to upper greenschist facies, with garnet-biotite (Figure 2.10(c)) and sericite assemblages and/or to proximal aluminous zones (alumino-silicates, quartz, sericite, sulfides) (Figure 2.10d), mainly developed in intermediate to felsic volcaniclastic units of transitional to calc-alkaline affinity (Mercier-Langevin et al. 2007b, 2007c; Dubé et al. 2014; Yergeau et al. 2015).

The largest ore deposits of the Noranda and Doyon-Bousquet-LaRonde mining camps were partly formed by replacement of altered host felsic volcaniclastic units over large distances (semi-concordant zones). This clearly distinguishes them from the "Noranda" model rather characterized by smaller lenses and by discordant alteration zones of limited lateral extent.

2.4.2.4. *VMS deposits – overview*

All the volcanic assemblages of the Abitibi greenstone belt are associated with VMS ore deposits and prospects, indicating a favorable period for this type of mineralization extending over more than 60 million years.

Generally speaking, the oldest deposits are slightly richer in Zn and Ag than the younger ore deposits, which are clearly enriched in gold, which suggests a link between the tectonic evolution of the belt and the nature of the associated volcanogenic deposits, as discussed in Mercier-Langevin et al. (2015).

Although the fertility of Archean tholeiitic rhyolites has been established some time ago, several VMS ore deposits of the Abitibi greenstone belt are directly associated with or hosted in intermediate to felsic volcanic units of transitional to calc-alkaline affinity according to the recent research, including some of the most important gold-rich VMS. Moreover, the ore deposits hosted in volcaniclastic sequences tend to have a larger size and to be associated with rather stratiform alteration zones largely exceeding the extent of mineralized zones.

The spatial distribution of VMS is uneven at the scale of the Abitibi and often shows a concentration in certain regions or camps (Figure 2.14). The potential of discoveries in the "mature" sectors remains high, particularly in depth where the alteration vectors and the geological context become first-order exploration guides.

Despite a favorable geology, other regions are under-represented in terms of discovered VMS (e.g. the Chibougamau-Chapais and Amos-Barraute camps, or the Stoughton-Roquemaure assemblage) (Figure 2.14) or host a single important ore deposit. Certain deposits could probably not be preserved because of the erosion level and the important deformation in several sectors, this deformation also being possibly responsible for the partial destruction and/or burial of deposits, complicating the exploration works.

2.4.3. *Ni-Cu-(PGE)-Cr-V deposits associated with mafic and ultramafic rocks*

The mafic and ultramafic intrusive rocks, as well as the ultramafic extrusive rocks (e.g. komatiites), are the main host rocks of orthomagmatic Ni-Cu and platinum group elements (PGE), Cr, and Fe-Ti-V mineralization. These rocks, generated from mantle magmas, are common in the Abitibi greenstone belt and are present in most of the volcanic assemblages (Figure 2.17). However, the main orthomagmatic deposits were formed during a relatively limited time interval, between ~2,735 and 2,692 Ma (Figure 2.5) (Houlé et al. 2020). At the Abitibi scale, the deposits can be associated with three large types of mafic and ultramafic systems: (1) komatiitic (Ni-Cu-PGE) intrusions and lavas, (2) gabbroic (Ni-Cu-PGE) intrusions and (3) differentiated layered intrusions (Fe-Ti-V, PGE and Cr) (Figures 2.18(a) and (b); section A2.4).

2.4.3.1. *Ni-Cu-(PGE) deposits of komatiitic rocks*

The Ni-Cu-PGE deposits associated with komatiitic rocks in the Archean and Proterozoic greenstone belts represent about one-fifth of the global Ni resources coming from magmatic deposits of Ni-Cu-(PGE). The main examples are found in the Archean greenstone belts of the Eastern Goldfields in Western Australia and in the Proterozoic terrains of the Circum-Superior belt in Canada (Thompson and Cape Smith-Raglan; Naldrett 2004). The komatiites are generally defined as being extrusive rocks of ultramafic composition, with a MgO concentration above 18 wt.%. The subvolcanic and intrusive equivalents can also constitute a significant portion of komatiitic successions (Figure 2.18(a)). Though the presence of spinifex textures (zones of platy or acicular olivine or pyroxene crystals) is generally a reliable criterion for the identification of a komatiitic composition, they are not always present. Houlé and Lesher (2011) and Houlé et al. (2017) present an exhaustive summary of komatiites of the Abitibi greenstone belt and associated deposits.

Many magmatic Ni-Cu-(PGE) deposits associated with komatiitic rocks have been discovered in the Abitibi greenstone belt (section A2.4). The deposits of this type represent the majority of nickel-bearing resources of the Abitibi. Two main styles of deposits are distinguished: "Kambalda" style (style I) and the "Mount Keith" style (style II). Style I deposits are found at the basis of komatiitic units and form stratiform lenses of massive sulfides (mainly pyrrhotite, pentlandite and chalcopyrite), matrix (*net-textured*) lenses or lenses disseminated in olivine mesocumulates overlain by a thin spinifex zone (Figure 2.18(c)). The mineralization in this style of deposit is rich in metals with lenses containing 1–3 Mt of ore at a Ni grade varying between 2 and 4% (e.g. the Marbridge ore deposit in Quebec). Style II consists of large stratabound accumulations of disseminated sulfides in olivine accumulates. Mineralization of this style generally forms very large mineralized bodies of low-grade material (e.g. >250 Mt at a <1% Ni, such as the Dumont deposit in Quebec).

The nature of the host rocks near the surface is one of the essential factors that control the emplacement of deposits associated with komatiites, irrespective of their style (Houlé et al. 2008). In general, a coherent and consolidated host (e.g. massive/pillowed volcanic rocks) forces lava to reach the surface and form flows, whereas a fragmentary and unconsolidated host (e.g. volcaniclastic or sedimentary rocks) facilitates the emplacement of the lava in the form of intrusions in the subsurface (Figure 2.18(a)).

The majority of Ni-Cu-(PGE) deposits associated with komatiites in the Abitibi are of style I (*Kambalda*) and several have been exploited (Houlé and Lesher 2011; Houlé et al. 2017). The Alexo deposit in Ontario (Figure 2.17), which is an excellent example of this mineralization style, contains lenses located in topographic paleo-depressions at the basis or near the basis of the host komatiitic unit (Figures 2.18(c) and 2.19(a)). The mineralized lenses are zoned with massive sulfides in the lower part (Figure 2.19(b)), a zone with net-textured sulfides (Figure 2.19(c)) in the central part, and disseminated and globular sulfides in the upper part. In most cases, paleo-depressions result from the thermomechanical erosion of the substrate or of the host rock by komatiitic magmas during their passage (flow, sill or dyke) (Houlé and Lesher 2011; Houlé et al. 2012). During this process, the magmas, which are undersaturated in S, locally incorporate S from the underlying substrate (e.g. graphitic and sulfide-bearing argilite or sulfide-bearing facies of iron formation), reach saturation in S and form a metal-rich and dense immiscible sulfide-bearing liquid at the origin of the Ni-Cu-(PGE) mineralization that accumulates in the lowermost part of komatiitic units (Houlé and Lesher 2011).

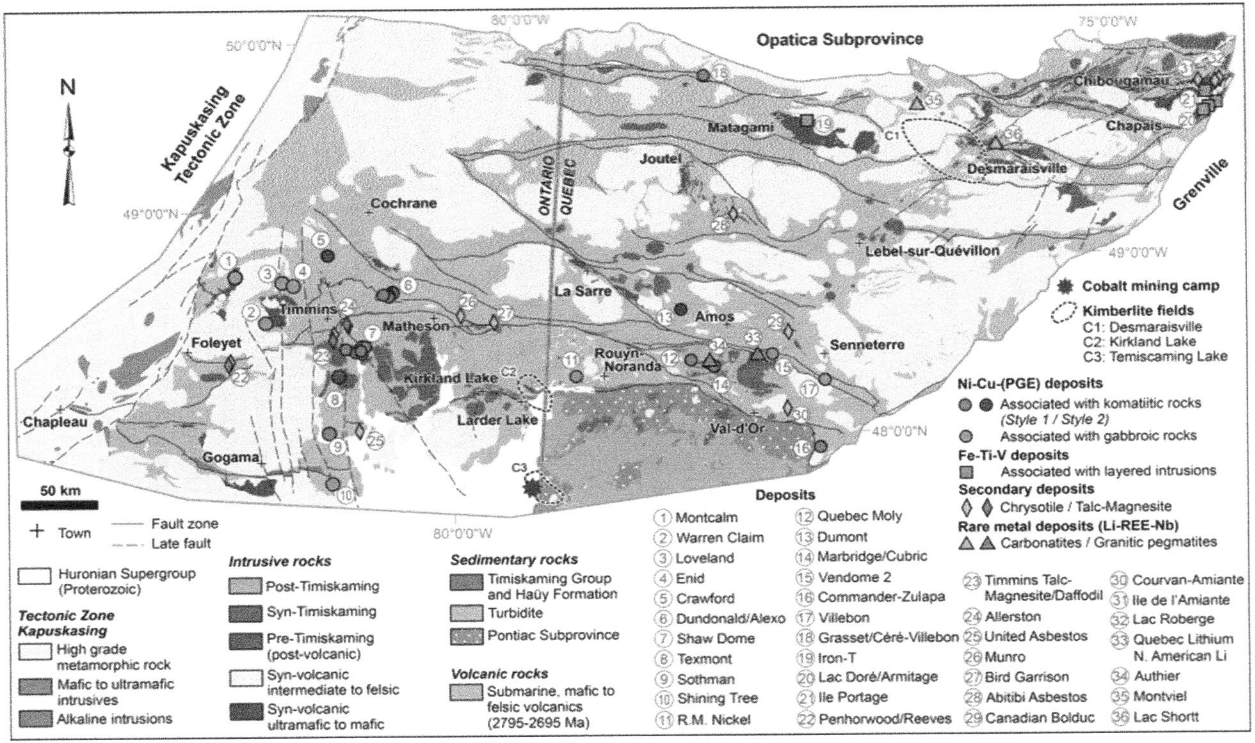

Figure 2.17. *Geological map of the Abitibi greenstone belt showing the distribution and location of Ni-Cu-(PGE)-Cr-V and affiliated deposits, rare metal and lithium deposits, and kimberlite fields. Geology from Dubé and Mercier-Langevin (2020), and deposit data from Houlé et al. (forthcoming) and Sappin et al. (forthcoming)*

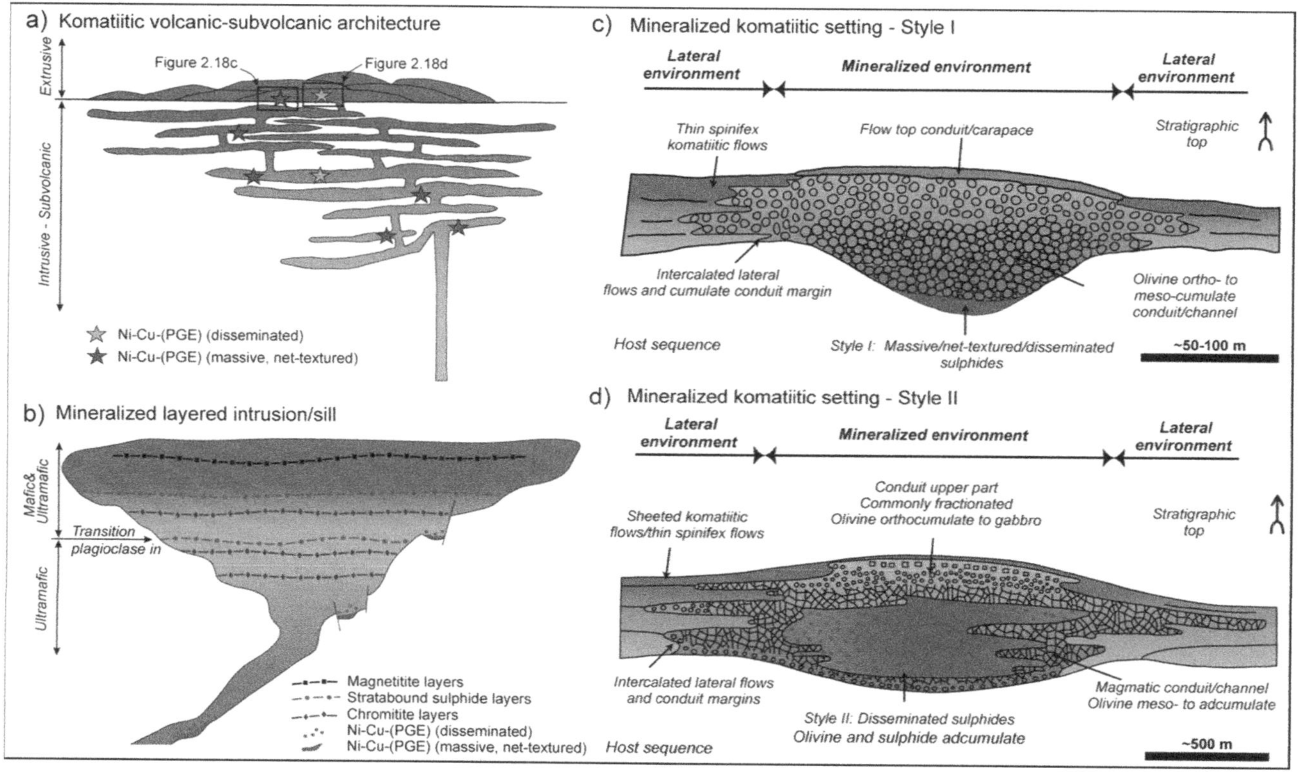

Figure 2.18. *Schematic diagrams showing the distribution of Ni-Cu-(PGE), Cr, and Fe-Ti-V mineralization in mafic-ultramafic systems. Diagrams modified from (a) Houlé et al. (2008); (b) Hoaston et al. (2006); (c and d) Hill (2001)*

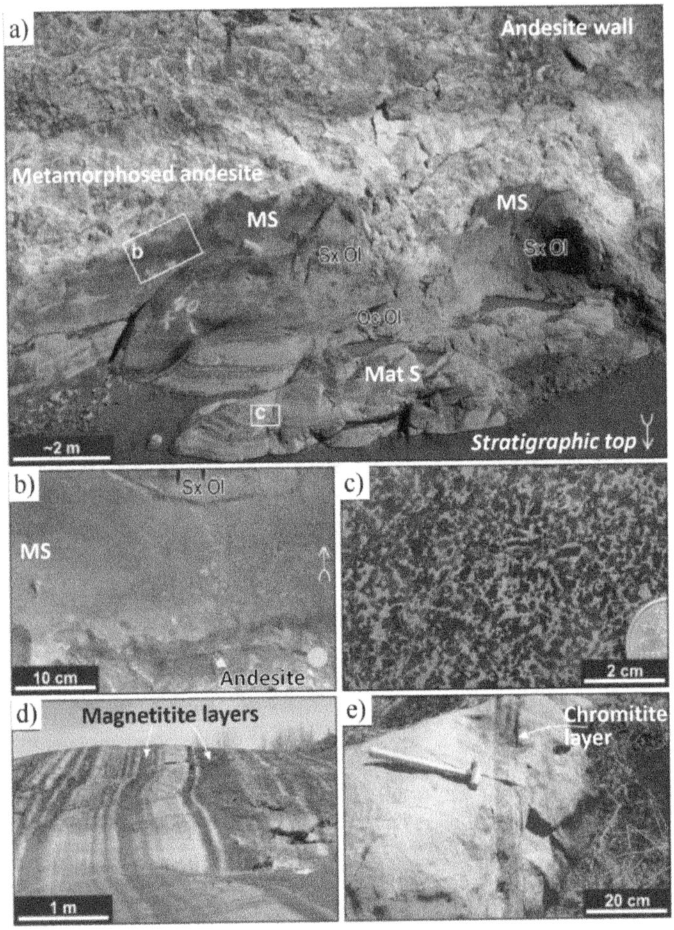

Figure 2.19. *Examples of Ni-Cu-PGE, Fe-V and Cr mineralization*

COMMENT ON FIGURE 2.19.– *(a) Accumulation of Ni-Cu-(PGE) sulfides at the basis of an erosion channel at the Alexo mine; (b) massive sulfides at the Alexo mine; (c) matrix sulfides at the Alexo mine; (d) levels of massive to semi-massive magnetitites in the Lac Doré complex, Chibougamau; (e) chromite horizon in the dunitic unit of Lava Lake in the Munro canton (from MacRae (1969)). MS = massive sulfides, Mat S = matrix sulfides, Sx = spinifex, Oc = orthocumulate, Ol = olivine.*

Style II (Mount Keith) deposits are less common than style I deposits in the Abitibi. The Texmont mine in Ontario (Figure 2.17), though only very briefly exploited, is a typical example of this style of deposit (section A2.4). The mineralization consists of disseminated sulfides in the central portion of a thick flow of komatiite that is essentially composed of olivine mesocumulates to adcumulates overlain by a series of thin flows with spinifex zones. The Dumont sill (Figure 2.17) is another example of this style of mineralization, though the deposit associated with this intrusion is still at the exploration stage. Unlike the Texmont mine, the Dumont komatiitic unit is differentiated. It is composed of a lower dunitic and peridotitic part and an upper gabbroic part. The mineralization of disseminated sulfides is stratiform and can be found in three main intervals (Duke 1986). The content of this deposit is relatively low compared to other deposits of this type (0.26% Ni for the Dumont deposit versus 0.6% Ni for the Mount Keith deposit in Australia), but it presents a total tonnage of the same order (Barnes 2006; Houlé et al. 2017) (section A2.4). One of the particularities of the Dumont deposit is the presence of many zones enriched in heazlewoodite (Ni_3S_2) and awaruite (Ni_3Fe), alloys of Ni easily recoverable by simple magnetic separation. The potential exploitation for this style of deposit, characterized by low contents but large tonnage, may lead to a significant increase in the relative contribution of the Abitibi to the world Ni resources (Houlé et al. 2020).

2.4.3.2. *Ni-Cu-(PGE) deposits associated with gabbros*

In the Abitibi greenstone belt, the Ni-Cu-(PGE) deposits associated with gabbroic units are divided into two large classes: (1) those associated with large gabbroic complexes (e.g. the Montcalm deposit) (Figure 2.17) and (2) those associated with gabbroic intrusions (e.g. the Lorraine deposit) (Figure 2.17) or with gabbroic sills (e.g. the Temagami Copper deposit) of smaller size, varying from a few hundred meters to a few kilometers in length. Generally speaking, these deposits are linked to later phases of mafic magmatism in the evolution of the belt (younger than ~2,705 Ma), with the exception of the Temagami deposit that is slightly older (~2,732 Ma) (Houlé et al. 2020; section A2.4).

Among the many Ni-Cu-(PGE) deposits associated with intrusions or gabbroic sills, only two are from ancient production sites, namely the Lorraine mine (1964–1968) and the Temagami Copper mine (1952–1972), both being located slightly south of the sector represented in Figure 2.17. In most cases, mineralization is enriched in Cu compared to Ni with Ni/Cu ratios that are generally ≤1. Moreover, in several cases, such as the Lorraine and Temagami Copper mines, sulfide lenses are located in the proximity of the base of the intrusion or host sill and are also found in the underlying host rocks (Franklin 1966; Descarreaux 1967). The mineralization of Ni-Cu-(PGE) deposits associated with large gabbroic complexes is however different and can be found in the lower part of intrusions, but generally at several hundred meters from the lower contact of the intrusive body. The Montcalm mine (Figure 2.17) exploited this style of mineralization. The gabbroic complex of Montcalm is composed of three main zones defining the internal intrusion stratigraphy with, from base to top: a pyroxenite zone, a gabbro-gabbro anorthositic zone and an iron-bearing gabbro zone. Secondary zones with gabbroic and peridotitic megabreccias are present locally within the iron-bearing gabbro zone (Barrie et al. 1990). The sulfide-bearing mineralization forms two lenses in the gabbro and gabbro anorthositic zone and consists of massive, matrix, disseminated and brecciated sulfides (Barrie et al. 1990). The common occurrence of gabbroic complexes in the belt renders the Abitibi an excellent target for Ni-Cu-(PGE) deposits, particularly those associated with late mafic magmatic phases.

2.4.3.3. *Fe-Ti-V, PGE and Cr deposits of layered intrusions*

The Cr-PGE, PGE and Fe-Ti-V deposits are commonly associated with large layered igneous complexes of mafic to ultramafic composition such as the Bushveld igneous complex in South Africa (Cawthorn et al. 2005). Orthomagmatic mineralization in those systems forms relatively thin (a few meters at most), stratiform layers that can however be relatively continuous laterally, sometimes over several kilometers. The mineralized layers can be located in the lower part (Cr-PGE), in the median part near the transition between ultramafic and mafic components (PGE) and in the upper part (Fe-Ti-V) of these layered complexes (Cawthorn et al. 2005) (Figure 2.18(b)).

The two most important layered complexes in the Abitibi greenstone belt are located in the Matagami and Chibougamau areas (Figure 2.17). These two layered complexes are of similar age (~2,728–2,725 Ma) (Mortensen 1993) and present several similarities including their predominantly gabbroic to anorthositic composition and the presence of Fe-Ti-V mineralization. The one located in Chibougamau represents a stratiform intrusive complex whose minimal thickness is approximately 5,000 m, which can be divided into (1) a lower series, essentially composed of anorthositic and gabbroic units, (2) a layered series, which consists predominantly of gabbroic, pyroxenitic units and of magnetitites and (3) an upper series, which comprises anorthositic gabbro and granophyre units (Daigneault and Allard 1990). The main mineralization is concentrated in the layered series that consists of a succession of massive to disseminated layers of titanium-bearing magnetite and ilmenite alternating with anorthositic to pyroxenitic silicated rocks, passing through anorthositic gabbros (Figure 2.19(d)). From an economic point of view, this layered complex is distinguished from other large stratiform complexes, such as Bushveld, by the absence of an ultramafic component and orthomagmatic mineralization such as stratiform chromite and PGE mineralization.

Stratiform PGE mineralized zones are unusual in the Abitibi greenstone belt and generally have low grades. They concentrate in mafic to ultramafic sills. Compared to other parts of the Superior Province, such as the McFaulds Lake (or Ring of Fire) area located in the Oxford-Stull domain in Ontario (Houlé et al. 2020), the Abitibi contains very little stratiform chromium-bearing mineralization. There are, however, locally, several prospects in the mafic to ultramafic sills, suggesting a potential for this mineralization style in some intrusive bodies of the belt. In the latter, chromite forms thin layers of chromitite (up to 5.5 cm thick; Figure 2.19(e)) alternating with dunitic units (e.g. the Amos sill (Hébert 1982) and Lava Lake in the Munro canton (MacRae 1969)).

2.4.3.4. *Secondary deposits (chrysotile, talc and magnesite)*

Several other types of mineral deposits associated with mafic and ultramafic rocks, though less important economically, are also present in the Abitibi greenstone belt, such as chrysotile deposits (e.g. Munro mine), talc deposits (e.g. the Penhorwood mine) and magnesite deposits (e.g. the Timmins Talc-Magnesite mine; section A2.4) (Figure 2.17) (Kretschmar and Kretschmar 1986). This secondary mineralization results from the alteration of ultramafic rocks in the oceanic environment or during exhumation, by hydrothermal fluids containing CO_2 (O'Hanley 1996). The deformation of the host rock also plays an important role in the genesis of these deposits, creating an increased secondary permeability that facilitates the circulation of fluids, leading to the transformation of primary mineralogy (olivine and pyroxene) in hydrated and/or carbonated secondary minerals (Duke 1996).

2.4.4. *Fe deposits associated with Algoma-type iron formations*

Banded iron formations (BIF) are sedimentary rocks that contain over 15% Fe (James 1954; Gross 1996). These rocks, which represent about 90–95% of the mined iron ore worldwide, are a common component of the Archean volcano-sedimentary sequences. Two main types of Precambrian iron formations are recognized depending on the nature of the sedimentary basin and on the associated rocks (Gross 1996). Those of the Superior Lake type, common in Paleoproterozoic terrains, originate in the continental platforms and are composed of thick units (up to several dozen meters), continuous over several dozen kilometers, forming in certain cases Fe deposits with very large tonnage (>1,000 Mt), such as those of the Labrador Trough in Eastern Canada. The Algoma type iron formations, common at the Archean, are generally of more limited thickness and extent, and are associated with clastic volcanic and sedimentary rocks (graywacke, turbidite and pelite), formed in the vicinity of volcanic centers (Gross 1996). They form deposits with tonnage varying from 1,000 t to <100 Mt. From a genetic point of view, the Algoma-type iron formations represent hydrolithic sediments (chemically precipitated in aqueous medium), formed by effusive and volcanogenic hydrothermal processes, which are commonly (temporally and spatially) linked to VMS. Such iron formations constitute marker horizons during exploration to locate potentially mineralized zones (e.g. VMS) resulting from syn-genetic hydrothermal processes.

The Algoma-type iron formations are common in the Abitibi greenstone belt (Gross 1996). Their ages are similar to those of surrounding volcanic and sedimentary rocks, that is, >2,680 Ma. The Abitibi iron formations are mainly composed of an alternation of millimeter- to meter-thick layers of Fe-rich minerals (magnetite, hematite, pyrite, pyrrhotite, and certain carbonates and silicates) and chert. Besides Fe, some lithofacies richer in sulfides can contain significant concentrations of Zn, Pb, Cu, Sn and Au when enriched through syn-genetic processes (Hannington and Scott 1988). Most of the Abitibi iron formations underwent regional metamorphism to greenschist facies, which enhanced the quality of the ore, recrystallization having facilitated the increase in granulometry (coarsening), simplifying the processing of the ore. Deformation also led to the thickening of mineralized layers by folding and/or by structural repetition. Though the Algoma-type iron formations may in some cases present a significant economic potential, their volume, far smaller than that of Lake Superior-type Proterozoic iron formations, limits the interest for Fe in the Abitibi sector, explaining the very limited number of mined deposits (Figure 2.13).

The Adams mine, in Ontario, remains today one of the most important Fe deposits of the Abitibi. Since its opening in 1962, until its closing in 1991, the mine produced a total of 27 Mt of Fe concentrate pellets (steel mixture of Fe, SiO_2, Al_2O_3, CaO and Mg) from approximately 102 Mt of ore at an average grade of 26% Fe (section A2.5). The iron formations exploited at the Adams mine are mainly composed of irregular lenses, totaling about 1,000–1,300 m of thickness, located in the hinge of a synform fold. These levels of iron formation contain mainly magnetite and chert intercalated with massive and volcaniclastic mafic volcanic rocks (Bowins and Crocket 2011).

In Quebec, the Adam River deposits (not to be confused with the Adams mine in Ontario) and Iron Hills, located to the west of the Matagami mining camp (Figure 2.14), have mineral resources estimated at 1508 Mt at 21.3% Fe (Grandillo and Bisaillon 2015; section A2.5). The two deposits are part of an east-west oriented assemblage composed of graywacke, graphitic shale, biotite schists and mafic volcanic rocks, in which banded iron formations are intercalated (Taner and

Chemam 2015). The mineralization is composed of levels containing 10–70% vol. of massive to disseminated magnetite (and/or hematite).

2.4.5. *Miscellaneous types of deposits*

Gold, Cu-Zn-(Au-Ag-Pb), Ni-Cu-(PGE), Cr and Fe-Ti-V deposits represent a very large proportion of mineral resources of the Abitibi greenstone belt. However, other types of deposits also contribute to the global mineral budget of the belt and illustrate its significant fertility and metallogenic diversity. Some of these deposits are briefly described below.

2.4.5.1. *Cu-(Au)-Mo deposits*

Several magmatic hydrothermal Cu-(Au)-Mo ore deposits, deposits and prospects (Figure 2.14) have been interpreted as being of porphyry (magmatic-hydrothermal) type in the Rouyn-Noranda, Val-d'Or and Chibougamau areas. These are generally closely associated with felsic, mainly syn-volcanic intrusions (Labbé et al. 2006). This type of mineralization is commonly associated with intrusive breccias produced during superimposed magmatic events. They all have alteration zones that vary in intensity and size (several dozen meters to kilometers) (Labbé et al. 2006). In terms of geometry and mineralogical assemblages, they are comparable to Phanerozoic porphyry systems. The Don Rouyn ore deposit, in the Noranda mining camp, is a good example of porphyry style Archean Cu-Mo-Au mineralization (Goldie et al. 1979). This ore deposit, exploited mainly as a source of silica for the Horne smelter, produced about 5 Mt of ore at a grade of 0.15% Cu (Lacroix 1998; section A2.5). It is contained in a silica-rich (SiO_2 >68 %) tonalite-trondhjemite associated with Powell syn-volcanic pluton (2,700 ± 1 Ma) (McNicoll et al. 2014). Chalcopyrite, molybdenite, bornite and pyrite mineralization is either in disseminated (intragranular) form (Figure 2.20(a)), of stockworks (Figure 2.20(b)), or in quartz veins. The associated alteration forms chlorite-carbonate, epidote-carbonate and white mica zones (Goldie et al. 1979). Located in the north-eastern part of the Abitibi belt, the Chibougamau region represents another sector favorable to magmatic-hydrothermal mineralization. It includes 16 mines having produced over 47 Mt at 1.73% Cu and 2.08 g/t Au (Labbé et al. 2006). These ore deposits have been described in section 2.4.1.2. The Cu-(Au-Mo) and Cu-Au mineralization of the Chibougamau region are associated with tonalitic to dioritic intrusions, slightly older (~2,715 Ma) than those of the southern part (~2,700 Ma) of the Abitibi. These mineralized zones are in many places inserted in, or intersected by ductile shears, proving their pre-deform timing. This mineralization is considered to be affiliated with porphyry-type deposits (Pilote and Guha 1995).

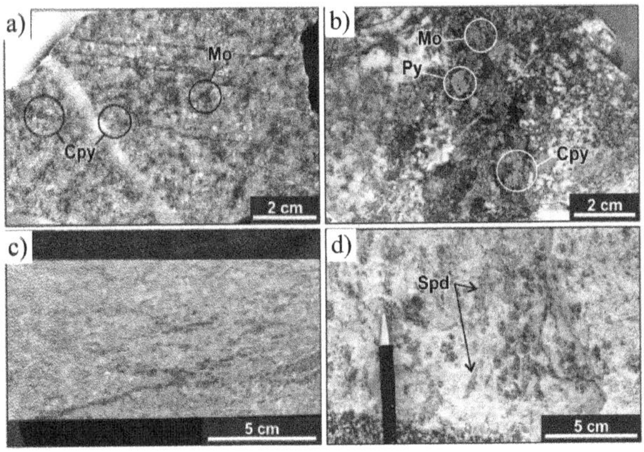

Figure 2.20. *Examples of porphyry-style mineralization, aluminosilicate-bearing alteration zones, and of Li-bearing pegmatite*

COMMENT ON FIGURE 2.20.– *(a) Tonalite-trondhjemite altered to chlorite ± epidote-albite with disseminated chalcopyrite ± molybdenite, Don Rouyn mine, Rouyn-Noranda ; (b) molybdenite, chalcopyrite and pyrite vein in chloritized tonalite-trondhjemite, Don Rouyn mine, Rouyn-Noranda; (c) dumortierite (pink) in a quartz ± feldspar and sericite matrix, Bevcon sector, Val-d'Or; (d) lithium-bearing pegmatite with large spodumene crystals, North American Lithium mine, Barraute. Cpy = chalcopyrite, Mo = molybdenite, Py = pyrite, Spd = spodumene.*

2.4.5.2. *Mo-Bi deposits*

Several Mo-Bi deposits present in the Abitibi greenstone belt (Figure 2.14) are associated with pegmatite systems or hydrothermal quartz veins hosted in late to post-tectonic intrusions (Taner et al. 1998). This type of mineralization was mainly documented in the southern part of the Abitibi, in the proximity of the Preissac-La Corne plutonic suite (Preissac, La Corne, Cadillac and Moly Hill deposits; section A2.5). In the case of the Preissac-La Corne plutonic suite, two generations of intrusions have been documented: (1) a first phase of gabbroic to granodioritic composition dated at about 2,675–2,671 Ma and (2) a second phase of monzogranite composition including pegmatites at about 2,655–2,630 Ma (Feng and Kerrich 1991; Mulja et al. 1995). The late intrusive phase associated with Mo-Bi mineralization is interpreted as being the product of partial melting of sedimentary sequences in depth (Feng and Kerrich 1992). Mo-Bi mineralization generally takes the form of molybdenite and native Bi or bismuthinite (Bi_2S_3) inside the millimeter- to meter-thick veins of quartz-muscovite-potassium feldspar with alteration envelopes varying from albite to potassium feldspar with disseminated molybdenite. The peraluminous nature and the abundance of volatile elements (e.g. fluorine) of certain phases of the Preissac-La Corne plutonic suite facilitated the development of rare-metal deposits (see section 2.4.5.4). Although the estimated reserves for Mo-Bi mineralization in this sector are about 7.2 Mt of ore at 0.2% Mo, only 19,000 t of concentrate have been produced (Taner et al. 1998) (section A2.5).

2.4.5.3. *Aluminosilicate deposits*

Rare deposits of aluminosilicates, an industrial mineral commonly used in the production of ceramics and as refractory material, are present in the Abitibi greenstone belt, including the Carpentier pyrophyllite deposit in the Amos area (≥2.4 Mt at ~60% pyrophyllite) (Assad 1990) (Figure 2.14; section A2.5). The deposit forms a semi-concordant lens of schist with pyrophyllite, quartz, chloritoid, muscovite and paragonite ± andalusite, diaspore, corundum, kyanite and garnet resulting from synvolcanic hydrothermal alteration (acid leaching) of rhyodacitic to rhyolitic volcaniclastic rocks of calc-alkaline affinity (Assad 1990) dated at ≥2,719 Ma (Deloro or Stoughton-Roquemaure; Pilote et al. 2019). A large zone of very intense synvolcanic alteration is located east of the town of Val-d'Or (Figure 2.14). It is characterized by a metamorphosed and deformed (schist) assemblage of quartz, andalusite, dumortierite, a borosilicate resulting from the late addition of B to andalusite (Taner and Martin 1993) (Figure 2.20(c)), kyanite, rutile and sericite (Taner 1993). Though this area has not been considered for the exploitation of aluminosilicates, the association with VMS deposits and copper and gold-bearing prospects makes it an exploration target for syn-volcanic and/or syn-magmatic mineralization.

2.4.5.4. *Rare metal deposits (Li-REE, -Nb-Ta-F)*

Rare metal deposits, sometimes defined as being critical or strategic due to their usefulness in various high technology applications, including Li, rare earth elements (REE), Nb, Ta, Zr, Th and Be, are associated with various types of host rocks (e.g. carbonatites, syenites, granites, pegmatites, metasomatic rocks, clays and nonconsolidated sedimentary rocks) formed in a variety of geological environments. However, most deposits are associated with carbonatites (e.g. Mountain Pass mine in the United States and Niobec mine in Canada) and pegmatites (e.g. Greenbushes mine in Western Australia). The Abitibi greenstone belt is the host of several prospects (Figure 2.17) (Ste-Croix and Doucet 2001; Boily and Gosselin 2004). These mineralized zones are scattered at the scale of the Abitibi (Figure 2.17). They are mainly associated with granitic pegmatites, carbonatite complexes and dykes, or more rarely with hyperalkaline intrusions.

In the Abitibi, the granitic pegmatites that host rare metal mineralization are rich in Li (locally with anomalous concentrations of Mo, Be, Nb and Ta) or more rarely in Nb-Ta ± U. They belong essentially to the Li, Cs and Ta (LCT) family of pegmatites defined by Cerný and Ercit (2005), and are associated with the Preissac-La Corne plutonic suite; 2,675–2,630 Ma (Feng and Kerrich 1991; Mulja et al. 1995; Ducharme et al. 1997; Pilote et al. 2019) (Figure 2.17). These pegmatites are formed from late magmatic liquids enriched in volatile elements. A series of lithium-bearing granitic pegmatites with spodumene ± lepidolite (Figure 2.20(d)) were exploited at the Quebec Lithium mine (Figure 2.17) between 1955 and 1965 for a total production of 0.9 Mt ore with a 1.4% Li_2O content (Boily et al. 1989) (section A2.5). More recently, the North American Lithium open-pit mine was put into production in the extensions of the Quebec Lithium deposit, with probable reserves estimated at 30 Mt of ore at a 1.12% Li_2O content (Stone and Selway 2009) (section A2.5).

Some rare metal prospects associated with carbonatites are present in the Abitibi greenstone belt. However, the age of most of these prospects is not known. The REE deposit associated with the Shortt Lake carbonatite complex (Figure 2.17) is currently the only known Archean deposit (2,691 Ma) (Morasse 1988; Joanisse 1994) of this type. The emplacement of this mineralized zone is due to the circulation of hydrothermal fluids released from REE-enriched carbonatitic magmas (Nadeau et al. 2014). This deposit contains up to 2.2% rare earth oxides (Quirion 1991). Finally, the hyperalkaline intrusions hosting rare earth prospects are not common in the Abitibi belt. Only a few mineralized zones, rich in REE ± Nb-Ag, have been discovered. They are associated with Neoarchean syenitic to granodioritic intrusions (Augland et al. 2016).

2.4.6. *Post-Archean deposits*

The Abitibi greenstone belt hosts several post-Archean deposits, including Co-Ag-Ni or rare metal deposits and diamond-bearing kimberlitic fields.

2.4.6.1. *Co-Ag-Ni-As vein systems*

The Cobalt mining camp (Figure 2.17), the most ancient in the Abitibi, produced 445 million ounces (~1,430 t) of Ag and significant quantities of Co and Ni since its discovery in 1903 (Marshall and Watkinson 2000). The deposits of this camp are part of a large network of subvertical and rectilinear veins with quartz, carbonates, native Ag and arsenides, located in the Cobalt Embayment, a basin of Paleoproterozoic siliciclastic sedimentary rocks (Huronian Supergroup: 2.45–2.22 Ga) forming an unconformity over the Archean bedrock of the Abitibi greenstone belt (Andrews et al. 1986a). The Co-Ag-Ni-As veins of the Cobalt mining camp are part of what is known as "five element-type deposits" (Kissin 1992). They are spatially associated with Nipissing diabase dykes (2.2 Ga; Andrews et al. 1986b), which intersect the Huronian sedimentary rocks, and are linked with the presence of alkaline, reducing and hyper-saline fluids (Kerrich et al. 1986).

2.4.6.2. *REE-Nb deposits*

Besides the Archean rare earth metal prospects, there is a Paleoproterozoic (1,894 Ma) REE and Nb deposit in the Abitibi (David et al. 2006), contained in the Montviel alkaline-carbonatitic complex (Figure 2.17). This deposit is located in the vicinity of the Waswanipi-Saguenay tectonic zone, considered as the continuation, into Archean rocks, of the Saguenay rift developed in the Grenville Province (Moorhead et al. 1999). The Montviel complex and its rare metal mineralization result from several injections of volatile element-rich magmas and fluids (Nadeau et al. 2015). The Montviel deposit has indicated mineral resources of 82.4 Mt at an average grade of 1.51% rare earth oxides (Belzile et al. 2015).

2.4.6.3. *Diamond-bearing kimberlites*

Although no diamond mine was ever put into production in the Abitibi greenstone belt, several diamond-bearing kimberlitic fields have been identified, including those of Desmaraisville (~1,104–1,100 Ma), which is associated with the tectonic zone of Waswanipi-Saguenay, and those of Kirkland Lake (~165–152 Ma) and Lake Temiscamingue (~154–125 Ma), which are located in the Lake Temiscamingue structural zone (Moorhead et al. 1999; Zurevinski et al. 2011) (Figure 2.17). The nature of the kimberlites in these fields varies, from fragmentary (volcaniclastic) to massive, and is largely comparable to other kimberlites of the Canadian Shield recently or currently exploited (e.g. Victor mine in Ontario and Renard mine in Quebec) (Moorhead et al. 1999; Kjarsgaard 2007; Zurevinski et al. 2011).

2.5. An evolving industry: technical and scientific challenges, and innovations in the mineral resources world

As illustrated in the previous sections, the Abitibi greenstone belt is characterized by the presence of many mines that were put into production during the last 120 years, including many major, or "world-class" deposits, conferring to it a rich and diversified metallogenic wealth. The Abitibi, both in Ontario and Quebec, remains to this day one of the main mining regions in Canada. The technological developments of the last few decades have greatly modified the way the industry operates, the industry being the driving force behind numerous innovations, both in exploration and underground and open-pit mining, including the responsible management of environmental impacts and the restoration of ancient mining sites. This section presents some examples of innovative and exemplary practices in the industry.

2.5.1. *Underground 4G LTE mobile network, the example of the LaRonde mining complex (Quebec)*

The LaRonde mining complex, which comprises the LaRonde Penna mine (Figure 2.8), the flagship exploitation of Agnico Eagle Mines Limited, became the deepest mine in the Americas in 2016, with a depth of over three kilometers, with resources located up to 3.4 km below the surface, the mineralized lenses still being open at depth. The challenges presented by a mining exploitation at this distance below the surface are enormous and the need for efficient and safe communication led the mine to help develop mobile communication technology. In 2018, Agnico Eagle Mines implemented a 4G LTE wireless network for all its underground operations at the LaRonde mining complex (see: www.abitibi.agnicoeagle.com). This network, more efficient and versatile than classic wireless internet access (Wi-Fi), offers better coverage and the possibility for workers to have telephone and video conversations and share data wherever they are in the mine. Consequently, there is an improvement in the workers' safety and the possibility to send and receive data in real time, offering various application possibilities from telemetry to geolocation of the workers, to remote control or automation of some vehicles and mining equipment. Though this technology is implemented at the surface, its underground deployment required research and development to adapt the LTE network to existing mining infrastructures.

2.5.2. *Use of electric vehicles in mining, the example of the Borden mine (Ontario)*

The use of fossil fuels generates particles and greenhouse gases (GHG) that are noxious for the environment and harmful to health. Consequently, a strong concentration of heavy vehicles and equipment in a closed environment, such as an underground mine, poses the risk of exposing the workers to noxious gases, in addition to being a major source of heat that must be managed. Ventilation (air quality and heat management) represents one of the main challenges in underground mines. This is the context in which the Borden mine, located near Chapleau in Ontario (Figure 2.8) and owned by Newmont Corporation, replaced the equipment using diesel fuel for a fleet of exclusively electric vehicles and equipment (Ross 2016) (see: www.newmont.com). The switch to electric equipment allowed the mine to reduce its CO_2 emissions by about 50%. This is reflected by long-term energy cost savings (Taylor and Lewis 2018). This innovation also contributes to better work conditions in terms of air quality, and to a decrease in ambient noise level. This is the reason why the Borden mine, inaugurated in the autumn of 2019, is perceived as a model of mining development (see: www.gold.org).

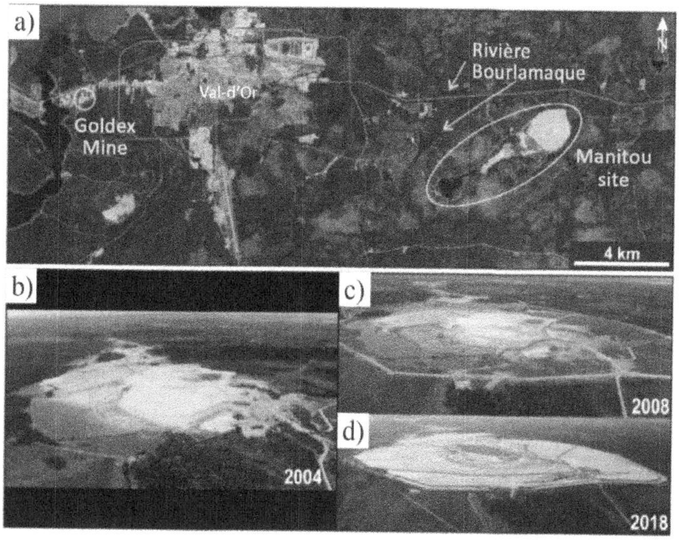

Figure 2.21. *Aerial views of the former Manitou mine site*

COMMENT ON FIGURE 2.21.– *(a) Location of Manitou project and Goldex mine in the vicinity of the city of Val-d'Or, Quebec (Google Earth image). Photos (b) to (d) (source: Quebec's Ministère des Ressources naturelles et des Forêts – MRNF) show the evolution of the restoration of the abandoned Manitou site from the moment it was taken over by the government in 2004 (b), to the start of containment works by MRNF and Agnico Eagle Mine Limited in 2008 (c), until complete covering and stabilization of the site with unreactive and uncontaminated residues from the Goldex mine, in 2018 (d).*

2.5.3. *Mine wastes used in the restoration of an abandoned mine site, the example of the Manitou project (Quebec)*

The exploitation of the Manitou Cu and Zn deposit, between 1942 and 1979, in Quebec (Figure 2.21(a)), left nearly 11 million tons of mine wastes generating acid mine drainage (Dumas 2019) (see: www.quebeccirculaire.org). For lack of adequate containment, a part of the leachate contaminated the surrounding environment with heavy metals (Figure 2.21b). Following the bankruptcy of the last owner of mining rights, Quebec's Ministère des Ressources naturelles et des Forêts (Ministry of Natural Resources and Forests) took charge of the site in 2003. With the wastes of the Manitou site being physically unstable and very reactive (oxidation of sulfides that drive a significant decrease of pH of the surface and underground waters and metal dissolution and dispersion), corrective measures were necessary. An innovative solution resulting from a collaboration between industry and the government was then proposed. The recommended approach involved the use of non-reactive wastes from the Goldex gold mine, exploited by Agnico Eagle Mines Limited, located less than 20 km west of the Manitou site (Figure 2.21(a)). Alkaline, non-leachable and cyanide-free waste of this mine proved to be a choice material for covering (Figure 2.21(c) and (d)) and neutralizing the acidity of interstitial waters of the abandoned Manitou site (Dumas 2019). For this purpose, many developments and infrastructures have been built, such as ditches, a polishing basin and the construction of dykes, parks and access roads, in order to, among others, limit the volume of water flowing through the site and slow down the discharge of contaminated material into the nearby river (Roy et al. 2018). A 23 km long pipeline was also built to carry the covering material from the Goldex mine to the site of the former Manitou mine.

This is an exemplary case of cost savings in the restoration of an abandoned mine site, while reducing the environmental footprint. The use of mine waste as covering material also reduces the use of surface materials such as sand, gravel and clay to a minimum. It also help lower the amount of space necessary to manage the waste from the Goldex mine, significantly limiting the impact of this exploitation that is located within the limits of the city of Val-d'Or. Cost sharing and the financial contribution of Agnico Eagle Mines Limited to this joint initiative, equal to the costs of building a new mine waste pad, represent savings estimated at 14.1 million dollars for the state, while contributing to the restoration and preservation of public land.

2.6. An exceptional metallogenic context: a brief discussion

There are still several knowledge gaps remaining in our understanding of the metallogeny of greenstone belts. The Abitibi greenstone belt is however one of the best places for addressing various questions that need to be answered. Full understanding of the source of metals, transport mechanisms, as well as concentration processes has not yet been achieved, which limits the efficiency of exploration models at all scales.

Remnants of Neoarchean crust are among the best mineralized parts of the Earth. When comparing the Abitibi to other Archean greenstone belts, it is clear that the surface area and the state of preservation are important characteristics to explain such mineral wealth. However, these cannot be considered the only factors responsible for the exceptional metallogeny of the Abitibi. The abundance of juvenile volcanic rocks seems to be one of the key factors in the abundance of synvolcanic mineralization (e.g. VMS) (Huston et al. 2014; Mercier-Langevin et al. 2014a). A higher thermal budget driven by higher mantle temperatures and mantle plumes, and deep structures linking the upper crust and the subcontinental lithospheric mantle in pericratonic context are also factors invoked to explain the presence of ortho-magmatic mineralization (e.g. komatitte-associated Ni-Cu-(PGE) deposits; Begg et al. 2010). The transition between mantle plume-dominated tectonics and tectonic environments with subduction followed by the accretion of multiple terranes is also considered as having possibly played a role in the metallogenic history of the belt (Kerrich et al. 2005), though the existence of modern-style plate tectonics in the Archean is still debated (Bédard 2018).

Metal-enriched source regions located in the upper mantle or lower crust have possibly been distributed heterogeneously, or have been tapped only under specific conditions during the geodynamic evolution of highly mineralized terranes (see Mercier-Langevin et al. (2012, 2014a) and the references therein). Repeated access to these fertile sources was proposed as an explanation for the superposition of various mineralization styles during the evolution of some regions (Hronsky et al. 2012). This may very well have been the case in the Abitibi greenstone belt, as indicated by the presence of various types of syngenetic and epigenetic gold and base metal deposits along specific corridors. For example, gold-rich VMS deposits seem to be more common, larger and/or richer in the regions where there are also significant quartz and carbonate vein-type (orogenic) gold deposits, as along the Larder Lake-Cadillac fault zone. The magmas associated with the formation of VMS deposits may have interacted with enriched material in the upper mantle or in the lower crust, which may have also served as reservoir during subsequent tectonic and magmatic episodes associated with the development of younger gold deposits, including orogenic gold deposits. This should however consider the significant age difference (several dozen million years) between the syn-genetic deposits and the orogenic deposits, and the formation of major tectonic episodes during this time interval (Mercier-Langevin et al. 2020). As discussed in Dubé and Mercier-Langevin (2020b), the orogenic gold resources in the Abitibi greenstone belt are concentrated along the Larder Lake-Cadillac and Destor-Porcupine fault zones, and this exceptional fertility is due to many factors. The presence of early to syn-Timiskaming alkaline/shoshonitic intrusions, the large-scale carbonatization and the presence of many world-class gold-bearing deposits indicate that these deformation zones are rooted in deep regions where gold-bearing metamorphic-hydrothermal fluids rich in H_2O-CO_2 have been concentrated. The structures may have drawn from one or several metal-enriched reservoirs of the upper mantle and/or of the lower crust. The early extension is responsible for the uplift (emergence) and the deposit of syn-Timiskaming conglomerates and the emplacement of alkaline to subalkaline intrusions. The presence of Timiskaming fluvial-alluvial subaerial conglomerates along these fault zones indicates a level of erosion favorable to the preservation of large gold-bearing deposits. These characteristics, combined with the presence of ultramafic volcanic rocks, distinguish the Larder Lake-Cadillac and Destor-Porcupine fault zones from several other faults of the belt.

Hence, even though the various geological and metallogenic processes active in the Archean are still not fully understood, it appears that certain modern geodynamic and metallogenic environments can be considered as possible analogues to Neoarchean belts, with the synchronous development of a spectrum of crustal, oceanic and continental conditions, driving the formation of various types of deposits. Such complex media, combined with the possible presence of mantle plumes in the Neoarchean, may explain the diversity of types of deposits in the terranes with exceptional metallogeny such as the Abitibi greenstone belt (see Mercier-Langevin et al. (2014a) and the references therein).

Steady supply of metals and minerals depends on the exploitation of known resources (e.g. using optimized extraction and recovery methods) and, critically, on the discovery of new deposits. To address the increasing demand for metals, new discoveries should take place both in established districts and in emerging camps and new regions, probably in remote geological environments and/or at a greater distance below the surface (Mercier-Langevin et al. 2020). To meet this need, research must continue in areas such as the Abitibi greenstone belt, where first-hand essential information is available and steadily generated in order to contribute to the improvement of exploration guides.

2.7. Conclusion

The Abitibi greenstone belt has an Archean geological history marked at first by several cycles of nearly continuous submarine volcanism and the associated intrusive magmatic activity, between ~2,795 and 2,695 Ma. Second, between 2,690 and ~2,640 Ma, several episodes of deformation, sedimentation and plutonism started in the northern part of the belt and propagated to the south in a few million years, illustrating the slight diachronism in the orogenic elements between the northern and southern parts of the belt. The nature and composition of the intrusive rocks and sedimentary sequences evolved depending on the tectonic context: (1) development of thick turbidite sequences and emplacement of pre- to syn-deformation multiphased intrusive rocks associated with the shortening (thrust and folding) of deformation episodes D_1 and D_2; (2) angular unconformity and deposition of alluvial and fluvial sediments, and emplacement of sub-alkaline to alkaline/shoshonitic, pre-D_3 intrusive bodies; (3) north-south shortening (D_3) and (4) late deformations and post-orogenic plutonism.

The metallogeny of the Abitibi greenstone belt has also evolved in response to the various episodes that marked the Archean geological history with the formation of various types of deposits, notably (1) VMS-type Cu-Zn ± Au-Ag-Pb, Ni-Cu-(PGE) and Fe associated with the volcanic activity, and Cu-(Au)-Mo, Cu-Au, Ni-Cu-(ÉGP) and Fe-Ti-V associated with syn-volcanic intrusive rocks; (2) gold and Cu-(Au)-Mo associated with pre- to syn-Timiskaming intrusions; (3) post-Timiskaming orogenic gold associated with large regional faults and with D_3 and (4) Li and Mo-Bi associated with post-orogenic plutons.

Due to this metallogenic wealth, the Abitibi is at the center of the Canadian mining and exploration activity, an important driver of the economic and social development, and also an exceptional laboratory for the study of metallogenic processes and the improvement of methods for the responsible exploration and exploitation of resources. It is therefore reasonable to expect that the Abitibi greenstone belt will continue to be an important source of learning for many years to come.

2.8. Acknowledgments

This contribution (RNCan number 20210068) is part of the Targeted Geoscientific Initiative Program of Natural Resources Canada and the Geological Survey of Canada (GSC). This chapter is the result of collaborative work of the

GSC, the Ministère des Ressources naturelles et des Forêts (MRNF, Géologie Québec) and the Ontario Geological Survey. This work was conducted in partnership with several universities and the industry. We thank many colleagues from these organizations for generously sharing their knowledge throughout the years.

We also thank many mining and exploration companies for giving us access to their properties, drill cores, data and expertise, in addition to providing us with technical, logistical and financial support in several cases. We thank F. Aucoin (GSC) for his help with some figures. Furthermore, we thank D. Fortin, M. Roy (Agnico Eagle Mines) and S. Turcotte (MRNF) for their comments on the section on challenges and innovations in the mining sector. This contribution was reviewed by N. Pinet (GSC).

2.9. References

Allard, G.O. (1976). Doré Lake complex and its importance to Chibougamau geology and metallogeny. Microfilm, Ministère des Richesses naturelles du Quebec, DP-368.

Andrews, A.J., Owsiacki, L., Kerrich, R., Strong, S. (1986a). The silver deposits at Cobalt and Gowganda, Ontario I: Geology, petrography, and whole-rock geochemistry. *Can. J. Earth Sci.*, 23(10), 1480–1506.

Andrews, A.J., Masliwec, A., Morris, W.A., Owsiacki, L., York, D. (1986b). The silver deposits at Cobalt and Gowganda, Ontario II: An experiment in age determinations employing radiometric and paleomagnetic measurements. *Can. J. Earth Sci.*, 23(10), 1507–1518.

Assad, J.R. (1990). Le gîte de pyrophyllite de Carpentier. Report, Ministère de l'Énergie et des Ressources naturelles, GM 50145.

Augland, L.E., David, J., Pilote, P., Leclerc, F., Goutier, J., Hammouche, H., Lafrance, I., Talla Takam, F., Deschênes, P.L., Guemache, M. (2016). Datations U-Pb dans les provinces de Churchill et du Supérieur effectuées au GEOTOP en 2012-2013. Report, Ministère de l'Énergie et des Ressources naturelles, RP 2015-01.

Ayer, J., Amelin, Y., Corfu, F., Kamo, S., Ketchum, J., Kwok, K., Trowell, N. (2002). Evolution of the southern Abitibi greenstone belt based on U-Pb geochronology: Autochtonous volcanic construction followed by plutonism, regional deformation and sedimentation. *Precambrian Res.*, 115(1–4), 63–95.

Barnes, S.J. (2006). Komatiite-hosted nickel sulfide deposits: Geology, geochemistry, and genesis. In *Nickel Deposits of the Yilgarn Craton: Geology, Geochemistry, and Geophysics Applied to Exploration*, Barnes, S.J. (ed.). Society of Economic Geologists, Littleton, Colorado.

Barrett, T.J., Cattalani, S., MacLean, W.H. (1991). Massive sulfide of the Noranda area, Quebec 1: The Horne mine. *Can. J. Earth Sci.*, 28(4), 465–488.

Barrie, C.T., Naldrett, A.J., Davis, D.W. (1990). Geochemical constraints on the genesis of the Montcalm gabbroic complex and Ni-Cu deposit, western Abitibi subprovince, Ontario. *Can. Miner.*, 28(3), 451–474.

Bateman, R., Ayer, J.A., Dubé, B. (2008). The Timmins-Porcupine gold camp, Ontario: Anatomy of an Archean greenstone belt and ontogeny of gold mineralization. *Econ. Geol.*, 103(6), 1285–1308.

Bécu, V., Dubé, B., Mercier-Langevin, P. (forthcoming). Gold deposits of the superior province – Database. Report, Geological Survey of Canada.

Bédard, J.H. (2018). Stagnant lids and mantle overturns: Implications for Archaean tectonics, magmagenesis, crustal growth, mantle evolution, and the start of plate tectonics. *Geosci. Front.*, 9(1), 19–49.

Begg, G.C., Hronsky, J.A.M., Arndt, N.T., Griffin, W.L., O'Reilly, S.Y., Hayward, N. (2010). Lithospheric, cratonic, and geodynamic setting of Ni-Cu-PGE sulfide deposits. *Econ. Geol.*, 105(6), 1057–1070.

Belzile, E., Marchand, R., Bouajila, A. (2015). NI 43-101 technical report – Montviel rare earth project, Quebec, Canada. Report, Géoméga Resources [Online]. Available at: https://geomega.ca/fr/montviel-ree-nb-2/ [Accessed 8 January 2021].

Bleeker, W. (1999). Structure, stratigraphy, and primary setting of the Kidd Creek volcanogenic massive sulfide deposit: A semi-quantitative reconstruction. In *The Giant Kidd Creek Volcanogenic Massive Sulfide Deposit, Western Abitibi Subprovince, Canada*, Hannington, M.D. and Barrie, C.T. (eds). Society of Economic Geologists, Littleton, Colorado.

Bleeker, W. (2015). Synorogenic gold mineralization in granite-greenstone terranes: The deep connection between extension, major faults, synorogenic clastic basins, magmatism, thrust inversion, and long-term preservation. In *Targeted Geoscience Initiative 4: Contributions to the Understanding of Precambrian Lode Gold Deposits and Implications for Exploration*, Dubé, B. and Mercier-Langevin, P. (eds). Geological Survey of Canada, Open File 7852, Ottawa, Ontario.

Boily, M. and Gosselin, C. (2004). Les principaux types de minéralisations en métaux rares (Y-Zr-Nb-Ta-Be-Li-ETR) du Quebec. Report, Ministère de l'Énergie et des Ressources naturelles, ET 2004-01.

Boily, M., Pilote, P., Raillon, H. (1989). La métallogénie des métaux de haute technologie en Abitibi-Témiscamingue. Report, Ministère de l'Énergie et des Ressources naturelles, MB 89-29.

Boily-Auclair, É., Mercier-Langevin, P., Ross, P.S., Pitre, D. (2020). Stratigraphic setting of the LZ5 and Ellison mineralized zones, LaRonde Zone 5 Project, Doyon-Bousquet-LaRonde mining camp, Abitibi, Quebec. In *Targeted Geoscience Initiative 5: Contributions to the Understanding of Canadian Gold Systems*, Mercier-Langevin, P., Lawley, C.J.M., Castonguay, S. (eds). Geological Survey of Canada, Open File 8712, Ottawa, Ontario.

Bowins, R. and Crocket, J. (2011). Tungsten in Archean banded iron formation of the former Sherman and Adams mines, Ontario; possible mineralization processes. *Can. J. Earth Sci.*, 48(1), 11–23.

Campbell, R.A. (2014). Controls on syenite-hosted gold mineralization in the Western Timmins camp. Master's Thesis, University of Western Ontario, London, Ontario.

Castonguay, S., Dubé, B., Mercier-Langevin, P., Wodicka, N. (2020). Geological setting and gold mineralization associated with the Sunday Lake and Lower Detour deformation zones, north-western Abitibi greenstone belt, Ontario and Quebec. In *Targeted Geoscience Initiative 5: Contributions to the Understanding of Canadian Gold Systems*, Mercier-Langevin, P., Lawley, C.J.M., Castonguay, S. (eds). Geological Survey of Canada, Open File 8712, Ottawa, Ontario.

Cawthorn, R.G., Barnes, S.J., Ballhaus, C., Malitch, K. (2005). Platinum group element, chromium, and vanadium deposits in mafic and ultramafic rocks. In *Economic Geologists: 100th Anniversary Volume*, Hedenquist, J.W., Thompson, J.F.H., Goldfarb, R.J., Richards, J.P. (eds). Society of Economic Geologists, Littleton, Colorado.

Cerný, P. and Ercit, T.S. (2005). The classification of granitic pegmatites revisited. *Canadian Mineral.*, 43(6), 2005–2026.

Couture, J.F., Pilote, P., Machado, N., Desrochers, J.P. (1994). Timing of gold mineralization in the Val-d'Or district, southern Abitibi belt: Evidence for two distinct mineralizing events. *Econ. Geol.*, 89(7), 1542–1551.

Daigneault, R. and Allard, G.O. (1990). Le Complexe du Lac Doré et son environnement géologique (Région de Chibougamau – sous-province de l'Abitibi). Report, Ministère de l'Énergie et des Ressources naturelles, MM 89-03.

Daigneault, R., Mueller, W.U., Chown, E.H. (2002). Oblique Archean subduction: Accretion and exhumation of an oceanic arc during dextral transpression, Southern Volcanic Zone, Abitibi Subprovince Canada. *Precambrian Res.*, 115(1–4), 261–290.

Daigneault, R., Mueller, W.U., Chown, E.H. (2004). Abitibi greenstone belt plate tectonics: The diachronous history of arc development, accretion and collision. *Developments in Precambrian Geology*, 12, 88–103.

David, J., Dion, C., Goutier, J., Roy, P., Bandyayera, D., Legault, M., Rhéaume, P. (2006). Datations U-Pb effectuées dans la Sous-province de l'Abitibi à la suite des travaux de 2004–2005. Report, Ministère de l'Énergie et des Ressources naturelles, RP 2006-04.

David, J., Davis, D.W., Dion, C., Goutier, J., Legault, M., Roy, P. (2007). Datations U-Pb effectuées dans la Sous-province de l'Abitibi en 2005-2006. Report, Ministère de l'Énergie et des Ressources naturelles.

Davis, W.J., Lacroix, S., Gariépy, C., Machado, N. (2000). Geochronology and radiogenic isotope geochemistry of plutonic rocks from the central Abitibi subprovince: Significance to the internal subdivision and plutono-tectonic evolution of the Abitibi belt. *Can. J. Earth Sci.*, 37(2–3), 117–133.

Davis, D.W., David, J., Dion, C., Goutier, J., Bandyayera, D., Rhéaume, P., Roy, P. (2005). Datations U-Pb effectuées en support aux travaux de cartographie géologique et de compilation géoscientifique du SGNO (2003-2004). Report, Ministère de l'Énergie et des Ressources naturelles, RP 2005-02.

De Souza, S., Dubé, B., McNicoll, V.J., Dupuis, C., Mercier-Langevin, P., Creaser, R.A., Kjarsgaard, I.M. (2017). Geology and hydrothermal alteration of the world-class Canadian Malartic gold deposit: Genesis of an Archean stockwork-disseminated gold deposit in the Abitibi greenstone belt, Quebec. In *Archean Base and Precious Metal Deposits, Southern Abitibi Greenstone Belt, Canada*, Monecke, T., Mercier-Langevin, P., Dubé, B. (eds). Society of Economic Geologists, Littleton, Colorado.

De Souza, S., Dubé, B., Mercier-Langevin, P., McNicoll, V., Dupuis, C., Kjarsgaard, I. (2019). Hydrothermal alteration mineralogy and geochemistry of the Archean world-class Canadian Malartic disseminated-stockwork gold deposit, southern Abitibi greenstone belt, Quebec, Canada. *Econ. Geol.*, 114(6), 1057–1094.

De Souza, S., Perrouty, S., Dubé, B., Mercier-Langevin, P., Linnen, R.L., Olivo, G.R. (2020). Metallogeny of the Neoarchean Malartic gold mining camp, Quebec, Canada. In *Geology of the World's Major Gold Deposits and Provinces*, Sillitoe, R.H., Goldfarb, R.J., Robert, F., Simmons, S.F. (eds). Society of Economic Geologists, Special Publication, Littleton, Colorado.

Debreil, J.A., Ross, P.S., Mercier-Langevin, P. (2018). The Matagami district, Abitibi greenstone belt, Canada: Volcanic controls on Archean volcanogenic massive sulfide deposits associated with voluminous felsic volcanism. *Econ. Geol.*, 113(4), 891–910. Ministère de l'Énergie et des Ressources naturelles, RP 2007-01.

Descarreaux, J. (1967). Géologie et géostatistique de la mine Lorraine. Master's thesis, Université du Quebec, Montreal.

Doucet, P., Mueller, W., Chartrand, F. (1998). Alteration and ore mineral characteristics of the Archean Coniagas massive sulfide deposit, Abitibi belt, Quebec. *Can. J. Earth Sci.*, 35(6), 620–636.

Dubé, B. and Gosselin, P. (2007). Greenstone-hosted quartz-carbonate vein deposits. In *Mineral Deposits of Canada: A Synthesis of Major Deposit Types, District Metallogeny, the Evolution of Geological Provinces and Exploration Methods*, Goodfellow, W.D. (ed.). Geological Association of Canada, Mineral Deposits Division, Special Publication, Saint-John's, Newfoundland.

Dubé, B. and Mercier-Langevin, P. (2020a). Gold in the Abitibi greenstone belt, Canada: A geological and historical overview. *Ressources Mines and Industry*, 1(2), 21–42.

Dubé, B. and Mercier-Langevin, P. (2020b). Gold deposits of the Archean Abitibi greenstone belt, Canada. In *Geology of the World's Major Gold Deposits and Provinces*, Sillitoe, R.H., Goldfarb, R.J., Robert, F., Simmons, S.F. (eds). Society of Economic Geologists, Special Publication, Littleton, Colorado.

Dubé, B., Gosselin, P., Mercier-Langevin, P., Hannington, M., Galley, A. (2007a). Gold-rich volcanogenic massive sulphide deposits. In *Mineral Deposits of Canada: A Synthesis of Major Deposit Types, District Metallogeny, the Evolution of Geological Provinces and Exploration Methods*, Goodfellow, W.D. (ed.). Geological Association of Canada, Mineral Deposits Division Special Publication, Saint-John's, Newfoundland.

Dubé, B., Mercier-Langevin, P., Hannington, M., Lafrance, B., Gosselin, G., Gosselin, P. (2007b). The LaRonde Penna world-class Au-rich volcanogenic massive sulfide deposit, Abitibi, Quebec: Mineralogy and geochemistry of alteration and implications for genesis and exploration. *Econ. Geol.*, 102(4), 633–666.

Dubé, B., Mercier-Langevin, P., Kjarsgaard, I., Hannington, M., Bécu, V., Côté, J., Moorhead, J., Legault, M., Bédard, N. (2014). The Bousquet 2-Dumagami world-class Archean Au-rich volcanogenic massive sulfide deposit, Abitibi, Quebec: Metamorphosed submarine advanced argillic alteration footprint and genesis. *Econ. Geol.*, 109(1), 121–166.

Dubé, B., Mercier-Langevin, P., Ayer, J., Pilote, J.L., Monecke, T. (2020). Gold deposits of the world-class Timmins-Porcupine camp, Abitibi greenstone belt, Canada. In *Geology of the World's Major Gold Deposits and Provinces*, Sillitoe, R.H., Goldfarb, R.J., Robert, F., Simmons, S.F. (eds). Society of Economic Geologists, Special Publication, Littleton, Colorado.

Ducharme, Y., Stevenson, R.K., Machado, N. (1997). Sm - Nd geochemistry and U - Pb geochronology of the Preissac and Lamotte leucogranites, Abitibi Subprovince. *Can. J. Earth Sci.*, 34(8), 1059–1071.

Duke, J.M. (1986). Petrology and economic geology of the Dumont sill: An Archean intrusion of komatiitic affinity in northwestern Quebec. Economic geology report 35, Geological Survey of Canada, Ottawa, Ontario.

Duke, J.M. (1996). Gîtes d'amiante dans des roches ultramafiques. In *Geology of Canadian Mineral Deposit Types*, Eckstrand, O.R., Sinclair, W.D., Thorpe, R.I. (eds). Geological Survey of Canada, Ottawa, Ontario.

Dumas, P. (2019). Restauration de l'ancien site minier Manitou : la force du partenariat entre l'État et l'entreprise privée. *Ressources Mines et Industrie*, 5(6), 30–35.

Fayard, Q., Mercier-Langevin, P., Daigneault, R., Perreault, S., Wodicka, N. (2020a). The B26 Cu-Zn-Ag-Au project, Brouillan volcanic complex, Abitibi greenstone belt, part 2: Hydrothermal alteration and mineralization. In *Targeted Geoscience Initiative 5: Contributions to the Understanding of Gold Deposit*, Mercier-Langevin, P., Lawley, C.J.M., Castonguay, S. (eds). Geological Survey of Canada, Open File 8712, Ottawa, Ontario.

Fayard, Q., Mercier-Langevin, P., Wodicka, N., Daigneault, R., Perreault, S. (2020b). The B26 Cu-Zn-Ag-Au project, Brouillan volcanic complex, Abitibi greenstone belt, part 1: Geological setting and geochronology. In *Targeted Geoscience Initiative 5: Contributions to the Understanding of Gold Deposit*, Mercier-Langevin, P., Lawley, C.J.M., Castonguay, S. (eds). Geological Survey of Canada, Open File 8712, Ottawa, Ontario.

Feick, K.E. (2016). An evaluation of the lithologies and geochemistry of the Upper Beaver deposit of the Kirkland Lake area. Master's Thesis, University of Western Ontario, London.

Feng, R. and Kerrich, R. (1991). Single zircon age constraints on the tectonic juxtaposition of the Ar-chean Abitibi greenstone belt and Pontiac subprovince, Quebec, Canada. *Geochim. Cosmochim. Acta*, 55(11), 3437–3441.

Feng, R. and Kerrich, R. (1992). Geochemical evolution of granitoids from the Archean Abitibi Southern Volcanic zone and the Pontiac Subprovince, Superior Province, Canada: Implications for tectonic history source regions. *Chem. Geol.*, 98(1–2), 23–70.

Franklin, J.M. (1966). The Pyrite zone of the Temagami mine of the Copperfields Mining Company, Temagami. Master's Thesis, Carleton University, Ottawa.

Franklin, J.M., Lydon, J.W., Sangster, D.F. (1981). Volcanic-associated massive sulfide deposits. In *Economic Geology 75th Anniversary Volume 1905-1980*, Skinner, B.J. (ed.). Society of Economic Geologists, Littleton, Colorado.

Gaboury, D. and Daigneault, R. (1999). Evolution from sea floor-related sulfide-rich to quartz vein-type gold mineralization during deep submarine volcanic construction: The Géant Dormant gold mine, Archean Abitibi belt, Canada. *Econ. Geol.*, 94(1), 3–22.

Galley, A.G. and Lafrance, B. (2014). Setting and evolution of the Archean synvolcanic Mooshla intrusive complex, Doyon-Bousquet-LaRonde mining camp, Abitibi greenstone belt: Emplacement history, petrogenesis, and implications for Au metallogenesis. *Econ. Geol.*, 109(1), 205–229.

Galley, A.G., Hannington, M.D., Jonasson, I.R. (2007). Volcanogenic massive sulphide deposits. In *Mineral Deposits of Canada: A Synthesis of Major Deposit Types, District Metallogeny, the Evolution of Geological Provinces and Exploration Methods*, Goodfellow, W.D. (ed.). Geological Association of Canada, Mineral Deposits Division, Special Publication, Saint-John's, Newfoundland.

Gibson, H.L. (1990). The mine sequence of the central Noranda volcanic complex: Geology, alteration, massive sulphide deposits and volcanological reconstruction. PhD Thesis, Carleton University, Ottawa.

Gibson, H.L. and Galley, A.G. (2007). Volcanogenic massive sulphide deposits of the Archean, Noranda district, Quebec. In *Mineral Deposits of Canada: A Synthesis of Major Deposit Types, District Metallogeny, the Evolution of Geological Provinces and Exploration Methods*, Goodfellow, W.D. (ed.). Geological Association of Canada, Mineral Deposits Division, Special Publication, Saint-John's, Newfoundland.

Gibson, H.L., Kerr, D.J., Cattalani, S. (2000). The Horne mine: Geology, history, influence on genetic models, and comparison to the Kidd Creek Mine. *Explor. Min. Geol.*, 9, 91–111.

Goldfarb, R.J., Baker, T., Dubé, B., Groves, D.I., Hart, C.J.R., Gosselin, P. (2005). Distribution, character, and genesis of gold deposits in metamorphic terranes. In *Economic Geologists: 100th Anniversary Volume*, Hedenquist, J.W., Thompson, J.F.H., Goldfarb, R.J., Richards, J.P. (eds). Society of Economic Geologists, Littleton, Colorado.

Goldie, R.J., Kotila, B., Seward, D. (1979). The Don Rouyn mine: An Archean porphyry copper deposit near Noranda. Quebec. *Econ. Geol.*, 74(7), 1680–1684.

Gosselin, G. (1998). Veines de quartz aurifères précoces à la zone Ouest de la mine Doyon, Canton de Bousquet, Preissac, Abitibi. Master's Thesis, Université du Quebec, Chicoutimi.

Goutier, J. and Melançon, M. (2010). Compilation géologique de la Sous-province de l'Abitibi (version préliminaire). Report, Ministère de l'Énergie et des Ressources naturelles, RP 2010-04.

Goutier, J., McNicoll, V., Dion, C., Lafrance, B., Legault, M., Ross, P.S., Mercier-Langevin, P., Cheng, L.Z., de Kemp, E., Ayer, J. (2009). L'impact du Plan cuivre et de l'IGC-III sur la géologie de l'Abitibi et du Groupe de Blake River. Report, Ministère de l'Énergie et des Ressources naturelles, MB 2009-09.

Goutier, J., Mercier-Langevin, P., McNicoll, V., Ayer, J. (2010). The Abitibi subprovince: Its evolution and its VMS deposits – An overview. *International Archean Symposium*, Abstracts of the Yilgarn-Superior Workshop, 37–41.

Grandillo, A. and Bisaillon, C. (2015). NI 43-101 technical report, Iron Hills, Adam River, and Orvilliers projects. Report, Ministère de l'Énergie et des Ressources naturelles, GM 69217.

Gross, G.A. (1996). Stratiform iron. In *Geology of Canadian Mineral Deposit Types*, Eckstrand, O.R., Sinclair, W.D., Thorpe, R.I. (eds). Geological Survey of Canada, Ottawa, Ontario.

Groves, D.I., Goldfarb, R.J., Robert, F., Hart, C.J.R. (2003). Gold deposits in metamorphic belts: Overview of current understanding, outstanding problems, future research, and exploration significance. *Econ. Geol.*, 98(1), 1–29.

Hannington, M.D. and Scott, S.D. (1988). Mineralogy and geochemistry of a hydrothermal silica-sulfide-sulfate spire in the caldera of Axial Seamount, Juan De Fuca Ridge. *Can. Mineral.*, 26, 603–625.

Hannington, M.D., Poulsen, K.H., Thompson, J.F.H., Sillitoe, R.H. (1999). Volcanogenic gold in the massive sulfide environment. In *Volcanic-Associated Massive Sulfide Deposits: Processes and Examples in Modern and Ancient Settings*, Barrie, C.T. and Hannington, M.D. (eds). Society of Economic Geologists Inc., Littleton, Colorado.

Hannington, M.D., de Ronde, C.E.J., Petersen, S. (2005). Sea-floor tectonics and submarine hydrothermal systems. In *Economic Geologists: 100th Anniversary Volume*, Hedenquist, J.W., Thompson, J.F.H., Goldfarb, R.J., Richards, J.P. (eds). Society of Economic Geologists, Littleton, Colorado.

Hébert, R. (1982). Géologie du complexe de filons-couches d'Amos, Quebec. Report, Ministère de l'Énergie et des Ressources naturelles, DP-786.

Hill, R.E.T. (2001). Komatiite volcanology, volcanological setting and primary geochemical properties of komatiite-associated nickel deposits. *Geochem-Explor. Env. A.*, 1, 365–381.

Hoaston, D.M., Jaireth, S., Jacques, A.L. (2006). Nickel sulfide deposits in Australia: Characteristics, resources, and potential. *Ore Geol. Rev.*, 29, 177–241.

Hodgson, C.J. (1993). Mesothermal lode-gold deposits. In *Mineral Deposits Modeling*, Kirkham, R.V., Sinclair, W.D., Thorpe, R.I., Duke, J.M. (eds). Geological Association of Canada, Saint-John's, Newfoundland.

Hodgson, C.J. and MacGeehan, P.J. (1982). A review of the geological characteristics of "gold-only" deposits in the Superior Province of the Canadian Shield. In *Geology of Canadian Gold Deposits*, Hodder, R.W. and Petruk, W. (eds). Canadian Institute of Mining and Metallurgy, Montreal.

Houlé, M.G. and Lesher, C.M. (2011). Komatiite-associated Ni-Cu-(PGE) deposits, Abitibi greenstone belt, Superior Province, Canada. In *Magmatic Ni-Cu and PGE Deposits: Geology, Geochemistry, and Genesis*, Li, C. and Ripley, E.M. (eds). Society of Economic Geologists, Littleton, Colorado.

Houlé, M.G., Gibson, H.L., Lesher, C.M., Davis, P.C., Cas, R.A.F., Beresford, S.W., Arntd, N.T. (2008). Komatiitic sills and multigenerational peperite at Dundonald Beach, Abitibi greenstone belt, Ontario: Volcanic architecture and nickel sulfide distribution. *Econ. Geol.*, 103(6), 1269–1284.

Houlé, M.G., Lesher, C.M., Davis, P.C. (2012). Thermomechanical erosion at the Alexo Mine, Abitibi greenstone belt, Ontario: Implications for the genesis of komatiite-associated Ni-Cu-(PGE) mineralization. *Mineralium Dep.*, 47(1/2), 105–128.

Houlé, M.G., Lesher, C.M., Préfontaine, S. (2017). Physical volcanology of komatiites and Ni-Cu-(PGE) deposits of the southern Abitibi greenstone belt. In *Archean Base and Precious Metal Deposits, Southern Abitibi Greenstone Belt, Canada*, Monecke, T., Mercier-Langevin, P., Dubé, B. (eds). Society of Economic Geologists, Littleton, Colorado.

Houlé, M.G., Lesher, C.M., Sappin, A.A., Bédard, M.P., Goutier, J., Yang, X.M. (2020). Overview of Ni-Cu-(PGE), Cr-(PGE), and Fe-Ti-V magmatic mineralization in the Superior Province: Insights on metallotects and metal endowment. In *Targeted Geoscience Initiative 5: Advances in the Understanding of Canadian Ni-Cu-PGE and Cr Ore Systems – Examples from the Mid-continent Rift, the Circum-Superior Belt, the Archean Superior Province, and Cordilleran Alaskan-type Intrusions*, Bleeker, W. and Houlé, M.G. (eds). Geological Survey of Canada, Open File 8722, Ottawa, Ontario.

Houlé, M.G. et al. (forthcoming). Ni-Cu-(PGE), Cr and Fe-Ti-V deposits of the Superior Province – Database. Report, Geological Survey of Canada, Ottawa, Ontario.

Hronsky, J.M.A., Groves, D.I., Loucks, R.R., Begg, G.C. (2012). A unified model for gold mineralisation in accretionary orogens and implications for regional-scale exploration targeting methods. *Mineralium Dep.*, 47(4), 339–358.

Huston, D.L. (2000). Gold in volcanic-hosted massive sulfide deposits: Distribution, genesis, and exploration. In *Gold in 2000*, Hagemann, S.G. and Brown, P.E. (eds). Society of Economic Geologists, Littleton, Colorado.

Huston, D.L., Champion, D.C., Cassidy, K.F. (2014). Tectonic controls on the endowment of Neoarchan cratons in volcanic-hosted massive sulfide deposits: Evidence from lead and neo-dymium isotopes. *Econ. Geol.*, 109(1), 11–26.

Ispolatov, V., Lafrance, B., Dubé, B., Creaser, R., Hamilton, M. (2008). Geologic and structural setting of gold mineralization in the Kirkland Lake-Larder Lake gold belt, Ontario. *Econ. Geol.*, 103(6), 1309–1340.

James, H.L. (1954). Sedimentary facies of iron-formation. *Econ. Geol.*, 49(3), 235–293.

Joanisse, A. (1994). Datation de la carbonatite du lac Shortt. PhD Thesis, Université du Quebec, Montreal.

Jolly, W.T. (1978). Metamorphic history of the Archean Abitibi belt. *Geological Survey of Canada*, 78–10, 63–77.

Katz, L.R., Kontak, D.J., Dubé, B., McNicoll, V., Creaser, R., Petrus, J.A. (2021). An Archean porphyry-type gold deposit: The Côté Gold Au(-Cu) deposit, Swayze greenstone belt, Superior Province, Ontario, Canada. *Econ. Geol.*, 116(1), 47–89.

Kerrich, R., Strong, D.F., Andrews, A.J., Owsiacki, L. (1986). The silver deposits at Cobalt and Gowganda, Ontario III: Hydrothermal regimes and source reservoirs – Evidence from H, O, C, and Sr isotopes and fluid inclusions. *Can. J. Earth Sci.*, 23(10), 1519–1550.

Kerrich, R., Goldfarb, R.J., Richards, J.P. (2005). Metallogenic provinces in an evolving geodynamic framework. In *Economic Geologists: 100th Anniversary Volume*, Hedenquist, J.W., Thompson, J.F.H., Goldfarb, R.J., Richards, J.P. (eds). Society of Economic Geologists, Littleton, Colorado.

Kissin, S.A. (1992). Five-element (Ni-Co-Ag-As-Bi) veins. *Geoscience Canada*, 19(3), 113–124.

Kitney, K., Olivo, G.R., Davis, D.W., Desrochers, J.P., Tessier, A. (2011). The Barry gold deposit, Abitibi Subprovince, Canada: A greenstone belt-hosted gold deposit coeval with Late Archean deformation and magmatism. *Econ. Geol.*, 106(7), 1129–1154.

Kjarsgaard, B.A. (2007). Kimberlite diamond deposits. In *Mineral Deposits of Canada: A Synthesis of Major Deposit Types, District Metallogeny, the Evolution of Geological Provinces, and Exploration Methods*, Goodfellow, W.D. (ed.). Geological Association of Canada, Mineral Deposits Division, Special Publication, Saint-John's, Newfoundland.

Kontak, D., Dubé, B., McNicoll, V., Creaser, R., Kyser, K. (2013). The Upper Beaver Au-Cu deposit, Kirkland Lake, Ontario, Canada: An Archean IOCG analogue or just an intrusion-related iron oxide copper gold deposit? In *Geological Association of Canada – Mineralogical Association of Canada (GAC-MAC), Winnipeg, May 22–24, 2013*. Abstract Volume, 36, 122, Saint-John's, Newfoundland.

Koopman, E.R., Hannington, M.D., Santaguida, F, Cameron, B.I. (1999). Petrology and geochemistry of proximal hydrothermal alteration in the mine rhyolite at Kidd Creek. In *The Giant Kidd Creek Volcanogenic Massive Sulphide Deposit, Western Abitibi Subprovince, Canada*, Hannington, M.D. and Barrie, C.T. (eds). Society of Economic Geologists, Economic Geology Monograph, Littleton, Colorado.

Kretschmar, U. and Kretschmar, D. (1986). Talc, magnesite and asbestos deposits in the Timmins-Kirk-land Lake area – Districts of Timiskaming and Cochrane. Report, Ontario Geological Survey, Study 28.

Krushnisky, A., Mercier-Langevin, P., Ross, P.S., Goutier, J., Pilote, C., Bernier, C. (2020). Geology and gold enrichment at the Horne 5 Archean volcanogenic massive sulphide deposit, Abitibi greenstone belt, Quebec: A synthesis. In *Targeted Geoscience Initiative 5: Contributions to the Understanding of Canadian Gold Systems*, Mercier-Langevin, P., Lawley, C.J.M., Castonguay, S. (eds). Geological Survey of Canada, Open File 8712, Ottawa, Ontario.

Krushnisky, A., Mercier-Langevin, P., Ross, P.-S., Goutier, J., McNicoll, V., Jackson, S.E., Monecke, T., Yang, Z., Pilote, C., Moore, L. (2022). The giant Archean Horne 5 volcanogenic massive sulfide deposit, Abitibi greenstone belt, Quebec: Geology and controls on gold enrichment. *Econ. Geol*, 118(2), 285–318.

Krushnisky, A., Mercier-Langevin, P., Meagher, D. (forthcoming). Volcanogenic massive sulfide deposits of the Superior Province, Canada – Compilation and database. Laurentian University, Sudbury.

Labbé, J.Y. and Couture, J.F. (1999) Contexte géologique des minéralisations métalliques, région d'Amos-Barraute. Report, Ministère de l'Énergie et des Ressources naturelles, ET 98-04.

Labbé, J.Y., Pilote, P., Lamothe, D. (2006). Évaluation du potentiel minéral pour les gîtes porphyriques de Cu-Au-Mo de l'Abitibi. Report, Ministère de l'Énergie et des Ressources naturelles, EP 2006-03.

Lacroix, S. (1998). Compilation et répartition des gisements polymétalliques à tonnage évalué dans la sous-province de l'Abitibi. Report, Ministère de l'Énergie et des Ressources naturelles, MB 98-06.

Lafrance, B., Mueller, W., Daigneault, R., Dupras, N. (2000). Evolution of a submerged composite arc volcano: Volcanology and geochemistry of the Normetal volcanic complex, Abitibi greenstone belt, Quebec, Canada. *Precambrian Res.*, 101(2–4), 277–311.

Lafrance, B., Davis, D.W., Goutier, J., Moorhead, J., Pilote, P., Mercier-Langevin, P., Dubé, B., Galley, A.G., Mueller, W.U. (2005). Nouvelles datations isotopiques dans la portion québécoise du Groupe de Blake River et des unités adjacentes. Report, Ministère de l'Énergie et des Ressources naturelles, RP 2005-01.

Larson, J.E. and Hutchinson, R.W. (1993). The Selbaie Zn-Cu-Ag deposits, Quebec, Canada: An example of evolution from subaqueous to subaerial volcanism and mineralization in an Archean caldera environment. *Econ. Geol.*, 88(6), 1460–1482.

Lauzière, K. (1989). Environnement géologique et minéralisation aurifère à la mine Bachelor, Desmaraisville, Quebec. Master's thesis, Université du Quebec, Chicoutimi.

Leclerc, F., Harris, L.B., Bédard, J.H., van Breemen, O., Goulet, N. (2012). Structural and stratigraphic controls on magmatic, volcanogenic, and shear zone-hosted mineralization in the Chapais-Chibougamau mining camp, Northeastern Abitibi, Canada. *Econ. Geol.*, 107(5), 963–989.

Legault, M., Gauthier, M., Jébrak, M., Davis, D.W., Baillargeon, F. (2002). Evolution of the subaqueous to near-emergent Joutel volcanic complex, Northern Volcanic Zone, Abitibi subprovince, Quebec, Canada. *Precambrian Res.*, 115(1–4), 187–221.

MacRae, N.D. (1969). Ultramafic intrusions of the Abitibi area, Ontario. *Can. J. Earth Sci.*, 6(2), 281–303.

Marshall, D. and Watkinson, D.H. (2000). The cobalt mining district: Silver sources, transport and deposition. *Explor. Mining Geol.*, 9(2), 81–90.

McNicoll, V., Goutier, J., Dubé, B., Mercier-Langevin, P., Ross, P.S., Dion, C., Monecke, T., Legault, M., Percival, J., Gibson, H. (2014). U-Pb geochronology of the Blake River Group, Abitibi greenstone belt, Quebec, and implications for base metal exploration. *Econ. Geol.*, 109(1), 27–59.

Mercier-Langevin, P., Dubé, B., Lafrance, B., Hannington, M.D., Galley, A., Moorhead, J., Gosselin, P. (2007a). Metallogeny of the Doyon-Bousquet-LaRonde mining camp, Abitibi greenstone belt, Quebec. In *Mineral Deposits of Canada: A Synthesis of Major Deposit Types, District Metallogeny, the Evolution of Geological Provinces and Exploration Methods*, Goodfellow, W.D. (ed.). Geological Association of Canada, Mineral Deposits Division, Special Publication, Saint-John's, Newfoundland.

Mercier-Langevin, P., Dubé, B., Hannington, M.D., Davis, D., Lafrance, B., Gosselin, G. (2007b). The LaRonde Penna Au-rich volcanogenic massive sulfide deposit, Abitibi greenstone belt, Quebec: Part I. Geology and geochronology. *Econ. Geol.*, 102(4), 585–609.

Mercier-Langevin, P., Dubé, B., Hannington, M.D., Richer-Laflèche, M., Gosselin, G. (2007c). The LaRonde Penna Au-rich volcanogenic massive sulfide deposit, Abitibi greenstone belt, Quebec: Part II. Lithogeochemistry and paleotectonic setting. *Econ. Geol.*, 102(4), 611–631.

Mercier-Langevin, P., Hannington, M.D., Dubé, B., Bécu, V. (2011). The gold content of volcanogenic massive sulfide deposits. *Miner. Deposita*, 46(5/6), 509–539.

Mercier-Langevin, P., Houlé, M.G., Dubé, B., Monecke, T., Hannington, M.D., Gibson, H.L., Goutier, J. (2012). A special issue on Archean magmatism, volcanism, and ore deposits: Part 1. Komatiite-associated Ni-Cu-(PGE) sulfide and greenstone-hosted Au deposits: Preface. *Econ. Geol.*, 107(5), 745–753.

Mercier-Langevin, P., Gibson, H.L., Hannington, M.D., Goutier, J., Monecke, T., Dubé, B., Houlé, M.G. (2014a). A special issue on Archean magmatism, volcanism, and ore deposits: Part 2. Volcanogenic massive sulfide deposits. *Econ. Geol.*, 109(1), 1–9.

Mercier-Langevin, P., Lafrance, B., Bécu, V., Dubé, B., Kjarsgaard, I., Guha, J. (2014b). The Lemoine auriferous volcanogenic massive sulfide deposit, Chibougamau camp, Abitibi greenstone belt, Quebec: Geology and genesis. *Econ. Geol.*, 109(1), 231–269.

Mercier-Langevin, P., Hannington, M.D., Dubé, B., Piercey, S.J., Peter, J.M., Pehrsson, S.J. (2015). Precious metal enrichment processes in volcanogenic massive sulphide deposits – A summary of key features, with an emphasis on TGI-4 research contributions. In *Targeted Geoscience Initiative 4: Contributions to the Understanding of Volcanogenic Massive Sulphide Deposit Genesis and Exploration Methods Development*, Peter, J.M. and Mercier-Langevin, P. (eds). Geological Survey of Canada, Open File 7853, Ottawa, Ontario.

Mercier-Langevin, P., Lawley, C.J.M., Castonguay, S., Dubé, B., Bleeker, W., Pinet, N., Bécu, V., Pilote, J.L., Jackson, S.E., Wodicka, N. et al. (2020). Targeted geoscience initiative 5, gold project: A summary of contributions to the understanding of Canadian gold systems. In *Targeted Geoscience Initiative 5: Contributions to the Understanding of Gold Deposit*, Mercier-Langevin, P., Lawley, C.J.M., Castonguay, S. (eds). Geological Survey of Canada, Open File 8712, Ottawa, Ontario.

Mercier-Langevin, P., Creaser, R.A., Dubé, B., Dubé, J., Kontak, D.J., Sutton, J., Côté-Mantha, O. (2021). Molybdenite Re-Os age of a gold-rich vein, Porphyry zone, Upper Beaver deposit, Abitibi greenstone belt, Ontario. Geological Survey of Canada, Open File 8789, Ottawa, Ontario.

Mines Agnico Eagle Limitée (2018). Mine LaRonde : un pas de plus vers le 4.0 [Online]. Available at: https://abitibi.agnicoeagle.com [Accessed 18 December 2020].

Monecke, T., Gibson, H., Dubé, B., Laurin, J., Hannington, M., Martin, L. (2008). Geology and volcanic setting of the Horne deposit, Rouyn-Noranda, Quebec: Initial results of a new research project. Report, Geological Survey of Canada, Current Research 2008-9.

Monecke, T., Mercier-Langevin, P., Dubé, B., Friemen, B.M. (2017). Archean base and precious metal deposits, southern Abitibi greenstone belt, Canada. *Reviews in Economic Geology*, 19, 7–49, Littleton, Colorado.

Moorhead, J., Beaumier, M., Lefebvre, D.L., Bernier, L., Martel, D. (1999). Kimberlites, linéaments et rifts crustaux au Quebec. Report, Ministère de l'Énergie et des Ressources naturelles, MB 99-35.

Moorhead, J., Tremblay, A., Pelz, P., Beaudoin, G. (2000). Géologie de la région de Val-d'Or, Sousprovince de l'Abitibi – Volcanologie physique et évolution métallogénique. Géologie de la Mine Louvicourt. Report, Ministère des Ressources naturelles, MB 2000-09, 47–64.

Morasse, S. (1988). Geological setting and evolution of the Lac Shortt gold deposit, Waswanipi, Quebec, Canada. Master's Thesis, Queen's University, Kingston.

Morasse, S., Wasteneys, H.A., Cormier, M., Helmstaedt, H., Mason, R. (1995). A pre-2686 Ma intrusion-related gold deposit at the Kiena mine, Val d'Or, Quebec, southerm Abitibi Subprovince. *Econ. Geol.*, 90(5), 1310–1321.

Mortensen, J.K. (1993). U-Pb geochronology of the eastern Abitibi Subprovince. Part 1: Chibougamau-Matagami-Joutel region. *Can. J. Earth Sci.*, 30(1), 11–28.

Mulja, T., Williams-Jones, A.E., Wood, S.A., Boily, M. (1995). The rare-element-enriched monzogranite-pegmatite-quartz vein systems in the Preissac-Lacorne batholith, Quebec. I: Geology and mineralogy. *Can. Mineral.*, 33(4), 793–815.

Nadeau, O., Stevenson, R., Jébrak, M. (2014). The Archean magmatic-hydrothermal system of Lac Shortt (Au-REE), Abitibi, Canada: Insights from carbonate fingerprinting. *Chem. Geol.*, 387, 144–156.

Nadeau, O., Cayer, A., Pelletier, M., Stevenson, R., Jébrak, M. (2015). The Paleoproterozoic Montviel carbonatite-hosted REE–Nb deposit, Abitibi, Canada: Geology, mineralogy, geochemistry and genesis. *Ore Geol. Rev.*, 67, 314–335.

Naldrett, A.J. (2004). *Magmatic Sulfide Deposits*. Springer, Berlin.

Newmont Corporation (2019). Newmont Goldcorp, dignitaries inaugurate Borden, Ontario's "Mine of the Future" [Online]. Available at: https://www.newmont.com [Accessed 30 October 2020].

O'Hanley, D.S. (1996). *Serpentinites: Records of Tectonic and Petrological History*. Oxford University Press, Oxford.

Oliver, J., Ayer, J., Dubé, B., Aubertin, R., Burson, M., Panneton, G., Friedman, R., Hamilton, M. (2012). Structure, stratigraphy, U-Pb geochronology and alteration characteristics of gold mineralization at the Detour Lake gold deposit, Ontario, Canada. *Explor. and Min. Geol.*, 20, 1–30.

Pacey, A., Wilkinson, J.J., Owens, J., Priest, D., Cooke, D.R., Millar, I.L. (2019). The anatomy of an alkalic porphyry Cu-Au system: Geology and alteration at Northparkes mines, New South Wales, Australia. *Econ. Geol.*, 114(3), 441–472.

Padilla, R.A. and Savard, M. (2018). Geology of the Windfall Lake gold project. *SEG 2018 Proceedings*, Keystone, Colorado, 22-25 September.

Percival, J.A., Skulski, T., Sanborn-Barrie, M., Stott, G.M., Leclair, A.D., Corkery, M.T., Boily, M. (2012). Geology and tectonic evolution of the Superior Province, Canada. In *Tectonic Styles in Canada: The Lithoprobe Perspective*, Percival, J.A., Cook, F.A., Clowes, R.M. (eds). Geological Association of Canada, Special Paper, Saint-John's, Newfoundland.

Perrouty, S., Gaillard, N., Piette-Lauzière, N., Mir, R., Bardoux, M., Olivo, G.R., Linnen, R.L., Bérubé, C.L., Lypaczewski, P., Guilmette, C. et al. (2017). Structural setting for Canadian Malartic style of gold mineralization in the Pontiac Subprovince, south of the Cadillac Larder Lake Deformation Zone, Quebec, Canada. *Ore Geol. Rev.*, 84, 185–201.

Piché, M., Guha, J., Daigneault, R. (1993). Stratigraphic and structural aspects of the volcanic rocks of the Matagami mining camp, Quebec: Implications for the Norita ore deposit. *Econ. Geol.*, 88(6), 1542–1558.

Pilote, P. and Guha, J. (1995). Metallogeny of the eastern extremity of the Abitibi belt. *Geological Survey of Canada*, 3143, 31–43.

Pilote, P., Dion, C., Cimon, J., Kirkham, R.V., Robert, F., Sinclair, W.D., Daigneault, R., Magnan, M. (1994). Les minéralisations de type Cu-Mo-Au-porphyrique et les gisements Cu-Au filoniens du lac Doré, Chibougamau : processus hydrothermaux distincts ou évolution d'un même évènement minéralisateur ? Report, Ministère de l'Énergie et des Ressources naturelles, DV 94-09.

Pilote, P., Robert, F., Sinclair, W.D., Kirkham, R.V., Daigneault, R. (1995). Porphyry-type mineralization in the Doré Lake Complex: Clark Lake and Merrill Island area. *Geological Survey of Canada*, 3143, 65–86.

Pilote, P., Marleau, J., David, J. (2019). Géologie de la région de Barraute, Sous-province de l'Abitibi, région de l'Abitibi-Témiscamingue, Quebec, Canada. Report, Ministère de l'Énergie et des Ressources naturelles [Online]. Available at: http://gq.mines.gouv.qc.ca/bulletins-geologiques/barraute-est/ [Accessed 15 December 2020].

Poulsen, K.H. (2017). The Larder Lake-Cadillac break and its gold districts. In *Economic Geologists: Archean Base and Precious Metal Deposits, Southern Abitibi Greenstone Belt*, Canada, Monecke, T., Mercier-Langevin, P., Dubé, B. (eds). Society of Economic Geologists, Littleton, Colorado.

Poulsen, K.H. and Hannington, M.D. (1996). Volcanic-associated massive sulphide gold. In *Geology of Canadian Mineral Deposit Types*, Eckstrand, O.R., Sinclair, W.D., Thorpe, R.I. (eds). Geological Survey of Canada, Ottawa, Ontario.

Powell, W.G., Carmichael, D.M., Hodgson, C.J. (1995). Conditions and timing of metamorphism in the southern Abitibi greenetone belt, Quebec. *Can. Jour. Earth Sci.*, 32(6), 787–805.

Préfontaine, S., Gibson, H.L., Houlé, M.G., Mercier-Langevin, P., Gamble A.D.P. (2008). Project Unit 08-022. Preliminary results of the alteration and mineralization study of the Potter mine, Munro Township, Abitibi greenstone belt. *Ontario Geological Survey*, 6226, 1–9.

Quebec Circulaire (2020). Restauration du site minier abandonné Manitou à l'aide des résidus de la mine Goldex [Online]. Available at: https://www.quebeccirculaire.org [Accessed 9 December 2020].

Quirion, D. (1991). Géologie de la mine d'or Lac Shortt. In *Lithotectonic Framework and Associated Mineralization of the Eastern Extremity of the Abitibi Greenstone Belt, Quebec (Field Trip 3)*, Guha, J., Chown, E.H., Daigneault, R. (eds). Geological Survey of Canada, Open File 2158, Ottawa, Ontario.

Riopel, J., Hubert, C., Cattalani, S., Barrett, T.J., Hoy, L.D. (1995). Gîtologie, structure et géochimie du gisement de sulfures massifs Mobrun (district de Rouyn-Noranda). Report, Ministère de l'Énergie et des Ressources naturelles, ET 92-02.

Robert, F. (1990). Structural setting and control of gold-quartz veins of the Val d'Or area, southeastern Abitibi Subprovince. In *Gold and Base-metal Mineralization in the Abitibi Subprovince, Canada, with Emphasis on the Quebec Segment*, Ho, S.E., Robert, F., Groves, D.I. (eds). Short Courses of the University of Western Australia.

Robert, F. (1994a). Timing relationships between Cu-Au mineralization, dykes, and shear zones in the Chibougamau camp, northeastern Abitibi Subprovince, Quebec. Geological Survey of Canada, Current Research, 1994-C, 287–294.

Robert, F. (1994b). Vein fields in gold districts: The example of Val d'Or, southeastern Abitibi subprovince, Quebec. Geological Survey of Canada, Current Research 1994-C, 295–302.

Robert, F. (2001). Syenite-associated disseminated gold deposits in the Abitibi greenstone belt, Canada. *Miner. Deposita*, 36(6), 503–516.

Robert, F. and Poulsen, K.H. (2001). Vein formation and deformation in greenstone gold deposits. In *Structural Controls on Ore Genesis*, Richards, J.P. and Tosdal, R.M. (eds). Society of Economic Geologists, Littleton, Colorado.

Robert, F., Poulsen, K.H., Cassidy, K.F., Hodgson, C.J. (2005). Gold metallogeny of the Superior and Yilgarn cratons. In *Economic Geologists: 100th Anniversary Volume*, Hedenquist, J.W., Thompson, J.F.H., Goldfarb, R.J., Richards, J.P. (eds). Society of Economic Geologists, Littleton, Colorado.

Ross, I. (2016). Goldcorp going electric with Chapleau gold mine. Northern Ontario Business [Online]. Available at: https://www.northernontariobusiness.com [Accessed 9 December 2020].

Ross, P.S., McNicoll, V.J., Debreil, J.A., Carr, P. (2014). Precise U-Pb geochronology of the Matagami mining camp, Abitibi greenstone belt, Quebec: Stratigraphic constraints and implications for volcanogenic massive sulfide exploration. *Econ. Geol.*, 109(1), 89–101.

Ross, P.S., Boulerice, A., Mercier-Langevin, P., McNicoll, V. (2020). Volcanology, chemo-stratigraphy, geochronology, hydrothermal alteration and VMS potential of the Lemoine member of the Waconichi Formation, Chibougamau district, Abitibi greenstone belt, Quebec. *Mineralium Dep.*, 55(1), 21–46.

Roy, M., Éthier, M.P., Turcotte, S. (2018). La restauration du site minier abandonné Manitou : bilan et résultats de dix années d'un partenariat novateur. *Résumés des conférences et des photoprésentations, Quebec Mines et Énergie 2018*. Ministère de l'Énergie et des Ressources naturelles, DV 2018-03.

Sappin, A.A. et al. (forthcoming). Critical and rare metal deposits of the Superior Province – Database. Report, Geological Survey of Canada.

Savoie, A., Trudel, P., Sauvé, P., Perrault, G. (1990). Géologie de la mine Doyon, Cadillac, Quebec. In *The Northwestern Quebec Polymetallic Belt: A Summary of 60 Years of Mining Exploration*, Rive, M., Verpaelst, P., Gagnon, Y., Lulin, J.M., Riverin, G., Simard, A. (eds). Canadian Institute of Mining and Metallurgy, Special Volume, Montreal.

Sillitoe, R.H. (2002). Some metallogenic features of gold and copper deposits related to alkaline rocks and consequences for exploration. *Miner. Deposita*, 37(1), 4–13.

Smith, J.P., Spooner, E.T.C., Broughton, D.W., Ploeger, F.R. (1993). Archean Au-Ag-(W) quartz vein/disseminated mineralisation within the Larder Lake-Cadillac Break, Kerr Addison-Ches-terville system, north east Ontario, Canada. Report, Ontario Geological Survey, 5831.

Ste-Croix, L. and Doucet, P. (2001). Potentiel en métaux rares dans les sous-provinces de l'Abitibi et du Pontiac. Report, Ministère de l'Énergie et des Ressources naturelles, PRO 2001-08.

Stone, M. and Selway, J. (2009). Independent technical report, Quebec Lithium Property. Report, Ministère de l'Énergie et des Ressources naturelles, GM 65011.

Stromberg, J.M., Barr, E., Van Loon, L.L., Gordon, R.A., Banerjee, N.R. (2019). Fingerprinting multiple gold mineralization at the Dome mine in Timmins, Ontario: Trace element and gold content of pyrite. *Ore Geol. Rev.*, 104, 603–619.

Taner, M.F. (1993). Gold abundance and an aluminosilicate-rich alteration zone in the Val D'Or mining district, Abitibi, Quebec: Examples of the fossil geothermal fields. *Ore Geol. Rev.*, 8(6), 451–475.

Taner, M.F. and Chemam, M. (2015). Algoma-type banded iron formation (BIF), Abitibi greenstone belt, Quebec, Canada. *Ore Geol. Rev.*, 70, 31–46.

Taner, M.F. and Martin, R.F. (1993). Significance of dumortierite in an aluminosilicate-rich alteration zone, Louvicourt, Quebec. *Can. Miner.*, 31(1), 137–146.

Taner, H., Williams-Jones, A.E., Wood, S.A. (1998). The nature, origin and physicochemical controls of hydrothermal Mo-Bi mineralization in the Cadillac deposit, Quebec, Canada. *Miner. Deposita*, 33(6), 579–590.

Taylor, S. and Lewis, B. (2018). First new all-electric mine dumps diesel; cuts costs, pollution [Online]. Available at: https://www.reuters.com [Accessed 30 October 2020].

Théberge, L., Daigneault, R., Labbé, J.Y., Brisson, T. (1999). Reconnaissance de l'héritage volcanogène d'une minéralisation de sulfures massifs en zone de forte déformation : le cas type du gisement de la mine Langlois, Lebel-sur-Quévillon, Abitibi. Report, Ministère de l'Énergie et des Ressources naturelles, MB 99-38.

Thurston, P.C., Ayer, J.A., Goutier, J., Hamilton, M.A. (2008). Depositional gaps in the Abitibi greenstone belt stratigraphy: A key to exploration for syngenetic mineralization. *Econ. Geol.*, 103(6), 1097–1134.

Tourigny, G., Brown, A.C., Hubert, C., Crépeau, R. (1989). Synvolcanic and syntectonic gold mineralization at the Bousquet mine, Abitibi greenstone belt, Quebec. *Econ. Geol.*, 84(7), 1875–1890.

Valliant, R.I., Barnett, R.L., Hodder, R.W. (1983). Aluminium silicate-bearing rock and its relation to gold mineralization; Bousquet mine, Bousquet township, Quebec. *CIM Bull.*, 76(850), 81–90.

Welch, M.J. (1995). Metal zoning, geochemistry and alteration of the Archean, Estrades Zn-Au massive sulfide deposit, Northwestern Quebec, Canada. Master's Thesis, Laurentian University, Sudbury.

Yergeau, D., Mercier-Langevin, P., Dubé, B., Malo, M., McNicoll, V.J., Jackson, S.E., Savoie, A., La Rochelle, F. (2015). The Archean Westwood Au deposit, southern Abitibi: Telescoped Aurich VMS and intrusion-related Au systems. In *Targeted Geoscience Initiative 4: Contributions to the Understanding of Precambrian Lode Gold Deposits and Implications for Exploration*, Dubé, B. and Mercier-Langevin, P. (eds). Geological Survey of Canada, Open File 7852, Ottawa, Ontario.

Yergeau, D., Mercier-Langevin, P., Dubé, B., Malo, M., Savoie, A. (2022a). The Westwood deposit, southern Abitibi greenstone belt, Canada: An Archean Au-rich polymetallic magmatic-hydrothermal system – Part 1. Volcanic architecture, deformation, and metamorphism. *Economic Geology*, 117(3), 545–575.

Yergeau, D., Mercier-Langevin, P., Dubé, B., McNicoll, V., Jackson, S., Malo, M., Savoie, A. (2022b). The Westwood deposit, southern Abitibi greenstone belt, Canada: An Archean Au-rich polymetallic magmatic-hydrothermal system – Part 2. Hydrothermal alteration, mineralization, and geological model. *Econ. Geol.*, 117(3), 577–608.

Zhang, J., Lin, S., Linnen, R., Martin, R. (2014). Structural setting of the Young-Davidson syenite-hosted gold deposit in the western Cadillac-Larder Lake deformation zone, Abitibi greenstone belt, Superior Province, Ontario. *Precambrian Res.*, 248, 39–59.

Zurevinski, S.E., Heaman, L.M., Creaser, R.A. (2011). The origin of Triassic/Jurassic kimberlite magmatism, Canada: Two mantle sources revealed from the Sr-Nd isotopic composition of groundmass perovskite. *Geochem. Geophys. Geosyst.*, 12(9). doi: 10.1029/2011GC003659.

3

The Unconformity-related Uranium Mineral System of the Athabasca Basin (Canada)

Patrick LEDRU[1], Antonio BENEDICTO[2], Guoxiang CHI[3], Charles KHAIRALLAH[4], Julien MERCADIER[1], Jonathan POH[5] and John ROBBINS[4]

[1] *GeoRessources, University of Lorraine, Vandœuvre-lès-Nancy, France*

[2] *GEOPS, CNRS, University of Paris-Saclay, Orsay, France*

[3] *University of Regina, Canada*

[4] *Orano Canada, Saskatoon, Canada*

[5] *Géosciences Rennes, CNRS, University of Rennes, France*

3.1. Introduction

The concept of mineral systems (Wyborn et al. 1994) emphasizes the importance of four interrelated steps in the formation and preservation of mineral deposits, that is, source, path, trap and preservation. Following further investigations, this concept has been reformulated considering critical processes that operate on a variety of scales to generate large ore deposits, including whole lithosphere architecture, transient favorable geodynamics, fertility and preservation of the primary depositional zone (Barnicoat 2007; McCuaig et al. 2010; McCuaig and Hronsky 2014). It has been applied to revisit the deposit framework and to suggest new exploration criteria at different scales, from the accretionary orogens (Hronsky et al. 2012) and cratons (e.g. the Yilgarn craton, McCuaig et al. 2010) to the uraniferous provinces in Africa (Eglinger et al. 2016) and Australia (Skirrow et al. 2009).

For a color version of all figures in this chapter, see www.iste.co.uk/decree/metallic2.zip.

Metallic Resources 2,

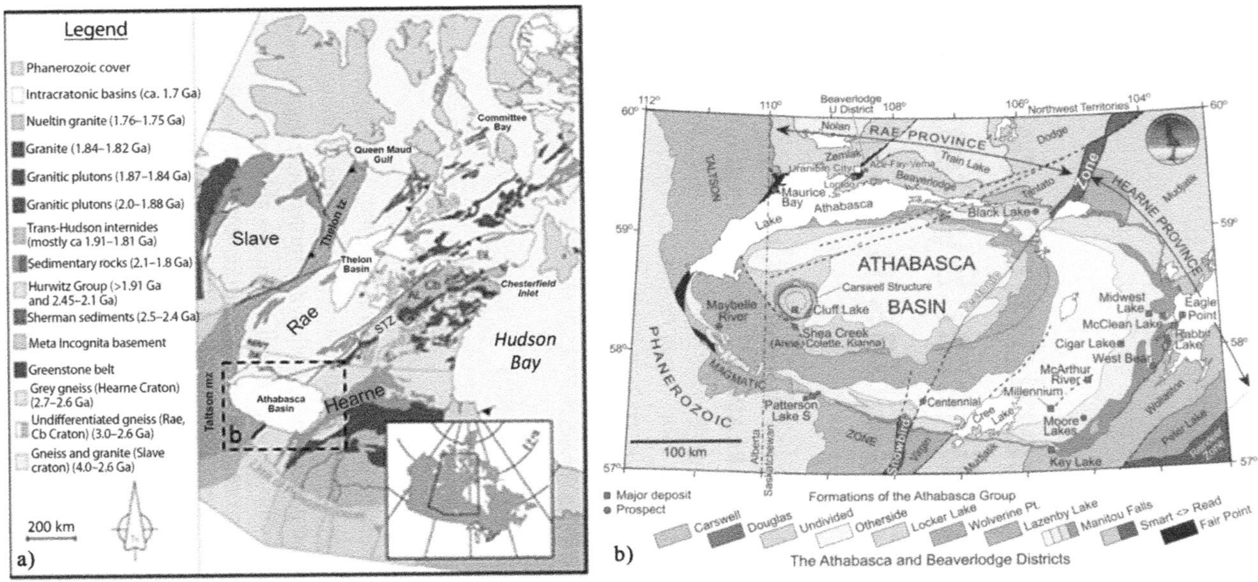

Figure 3.1. *Geological map of the Canadian Shield (a: modified from Pehrsson et al. 2013) and the Athabasca Basin (b: Jefferson et al. 2007)*

Unconformity-related uranium deposits of the Athabasca Basin (Canada) (Figure 3.1) constitute a major metallogenic province on earth that has produced more than 500,000 t of uranium ore since the first discovery in late 1960s. The major deposits have been discovered in the vicinity of the unconformity that separates the Archean-Paleoproterozic metamorphic basement of the Hearne and Rae cratons from the almost undeformed, mainly siliciclastic rocks of the Paleo- to Mesoproterozic Athabasca Basin. Since the pioneer work of Hoeve and Sibbald (1978), a diagenetic-hydrothermal model is considered as the main process at the origin of this type of deposit. This model invokes the percolation of fluids from the basin toward the basement, the uranium mineralization resulting from the interaction between these oxidizing diagenetic fluids with the methane-bearing reducing fluids from the basement (Raffensperger and Carven 1995). Thus, the main traps of the mineralizations are chemical and structural, located at the unconformity and within reactivated basement faults. The source of uranium is debated, with both basement rocks that floor the basin and uranium-bearing clastic components in platform-cover sandstones that fill the basin being considered as possible sources (IAEA 2018).

Following the discovery of outcropping deposits by airborne and ground radiometric surveys, the first-order metallotectused for exploration has been the intersection between basement-hosted graphitic shear zones, interpreted from magnetic and electromagnetic geophysical surveys, and the unconformity at the base of the Athabasca Basin. This metallotect is at the origin of the discovery of the classical unconformity-related uranium deposits including the world class McArthur River and Cigar Lake deposits (Bruneton 1993; McGill et al. 1993; Jefferson et al. 2007; IAEA 2018). The association of the uranium deposits with basement-hosted fault systems, such as in the Wollaston Mudjatik Transition Zone (Annesley et al. 2005) and the Patterson Lake corridor (Johnstone et al. 2021), clearly indicates a structural control of fluid circulation and of the localization and geometry of the main orebodies. Geological elements involved in the metallogenic process, that is, host rocks, associated structures, alterations, geometries of the traps, grades and tonnages, have been synthetized by considering end-members based on the ore models of four "large to giant deposits": Cigar Lake, McArthur River, Eagle Point and Millennium (IAEA 2018). However, as recently summarized by the Footprint Project (Lesher et al. 2017), around 20 individual footprints and vectors have been identified for McArthur River and around 30 at Millennium. Some of the footprints/vectors are similar at both sites, but others are different, highlighting multiple factors involved in the mineralizing processes in the Athabasca Basin. Moreover, discoveries of large to giant basement-hosted deposits like the Arrow and Triple R deposits in the Patterson Lake corridor (Hillacre et al. 2017, 2021; Card 2020; Potter et al. 2020; Johnstone et al. 2021; Rabiei et al. 2021), Centennial (Reid et al. 2014) or Gryphon (Liskowich 2018), respectively, in West, Central and East

Athabasca, are not located at the intersection between basement-hosted graphitic shear zones and the unconformity, illustrating the fact that this intersection is not the only "ideal target". Thus, the identification of some footprints may be more relevant than a full record of all footprints that are considered to be indicative of the targeted ore depositmodel. The application of the mineral systems concept and a review of the critical elements as recommended by McCuaig and Hronsky (2014) provide an opportunity to revisit some of the main concepts that are currently applied for guiding exploration in the Athabasca Basin (Figure 3.2). The first critical element to be considered is the architecture of the geological domain, from the lithospheric to the upper crustal scale. The variety and complexity of the metamorphosed basement has been revealed by regular drilling along graphitic bearing shear zones. As structural and petrological studies identify the structural control and the importance of these inherited fault systems in the initiation and sustained development of fluid-rock interaction between the basin and its basement, the tectonic processes at the origin of these structures are of fundamental importance in order to understand their possible role as lithospheric boundaries and fluid-thermal conduits prior, during and after the deposition of the Athabasca Basin.

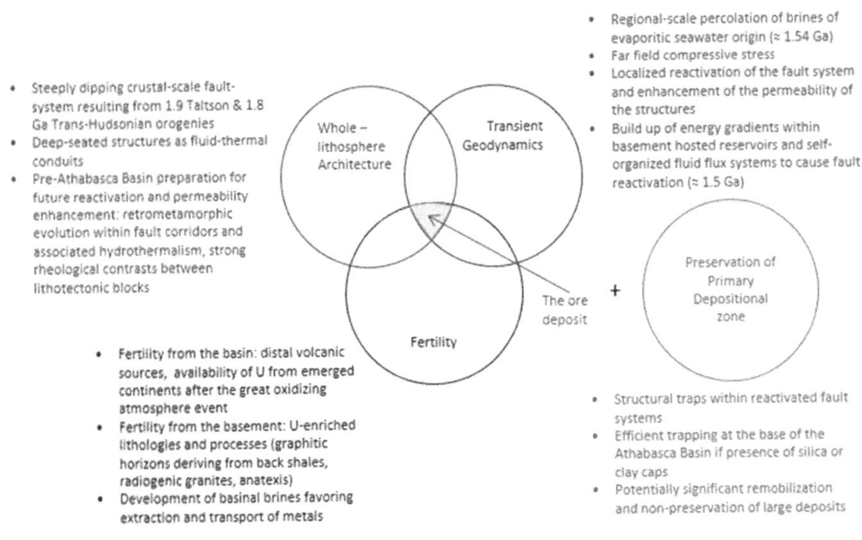

Figure 3.2. *The mineral system concept (McCuaig and Hronsky 2014) applied to the unconformity-related uranium deposits: review of the main critical elements*

A major discovery has been the characterization of the brine signature of fluid inclusions in quartz, enriched in Ca and Mg, in the vicinity of the uranium deposits

(Pagel 1975; Derome et al. 2005; Richard et al. 2011, 2012; Mercadier et al. 2012). The role of these brines is considered as a necessary condition for the formation of the unconformity-related giant deposits as they are exceptionally rich in uranium. Fluid flow systems have been summarized and schematized as ingress and egress flows corresponding, respectively, to descending oxidized fluids from the basin penetrating into the basement and to ascending reducing fluids from the basement reaching the unconformity (Jefferson et al. 2007). If the unconformity obviously has a major role as a regional and crustal scale exchange interface, then the dynamic conditions or processes leading to the coincidence between the unconformity, deposition of ore, intense dissolution and alteration, and reactivation of the basement-hosted fault zone (mainly under compressional stress regime), remain poorly understood. In this context, the characterization of the transient geodynamic conditions, which may have controlled the driving forces of fluid flow, the permeability of conduits, and connectivity between basinal fluids and basement-hosted reservoirs, may be considered as a second critical element in the mineral system. The ambiguity in distinguishing these ore-forming processes – which may correspond to transient episodes, from those related to a steady-state fluid circulation responsible for widespread regional alteration – may challenge the pertinence of some pathfinders.

In this chapter, we focus on the architecture of the Athabasca Basin and its substratum at the regional scale and on the fluid flow systems in light of recent surveys and scientific advances that complement the results of the EXTECH project (Jefferson et al. 2007) as well as the review performed by the IAEA (2018). Implications of the mineral systems concept applied to the exploration of unconformity-related uranium deposits are discussed in terms of project generation, targeting and modeling.

3.2. Defining the critical elements of the unconformity-related uranium mineral system of the Athabasca Basin

3.2.1. *The architecture of the Athabasca Basin and its basement at the regional scale*

3.2.1.1. *The tectonic evolution of the Athabasca basement*

The Trans-Hudson Orogen (THO, 1.84–1.78 Ga; Hoffman 1988; Alexandre et al. 2009; Jeanneret et al. 2017) and the Taltson/Thelon orogen (1.9–1.8 Ga; Card et al. 2014) are the main Paleoproterozoic orogens that contributed to the assembly of the Canadian Shield, a part of Laurentia, between 2.0 and 1.8 Ga (Hoffman 1988; Eglington et al. 2013), marking the collision between the eastern Churchill Province

(Hearne) and the Superior Province to the east, and between the western Churchill Province (Rae) and the Slave Craton to the west (Figure 3.1). The Hearne Craton is mainly constituted by the Mudjatik domain, composed of Archean granulitic tonalitic-trondhjemitic gneiss domes and the Wollaston domain composed of Paleoproterozoic meta-sedimentary gneisses (Annesley et al. 2005). According to Tran et al. (2008), most of the Wollaston group was deposited in a back-arc to foreland basin setting, contemporaneous with the build-up of nearby volcanic arcs that could be the source of detrital zircons ranging from 1920 to 1880 Ma.

These domains were strongly imbricated during orogenic processes marked by vertically-accretive growth histories and the development of steeply dipping anastomosed shear zones that extend over several hundred kilometers along the strike, mainly the Wollaston-Mudjatik Transition Zone (Figure 3.3) and the Snowbird Tectonic Zone/Virgin River shear zone (Figure 3.1). Discrete geophysical signatures, mainly derived from the magnetic field and electromagnetic conductors, reveal the contrast between dominantly linear structures for the Wollaston domain and the dome shape of the main Archean gneisses of the Mudjatik domain (Figure 3.3).

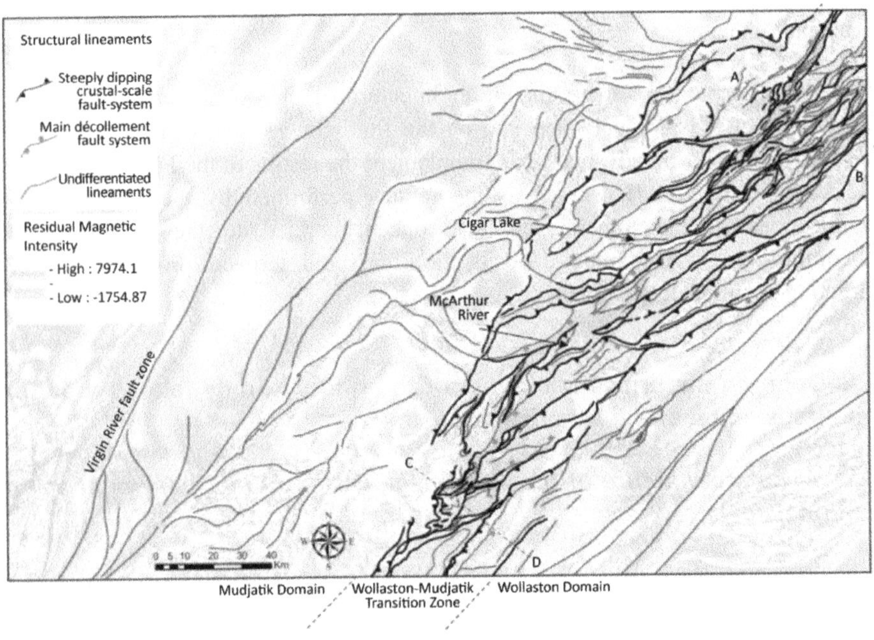

Figure 3.3. *Interpretative structural map of the magnetic linear gradients and electromagnetic conductors of the Wollaston Mudjatik transition zone. See Figure 3.5 for AB, CD cross sections*

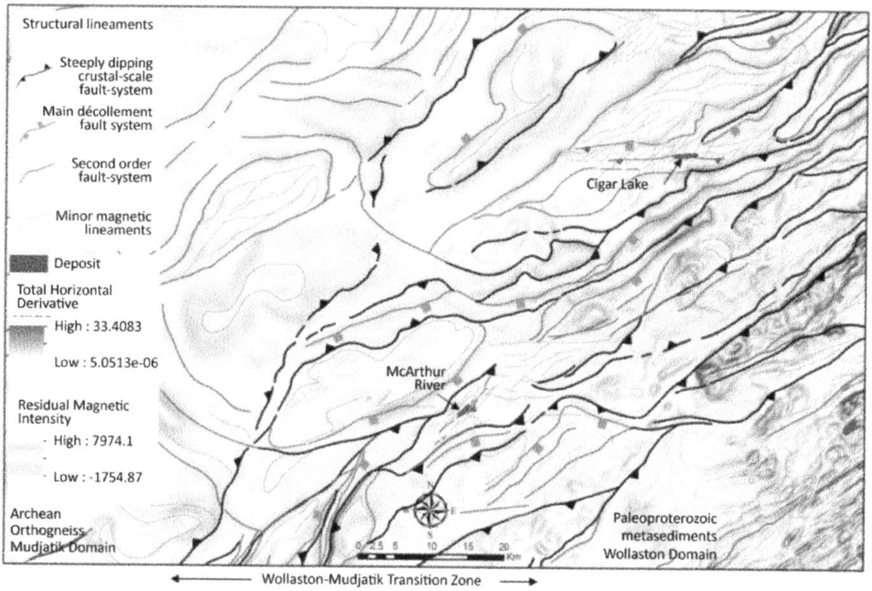

Figure 3.4. *Interpretative structural map of the magnetic linear gradients and electromagnetic conductors of the Wollaston Mudjatik transition zone focused on the median part, from Cigar Lake to McArthur River*

The nature of these contacts has been disputed since the pioneer articles from Gibb et al. (1983) and Hoffman (1988). As far as oceanic crust or eclogite bearing terranes are absent or poorly exposed, there have been debates, for example, on the significance of the Snowbird Tectonic Zone, considered as a suture between the Hearne and Rae cratons (Hoffman 1988; Berman et al 2007) or as an intercontinental fault zone (Hanmer et al. 1995; Regan et al. 2017). Geophysical surveys performed in the framework of the Lithoprobe project and recently by the Geological Survey of Canada provide new constraints on some of these contacts and reveal deeply rooted frontiers at the scale of the lithosphere. This was illustrated by the magneto telluric (MT) profiles performed in the southern extension of the Trans-Hudson Orogen by Jones et al. (2005) and recently for the Taltson orogeny to the west (Potter et al. 2020; Tschirhart et al. 2020). Based on these arguments and on the presence of carbonatites dykes within the associated shear zones of the Patterson Lake corridor, Card (2020) concluded that crustal-scale fault zones with access to the mantle should be considered a key parameter in the exploration model for Proterozoic unconformity uranium deposits. As it is suspected and well documented by MT networks in Australia (Thiel and Heinson 2010; Motta et al. 2019), it is likely that the frontiers between Archean blocks, corresponding to accretionary zones during the Paleoproterozoic Era, have probably preserved some specific conductive properties that are still marked in the structure of the lithosphere.

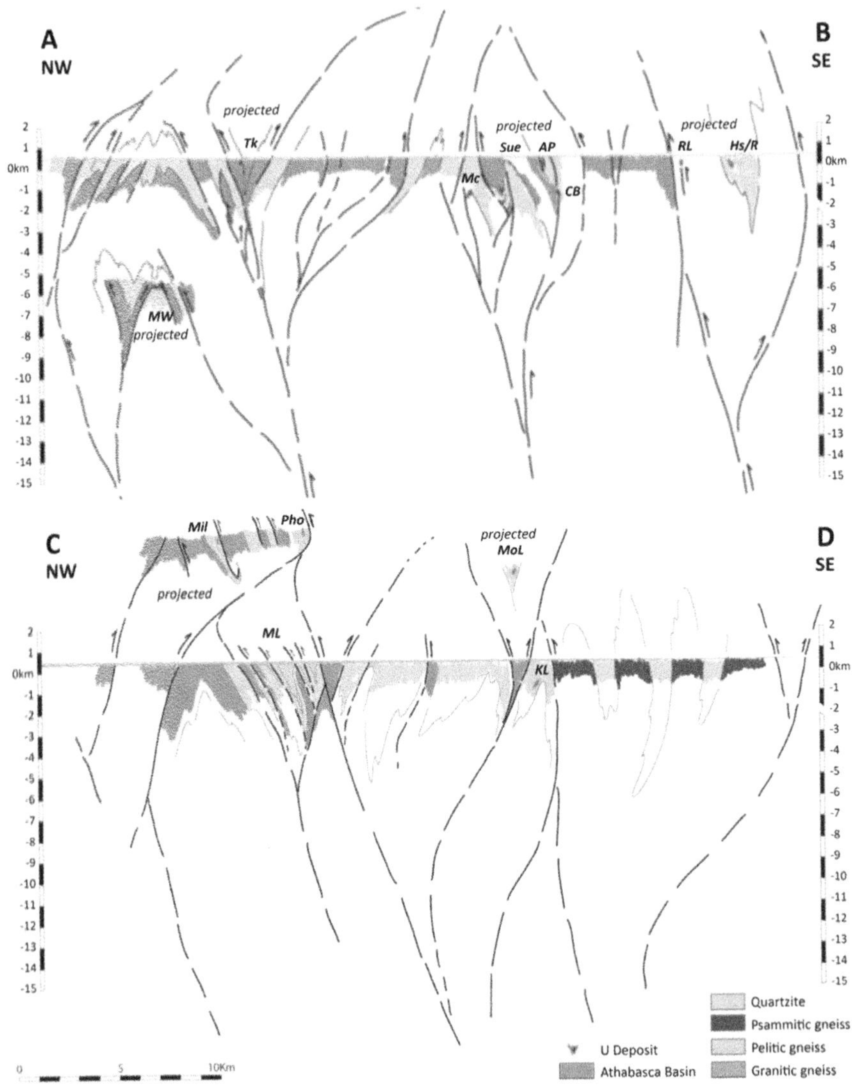

Figure 3.5. *Interpretative cross sections through the Wollaston Mudjatik transition zone based on the structural map and information from oriented cores and geophysical modeling. See location in Figure 3.3*

COMMENT ON FIGURE 3.5.– *Abbreviations: MW: Midwest, Tk: Tamarak, Mcl: McClean, EP: Eagle Point, CB: Collins Bay, RL: Rabbit Lake, Hs/R: Horseshoe/ Raven, MoL: Moore Lake, Pho: Phoenix, Mil: Millennium, ML: Martin Lake.*

In order to constrain this critical element (i.e. basement structural framework) and to define some markers that can be used for understanding the structural control of the mineralization and identify the role of the main lineaments, several multidisciplinary actions have been undertaken at different scales.

A new interpretation of the magnetic linear gradients and electromagnetic conductors has been completed (Figure 3.3). As already noticed by Condor Consulting Inc. (2013) and Witherly (2014), by doing such interpretation, structures in the gneissic basement appear more as curvilinear than by considering superposed and transecting fault systems as interpreted from a classical polyphased tectonic history (Annesley et al. 2005). A detailed analysis of these basement structures, by referring to geological maps and structural data (mainly foliation, stretching lineation and shear zones) that have been determined and measured on oriented cores of the exploration projects, gives an added value to such magnetic gradient interpretation. Taking into account the observation that post-unconformity reactivation displacement never exceeds 200 m, the observation that the Archean basement slices are thrusted over Paleoproterozoic metasediments, that is, in an upper geometrical position, implies that large displacements at the crustal scale, which have been considered as first-order structures in the basement, are related to the Trans-Hudson Orogen (Figures 3.3 and 3.4). Based on this first filter, other basement magnetic lineaments have been interpreted as second-order fault systems, displaying an anastomosed pattern. The fact that the main foliation and lineation within the Archean slices are in geometrical continuity with the foliation of the Paleoproterozoic metasediments and that there is no metamorphic break between these terranes at the scale of these thrust zones, suggests that the deep crustal structures were formed during the Trans-Hudson Orogen and before the Athabasca Basin.

The regional structural map of the pre-Athabasca fault system for the eastern part of the basin resulting from this interpretation enables the construction of crustal-scale structural cross sections (Figure 3.5). Based on this interpretation, a revision of the transition zone between the Wollaston and Mudjatik domains is proposed; it compares similar orogenic context elsewhere developed during this Archean-Proterozoic transition period (Gapais et al. 2014; Gapais 2018). The regional strain field, characterized by subvertical anastomosing fault systems and foliation showing mainly down dip extension lineation, is interpreted as resulting from a major compressional event characterized by horizontal shortening and vertical stretching of a weak lithosphere. In this model, structures were originally formed vertically and variation in orientation and local superposition of structures are interpreted as the result of progressive deformation. This model is an alternative to a general tectonic evolution as proposed by Tran (2001), Annesley et al. (2005) and Jeanneret et al. (2016) that invokes several stages of collision, including early stages of crustal thickening caused by crustal scale thrusts comparable to modern collision orogens. Several arguments can be discussed:

– Although it is possible to describe superposed foliations and refolding patterns at the scale of exploration projects, it is very difficult to identify a clear reproducible chronology, geometry and kinematics of this pre-Athabasca tectonic evolution from one project to another and in particular to characterize an early phase related to thick skinned tectonics. Annesley et al. (2005) describe at the scale of the Wollaston-Mudjatik transition zone two phases related to syn-collisional stages of deformation and two late oblique collision phases, that is, a sinistral transpressional tectonics marked by a NE–SW shearing and NW–SE shortening (i.e. regional northeast-trending structure) at 1,820–1,805 Ma followed by dextral strike slip under NE–SW shortening at 1,805–1,775 Ma. In the north-eastern termination of this transition zone, Jeanneret et al. (2017) identified two superposed tectonic phases related to collision under N–S shortening at 1,840–1,813 Ma and 1,813–1,770 Ma, considering that the second phase is mainly developed under a sinistral transpressional tectonic regime during the final oblique collision of the Trans-Hudson Orogen. In the south-western termination of this transition zone, Tran (2001) and Tran et al. (2000) summarized the tectonic evolution of the Cree area by describing fives phases of deformation, considering that the regional northeast-trending structure was developed in axial plane of open to tight F3 folds refolding the main foliation resulting from collision tectonics, with no footprint of any transcurrent tectonics. This structural complexity is a strong argument allowing us to consider that the variations are related to a local deviation of stresses and progressive accommodation of the deformation under moderate to low average strain rates of horizontal shortening in a weak lithosphere rather than to superposed tectonic phases marking changes in the far-field stress field.

– Structural gradients are mainly identified along the basement-cover contact, which is interpreted as a basal décollement (Tran 2001). However, these gradients are mainly related to the rheological contrasts between these terranes and the normal stratigraphic superposition cannot justify large thrusts. In other areas, over-thrusting of basement complex on the Wollaston domain is recognized and is equivalent to the first-order structures related to the Trans-Hudson Orogen as described above. Stereograms and maps show that the structures are steeply dipping and that the lineation is generally downdip (Tran 2001). It is also noted that there is evidence for both northwest and southeast-directed displacement in the form of conjugate sets which form pop-up structures. This geometry is interpreted by Tran (2001) as related to the F3 folding of an originally flat lying S1-2 foliation.

– The mineral assemblages and pressure–temperature (P-T) conditions remain fairly constant throughout a section of the Trans-Hudson Orogen, characterized by peak conditions at 0.6–0.9 GPa and 750–825°C in the SE basement (Cree Lake, Fraser Lake) (Annesley et al. 2005; McKechnie et al. 2012) and at 1 GPa and 750–800°C in the NE basement (Wollaston Lake), as well as a regional foliation developed under upper amphibolite facies, marked by incipient melting at 0.4–0.6 GPa and 700–800°C (Jeanneret 2016).

– Evidence of high-pressure or oceanic relics and of inverse metamorphism associated with thrust systems, considered as key markers of modern collision tectonics, are lacking.

The regime at the origin of the crustal thickening and tectonic imbrication of the different lithotectonic units has been discussed by Tran et al. (2000). They considered that the earliest Trans-Hudson tectonic events involved crustal shortening, west-verging fold/thrust stacking marked by both thin-skinned and thick-skinned tectonics and regionally extensive initially flat-lying foliation. Our interpretation differs from that of Tran et al. (2000) as we believe that the structures were originally formed in a sub-vertical position caused by a bulk homogenous crustal thickening process, rehabilitating Lewry and Sibbald's (1980) model which considered the extrusion of the Mudjatik migmatitic domain as the main driving force for the construction of this part of the Canadian Shield. The presence of local flat lying shear zones and tangential tectonics is not ruled out from this new interpretation but is considered as related to the partitioning of deformation and boudinage around large Archean domes.

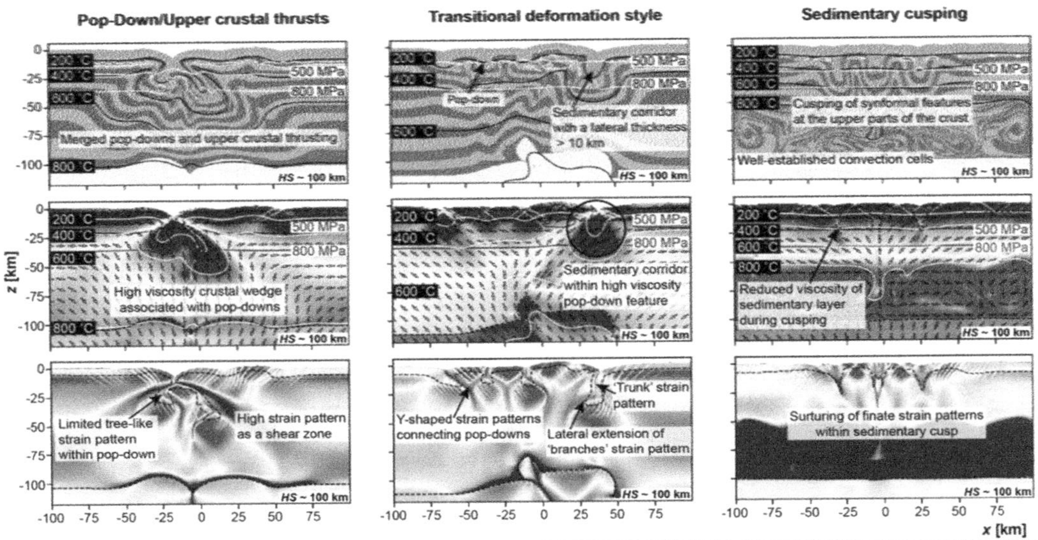

Model Initial Conditions (Moho Temperature, Strain-rate)	T_{Moho} = 750°C, $\dot{\varepsilon}_{BG}$ = 1.e-14 s^{-1}	T_{Moho} = 650°C, $\dot{\varepsilon}_{BG}$ = 5.e-15 s^{-1}	T_{Moho} = 750°C, $\dot{\varepsilon}_{BG}$ = 1.e-16 s^{-1}
Topographic Range	9.66 km	7.18 km	2 km
Maximum *P-T* in metasediments at *HS* ~ 100 km	*P* = 886 MPa, *T* = 236°C	*P* = 798 MPa, *T* = 216°C	*P* = 886 MPa, *T* = 236°C

Figure 3.6. *Representative snapshots of the main deformation styles and relevant indicators at a horizontal shortened state (HS ~100 km): morphology (top), viscosity (center) and accumulated strain (bottom)*

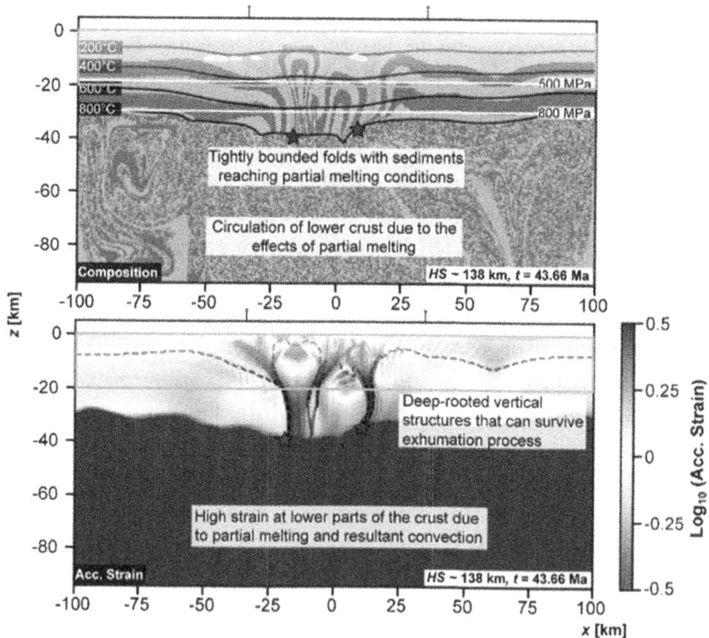

Model Initial Conditions (Moho T, Strain rate)	T_{Moho} = 750°C, $\dot{\varepsilon}_{BG}$ = 5.e-16 s^{-1}
Topographic Range	3.33 km
Maximum *P-T* in metasediments at HS ~ 138 km	*P* ≥ 800 MPa, *T* ≥ 700°C

Figure 3.7. *Reference model with the lithosphere at a higher horizontal shortened state (HS ~ 139 km), showing the morphological (top) accumulated strain plots (bottom)*

The scenarios of subvertical structure development under a compressional regime have been tested with a series of lithospheric-scaled thermo-mechanical numerical models over a wide range of Moho temperatures and shortening rates (Figure 3.6; Poh et al. 2020). Two end members, pop-down/upper crustal thrusts and sedimentary cusping have been considered and reflect different types of orogens (Chardon et al. 2009; Gapais et al. 2014; Gapais 2018). The initial geological model that mimics a simplified Precambrian terrain consists of a rectangular 30 km thick crust that is subdivided into two homogeneous components: a 2.5 km sedimentary cover, representing the Wollaston metasedimentary sequence, on top of a 27.5 km granitic crust, corresponding to the Mudjatik domain. The resultant deformation styles are characterized for a horizontal shortening of 100 km and constrained by four types of geological observations from the Trans-Hudson Orogen (Figure 3.6): (i) structural evolution, (ii) associated finite strain patterns, (iii) evolution of topography and (iv) pressure–temperature–time evolution of the sedimentary rocks (i.e. P-T-t paths). Each deformation style resulting from this numerical perspective has then been compared to the vertical structures of the Wollaston-Mudjatik Transition Zone as described above and P-T estimates from petrological studies (Annesley et al. 2005; McKechnie et al. 2012; Jeanneret et al. 2016). It is concluded that sedimentary cusping, occurring from moderate to low average strain rates of horizontal shortening with the initial conditions ($T_{\mathrm{Moho}} = 750°\mathrm{C}$, $\dot{\varepsilon}_{\mathrm{BG}} = 5 \times 10^{-16}\ \mathrm{s}^{-1}$), is the most likely deformation style that fulfills the dominantly vertical structures comparable to the structural cross sections. The peak P-T conditions found within metamorphic assemblages are reached for this sedimentary cusping tectonics for a horizontal shortening of 138 km (Figure 3.7). In these conditions, heterogeneous thickening related to thick-skinned tectonics and generalized crustal scale thrusts are not requested to explain the thermal evolution of the Trans-Hudson Orogen.

This tectonic evolution has been sealed by the deposition of Paleoproterozoic to Mesoproterozoic basins (Athabasca, Thelon, Elu and Hornby Bay) mainly composed of detrital sediments that rest unconformable over these deeply eroded domains (Figure 3.1). The youngest ages for the late Trans-Hudson tectonic evolution, that is, prior to the deposition of the Athabasca Basin, is given at 1,742 ± 25 Ma by 40Ar/39Ar dating (Alexandre et al. 2009), at 1,718 ± 12 Ma by U-Th-Pb Concordia diagrams on monazite grains from the Wollaston Lake district (Jeanneret et al. 2017) and at 1,713 ± 10 Ma by 207Pb/206Pb on rutile (Adlakha and Hattori 2021). The oldest zircon from reworked tuff layers in the middle-lower part of the Athabasca Basin stratigraphy is dated at 1,644 ± 13 Ma (U/Pb; Rainbird et al. 2007). These ages indicate a rather restricted bracket at about 1,700 Ma that raises the question of the exhumation processes. Metamorphic mineral assemblages exposed under the unconformity corresponds to a maximum burial of about 15 km depth, which means that these rocks were brought from a depth of 15 km to the surface and suffered a weathering marked by a paleoregolith in less than 50 Ma. So far, no regional scale extensional tectonics is described, and erosion and denudation appear to be the main process responsible for this exhumation.

3.2.1.2. *Perspectives for further research to guide exploration*

Regarding the Athabasca uranium mineral system, this review of the architecture at the regional scale highlights critical elements and justifies further investigation in terms of project generation:

– Although translating crustal scale investigation by geophysical imagery into an effective exploration targeting system is challenging (McCuaig et al. 2010), the role of the long lasting tectonic history of deeply rooted structures is critical in the definition of metallogenic provinces. These structures are responsible for crustal segmentation and are natural links between deep and shallow processes. They guided in many cases the intrusion of magmatic bodies extracted from the mantle or generated during crustal melting and have recorded significant retrometamorphic evolution and hydrothermal alteration during the late orogenic evolution, leading to the segmentation of the crust. Understanding their significance and their extension is certainly an opportunity to focus exploration on zones of interest at the regional scale. Thus, by referring to geophysical imagery, Card (2020) considers that protracted reactivation of deep-seated structures and their subsidiaries was a fundamental control on uranium mineralization in the southwestern Athabasca Basin. Based on the MT profiles realized through the southern extension of the Trans-Hudson Orogen (Jones et al. 2005), on the structural analysis and resulting cross sections presented in this chapter and on numerical modeling (Poh et al. 2020), a similar interpretation is proposed for the Wollaston-Mudjatik transition zone that concentrates the main large to giant uranium showings or deposits (Ledru et al. 2019).

– The reconciliation between project scale geological information, mainly interpreted from structural and mineralogical datasets, from drill hole and field surveys, as well as geophysical surveys is critical for building reliable 3D reference models and efficient targeting. The qualification and quantification of parameters used to build these models must be performed according to the advances of geological knowledge. Thus, a change of paradigm from a modern-style fold and thrust orogen to an ancient-style orogen generating deep-rooted vertical structures may significantly change results of direct and inverse modeling of geophysical surveys, as modeling is directly influenced by the a priori geological model of the crust. However, it should be pointed out that although we contrast two end-member structural styles and we prefer the deep-rooted vertical one, the actual geodynamic mechanism causing such deformation still needs further study.

– The reactivation of the inherited pre-Athabasca fault system during and after the deposition of the basins is well constrained by the extension of basement faults into the basin, by some syn-sedimentary structures (faults, rotated bedding) and by the offset of the unconformity along some faults. However, the role of the reactivation of these faults remains unclear; it is considered to be localized brittle deformation facilitating circulation of the fluid flow or considered to be linked to transient favorable geodynamics that enable the focused fluid expulsion through transient exit pathways, using pre-existing weaknesses in the rock mass and the reactivated fault network. This discussion is open as it is difficult to date the polyphased evolution of the basement faults outside of the basin or when their reactivation has not left any footprint within the basin. Thus, there is an uncertainty about the frontier between late retrograde metamorphism and ductile-brittle transition related to the late Trans-Hudson or Talston Orogens and mylonitic to cataclastic structures developed during and after the deposition of the Athabasca Basin. Second, if the structural control of ore deposits is well characterized along the Sue Trend (Tourigny et al. 2007; Benedicto et al. 2021a), at McArthur River mine (Harvey 2020) or Cigar Lake (Martz et al. 2017; Harvey 2018), then most studies would conclude that the role of the pre-existing faults is mainly mechanical, localizing brittle reactivation and traps, mainly in jogs and other subsidiary structures. However, recent detailed analysis of microstructures and their relation with successive deformation events, mainly in basement-hosted deposits, show a zonation of the hydrothermal alteration and mineralization between the core and damage zones of the fault, directly controlled by the reactivation of the fault (Grare et al. 2020; Benedicto et al. 2021b; Hillacre et al. 2021). This is in particular the case of basement-hosted uranium deposits along the Patterson Lake conductive trend, where Hillacre et al. (2021) underline the control of the fluid flow by previously established and newly formed subsidiary fractures, cataclasis zones and shear and mylonite zones, considering that these structures have a long lasting history. The development of fluid inclusion assemblages indicating fluid boiling in many uranium deposits, both in the eastern and western Athabasca Basin, also suggests episodic reactivation of the basement faults that caused dramatic drop in fluid pressure and subsequent fluid flow (Wang et al. 2018; Rabiei et al. 2021). Thus, the characterization of the successive stages of deformation, in relation with hydrothermal alteration in the basement and in the basin, is a major target for structural studies.

In a mineral system approach, the characterization of the main triggering event that enhances the permeability of the basement and enables the creation and the sustainability of large-scale fluid flow is key for constraining the requested conditions to form giant deposits and is discussed in the next section, which focuses on the fluid flow system.

3.2.2. *The fluid flow system*

3.2.2.1. *Hydrothermal alteration and mineralization in their structural context*

Various generalized models are envisaged for the fluid flow system that embrace the origin of uranium and the respective role of fluids from the basin and from the basement (Figure 3.8). As the crustal scale architecture and related structures resulting from the tectonic history are considered as critical elements in understanding the unconformity-related uranium mineral system, transient geodynamics and associated fluid flow is an innovative approach to consider these generalized models (Figure 3.2). If the structural control of the fluid flow is generally accepted, the primary attributes of fault zones (i.e. their structure, mechanics and fluid flow properties; Faulkner et al. 2010) and their roles in fluid flow are rarely considered as a self-organized system (Hronsky 2011; McCuaig and Hronsky 2014).

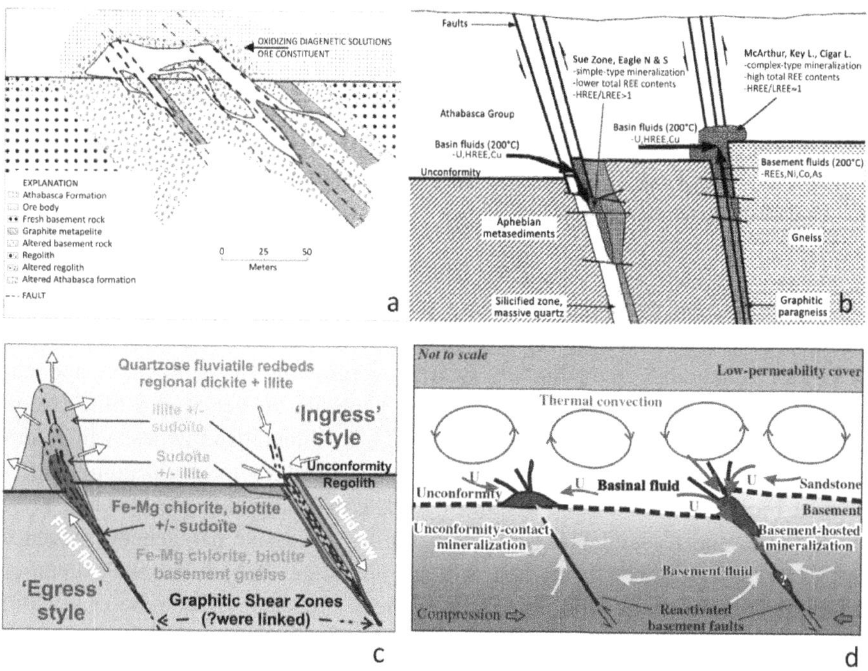

Figure 3.8. *Evolution of the conceptual models for the fluid flow at the origin of the uranium deposits in the Athabasca Basin*

COMMENT ON FIGURE 3.8.– *(a) Basin fluids are the only source of fluids (Hoeve and Sibbald 1978); (b) circulation of basin- and basement-derived fluids (Fayek and Kyser 1997); (c) "egress-" versus "ingress-" style alteration zones, reducing fluids from the basement; oxidizing fluids from the basin (Jefferson et al. 2007); (d) coexistence of thermal convection and compressive deformation-driven fluid flow (Li et al. 2021).*

The interaction between fluid overpressure and transient tectonic activity along reactivated fault systems (core and damage zones) is here discussed in order to evaluate how giant deposits can be generated in some very specific areas. Mineralogical, geochemical and geochronological data, relationships between pre- and syn-mineralization hydrothermal alteration and deformation episodes and new insights from fluid flow modeling are converging to propose some new concepts that can have significant impact on exploration at the regional scale.

All uranium occurrences are found within hydrothermal alteration halos that are the main footprint used for exploration. An attempt to define end members of the ore deposit models of the Athabasca Basin (IAEA 2018) has classified the alteration footprint to large (Cigar Lake) or tight/narrow (McArthur River, Eagle Point, Millennium) sandstone halos according to their size, and to illite and/or kaolinite, dravite, desilicification in the sandstone, and quite diversified mineral assemblages in the basement (massive hematite clay, chlorite-dravite-illite, chlorite, sericite, quartz-carbonate veins) (Figure 3.9).

As summarized by IAEA (2018), sediments of the Athabasca Basin record two types of regional scale alteration, basin-wide pre-ore diagenetic sandstone alteration and alteration halos associated with uranium mineralization. Diagenesis is preserved in barren zones and is marked by quartz overgrowths, hematitization and clay mineral transformation and conversion from kaolinite to dickite and illite (Hoeve and Quirt 1984; Quirt 2001; Chu et al. 2015). In mineralized zones, while some authors consider a major hydrothermal event related to mineralization with no marked break between diagenesis and hydrothermal alteration (Kotzer and Kyser 1995; Fayek and Kyser 1997; Reid et al. 2014), successive paragenetic sequences are attributed by others to at least one pre-ore stage of hydrothermal alteration, superposed to diagenesis and preceding the main ore stage (Alexandre et al. 2005, 2009; Derome et al. 2005; Cloutier et al. 2009, 2010; Ng et al. 2013; Kaczowka 2017, 2021). If it is difficult to make a distinction between the diagenetic evolution and the hydrothermal alteration. It is also important to consider the extent of this pre-ore stage of hydrothermal alteration and how much it is related to the ore stage so that it can be considered as a footprint of uranium deposits. It is likely that the onset of hydrothermal alteration related to mineralization corresponds to some changes in the dynamic of the basin. In addition to the mineralogical evolution, several lines of evidence enable the characterization of this change: the imprint of brine percolation as determined by fluid inclusion, the onset of the interaction between fluid circulation in the basin and the basement, the chronological constrains, the reactivation of the inherited Trans-Hudson fault system and the concomitance with hydrothermal alteration.

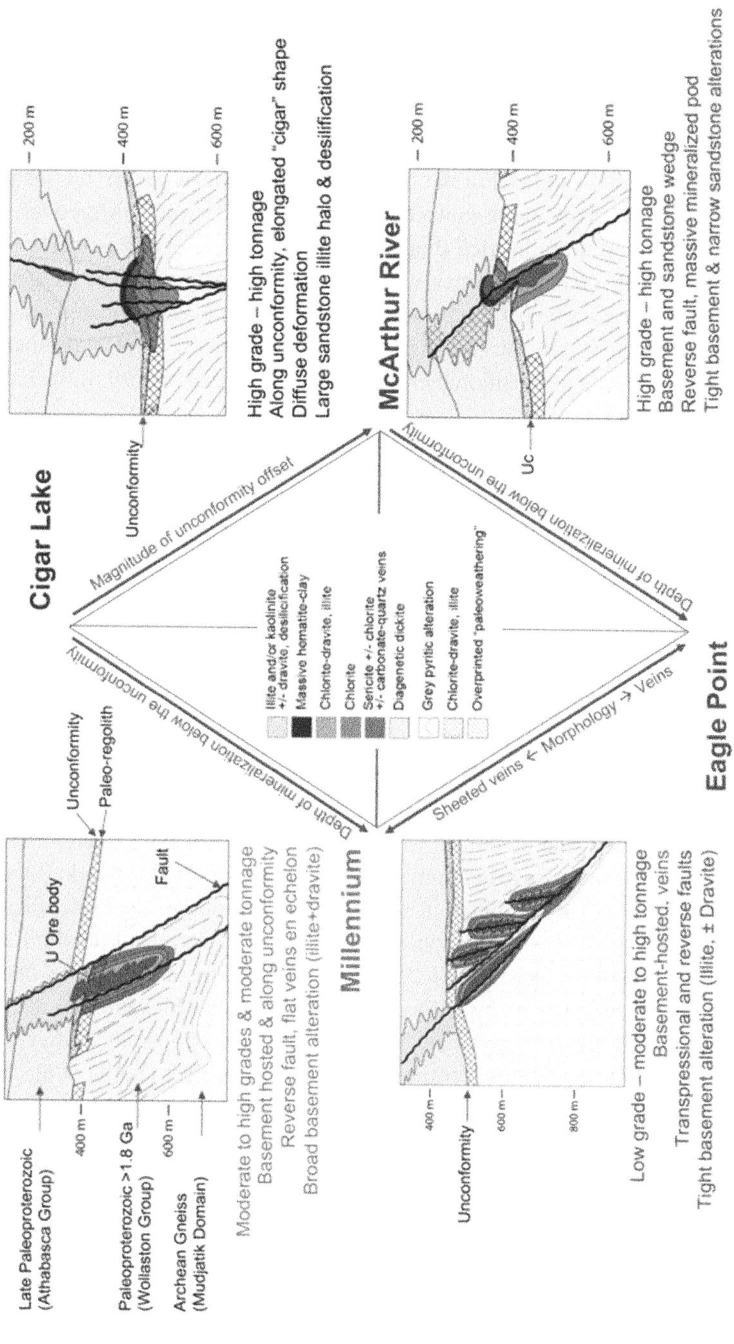

Figure 3.9. *Lithostructural classification diagram with end member summary characteristics (IAEA 2018)*

The brine origin of the mineralizing fluids has been determined from fluid inclusions (Richard et al. 2011, 2012) and boron isotope signatures in tourmaline associated with mineralization (Mercadier et al. 2012). Based on the mainly detrital litho-stratigraphy of the Athabasca Basin, these brines could only be generated by evaporitic seawater during the deposition of the Carswell Formation (Hendry and Weathley 1985; Chi et al. 2018), whose exact age is unknown but may be close to that of the underlying Douglas Formation (1,541 ± 13 Ma; Creaser and Stasiuk 2007), based on their transitional contact (Ramaekers et al. 2007). Moreover, Chi et al. (2019a) show that uranium-rich diagenetic fluids are preserved as inclusion in the quartz overgrowth in barren zones within pre-Douglas formations, suggesting that the downward percolation of the brines occurred within unconsolidated and still permeable sandstones. It can thus be concluded that diagenesis and hydrothermal alteration occurred during a transitional episode during the long history of the basin.

Thermally-driven free convection has been suggested to be the main force driving brines toward the basal Athabasca sandstones and basement rocks (Raffensperger and Garven 1995; Boiron et al. 2010). It is likely that these oxidized and acidic brines (Richard et al. 2012) deeply penetrated the underlying basement by analogy with the observed presence of fluid at depth in the continental crust (Emmermann and Lauterjung 1997; Cathelineau and Boiron 2010; Bons et al. 2014) and synthetized by Boiron et al. (2010) for unconformity-related deposits. Extensive exploration drilling has demonstrated the generalized presence of a pluridecametric alteration profile at the top of the basement, interpreted as a paleoregolith and/or the result of long term fluid/rock interaction between the fluids from the basin and the basement. The permeability of this profile is marked by a decompressed highly fractured zone, friable oxidized lithologies and a general drop of resistivity recorded by down-hole surveys. Development of large oxidized zones is also observed downdip within some fault core and damage zones reaching several hundred meters below the unconformity (Mercadier et al. 2010). The extension of the mineralization several hundred meters below the unconformity is well characterized for the Rabbit Lake/Eagle Point, Dawn Lake and McArthur River deposits where three distinct paragenetic stages, pre-, syn- and post-ore, are distinguished (Alexandre et al 2005; Cloutier et al. 2011). The discoveries of the Millennium deposit (Cloutier et al. 2009), the Gryphon mineralization on the Wheeler River project (Sorba et al. 2016) and the Arrow deposit along the Patterson lake conductive trend (Hillacre et al. 2021) have confirmed and documented how deep fluids originally from the basin have been penetrating and exchanging with basement lithologies (Rabiei et al. 2021).

The percolation of these brines in the basement is also recorded by the stable isotope geochemistry (Kotzer and Kyser 1995) and the geochemical signatures of fluid inclusions. Thus, analyses of halogen concentration and stable chlorine isotope composition of fluid inclusions from hydrothermal quartz and carbonate veins in the basement of the Athabasca Basin illustrate the interaction between meteoric water and/or seawater, evaporites and basement rocks (Richard et al. 2011), chlorine leaching of biotite during their retrogression to chlorite being considered as a minor source of salinity for both NaCl-rich and $CaCl_2$-rich brines involved in the uranium mineralizing events. In the case of the Cigar Lake deposit, Martz et al. (2019a) show that the circulation of the basinal brines into these basement structures was enhanced and structurally controlled by the microfracture network, as demonstrated in other localities (Mercadier et al. 2010). Thus, the similar distributions of micro-fracture surface densities obtained for Ca-dominated fluid inclusion planes attributed to pre-Athabasca alteration stage and brine inclusion planes indicate that this brine circulation was greatly influenced by the microfracture network inherited from Trans-Hudson exhumation events (Mercadier et al. 2010; Martz et al. 2019a). These structures favored intense interaction between the brines and basement rocks that induced the formation of illite-sudoite alteration halos and scavenged Mn, K, Ba, Sr and especially U, and evolved from the original NaCl-dominant composition toward a higher salinity $CaCl_2$-rich composition. This evolution from the original NaCl-dominant composition toward a higher salinity $CaCl_2$-rich composition is considered as a marker of enriched uranium structures. It should however be noted that Ca-enriched brines do not necessarily indicate interaction with basement rocks, as such brines could be developed within the basin (Chu and Chi 2016; Chi et al. 2019a) and may be dominantly derived from the basin even when they are located in the basement (Rabiei et al. 2021). In the case of the Key Lake uranium deposit, Wilson and Kyser (1987) have shown that differences between the $\delta^{18}O$ values in fluids that formed basement-hosted deposits result from the interaction of basinal fluids with various lithologies. They concluded that if the similarity in H and O isotopic compositions of both basin- and basement-hosted deposits with the basinal brines is consistent with this fluid being dominant in the formation of the deposits, then a basement-derived fluid is also involved; this is particularly evident thanks to the compositions of fluids in basement-hosted deposits. According to Richard et al. (2012), if these giant deposits were formed from these exceptionally uranium-rich acidic brines, the onset of the interaction between fluid circulation in the basin and the basement is then a milestone and a transitional episode in the metallogeny of the Athabasca Basin.

Up to now, dominantly C-O-H(-N) fluid inclusions trapped within ductile-brittle graphitic shear zones have been considered to be associated with the retrograde metamorphism during the exhumation of the Trans-Hudson Orogen (Martz et al. 2017). However, the high geothermal gradient recorded during the formation of the ore deposits cannot exclude that deeply rooted fault systems could operate as conduits to bring fluids and heat toward the upper parts of the crust, even after the deposition of the Athabasca Basin. Possible deeply rooted basement fluids mixed with fluids originating from the basin would certainly lead to a major revision of the current model (Rabiei et al. 2021). Investigating gradients downdip in the main reactivated shear zones recorded by mineralogy and fluid inclusions could be one way to test the hypothesis that brittle faults marking a significant offset of the unconformity could be rooted at depth and have a ductile equivalent enabling metamorphic fluids to be generated. In such a scheme, the role of active faulting and shearing processes would be a trigger of the initiation of a self-organized hydrothermal system (Micklethwaite et al. 2010; Hronsky 2011) and the brittle-ductile transition could be considered as a threshold barrier.

Chi et al. (2018) have compiled a large number of U-Pb ages of uraninite and some Ar-Ar ages of illite alteration related to U mineralization obtained from U deposits in the Athabasca Basin, ranging from ca. 1,590 to 60 Ma. Many of these ages likely reflect U remobilization and it is considered that the maximum U-Pb ages obtained from uraninite in the largest U deposits (e.g. McArthur River and Cigar Lake) are close to the ca. 1,540 Ma age assumed for the development of basinal brines during the deposition of the Carswell Formation (Chi et al. 2018). The conventional belief is that uraninite precipitated over very long periods of time (Hoeve and Quirt 1987), perhaps hundreds of millions of years (Kyser et al. 2000) and this is reflected in most of the schematic representation of the hydrothermal systems. However, recent findings and the comparison with their modern counterpart exploited as geothermal reservoirs, suggest that the life of such hydrothermal systems might be much shorter (a few tens of thousands of years) and more chaotic than previously assumed (McCuaig and Hronsky 2014; Cox 2016; Cernuschi et al. 2018; Moncada et al. 2019). Moreover, it seems that such pulsations could correspond to modifications of the crustal stress regime marked by highly localized deformation and enhancement of the permeability of the geological systems (Manga et al. 2012). Although no geochronological constraints exist on the time required to form the Athabasca deposits, first-order approximation from Raffensperger and Garven (1995), Richard et al. (2012) and Dargent et al. (2015) suggest that 0.1–1 Myr is a viable estimate for the duration of the mineralizing event. If this change of paradigm is considered, the mineralization event could then correspond to a transient event, limited in time and space.

The structural control for focusing mineralizing fluids and mineralization is well established as ore formation is spatially and temporally focused by the reactivation of the pre-Athabasca fault system (Hoeve et al. 1980; Kotzer and Kyser 1995; Baudemont and Paquet 1996; Fayek and Kyser 1997; Thomas et al. 2000; Jefferson et al. 2007). The general concept summarized in IAEA (2018) is that deposits were formed in fault zones and along the unconformity where reduced fluids from the basement and oxidized fluids from the basin were mixed. Structural traps opened during active faulting and were repeatedly brecciated (Hoeve and Quirt 1987; Kyser et al. 2000). In more detail, the alteration halos considered to be related to the deposits are always developed in the vicinity of a fault system that crosscuts the unconformity. Most of these faults correspond to reactivated-inherited fault zones of the basement and have an associated reverse offset of the unconformity, up to several hundred meters along the Virgin River shear zone for example and below 100 m in the Wollaston-Mudjatik Transition Zone (Figure 3.10). These faults are often graphite-rich, and cohesive to non-cohesive breccias are the most obvious expression of this reactivation.

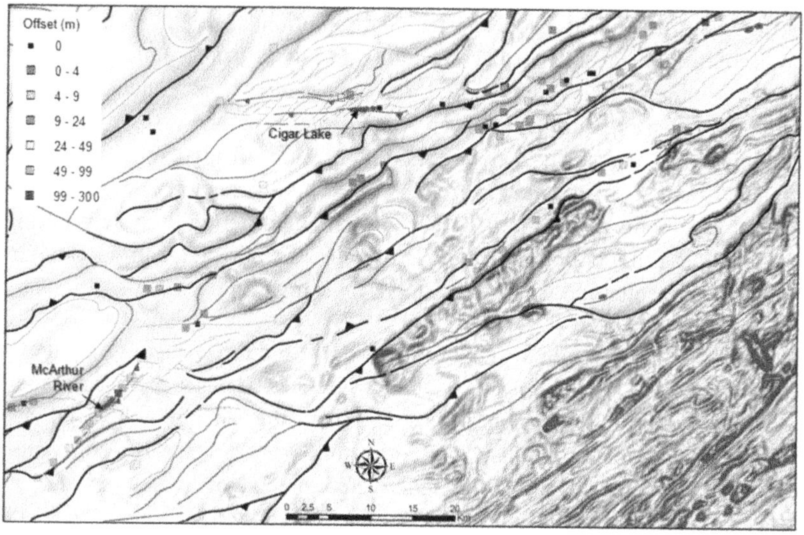

Figure 3.10. *Calculated offset of the unconformity along the first- and second-order fault systems*

A detailed study has been done along the Cigar North conductor where the post-Athabasca reactivation is attested by an offset up to 30 m of the unconformity by a ductile-brittle shear zone (Figures 3.11 and 3.12). Along with structural logging on core, microscopic observations show that the basement rocks from the fault cores and damage zones have recorded deformation patterns ranging from ductile to brittle, some ultramylonitic textures being suspected to correspond to pseudotachylite. Hydrothermal alteration and mineralization phases happened on the later brittle stages and correspond to: a first dissolution-precipitation event affecting the fault rocks, with the development of stylolites and veins of macrocrystalline undeformed quartz (Figure 3.11); the precipitation of sudoite associated with dissolution and initiation of the cataclastic stage; the precipitation of microcrystalline illite associated with dissolution and the main cataclastic stage (Figure 3.12); and an extensive carbonate veining. Relations between the alteration phases and micro-structures illustrate how quartz dissolution in the shear zone along with brittle reactivation of the structures created an open space for fluids to percolate. A first silicification event took place in conditions set below the quartz plasticity domain. Sudoite replaced primary minerals and hydrothermal quartz during the first brittle event associated with a dissolution stage. Illite, also associated with a dissolution event, then replaced sudoite and other primary minerals during a second and main brittle event. Based on this study, it is concluded that the interaction between pressure solution creep and fracturing processes were the main deformation mechanisms active along this type of fault, making it behave in both a brittle and viscous manner (Gratier et al. 1999). This interaction is considered as a critical factor to constrain the permeability of this inherited fault system during its reactivation and fluid flow. The role of graphite as lubricant and as catalytic is probably a major factor and is presently under investigation.

A comparison with the deformation observed in the direct vicinity of the Cigar Lake deposit about 1 km south shows some differences. It is generally accepted that there is no offset at the unconformity and that the "bump" observed at the unconformity, underlying the deposit, corresponds to a Trans-Hudson paleohill (Bruneton 1987). Consequently, the reactivated brecciated structures and semi-brittle faults, which served as the main conduit for fluid circulation, are considered to be pre-Athabasca (Harvey 2018).

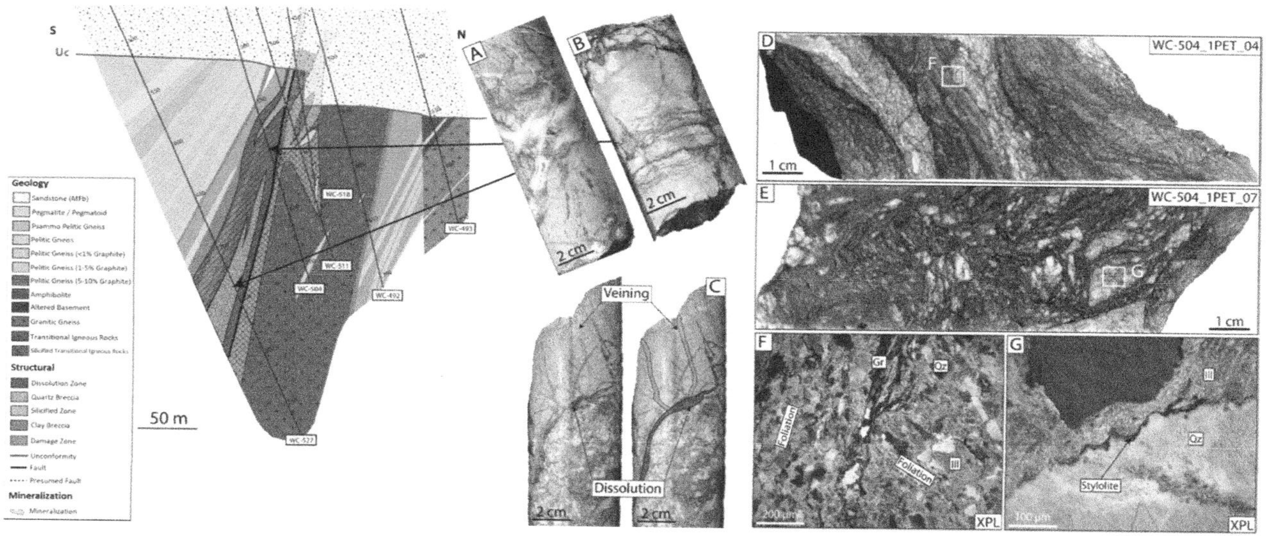

Figure 3.11. *Macro- and microstructures across a reactivated Trans-Hudson fault zone. (A) Polyphased quartz breccia; (B) pressure solution in silicified gneiss; (C) dissolution and veining in silicified gneiss; (D and F) pressure solution creep in metasediment ; (E and G) stylolithe and dissolution*

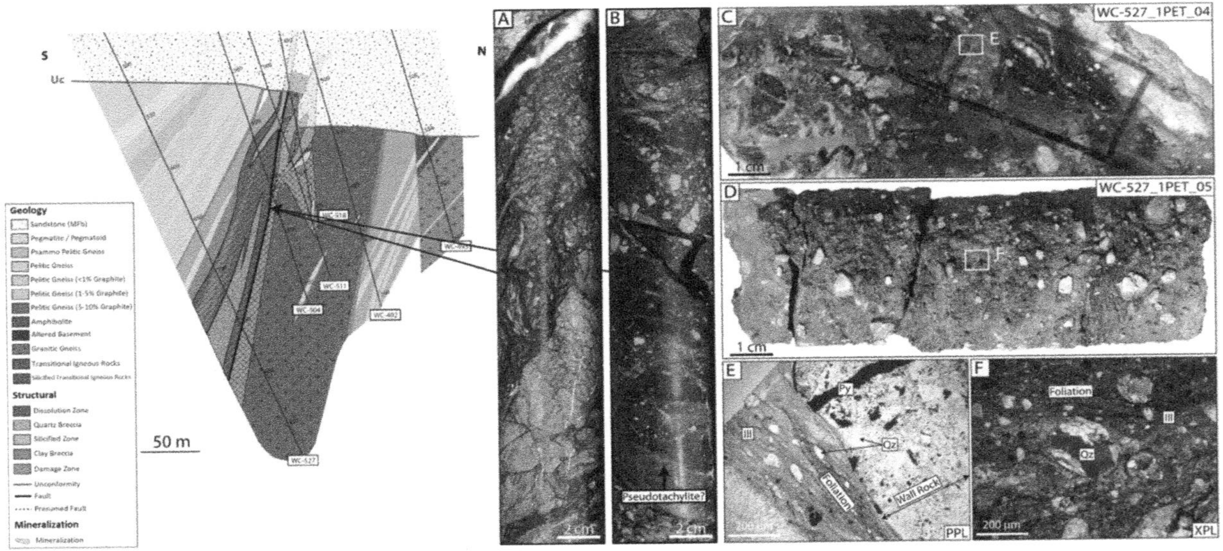

Figure 3.12. *Macro- and microstructures across a reactivated Trans-Hudson fault zone. (A and B) Graphitic cataclastic breccia; (C and E) polyphased quartz breccia and sulfide; (D and F) cataclastic flow in the graphitic zone*

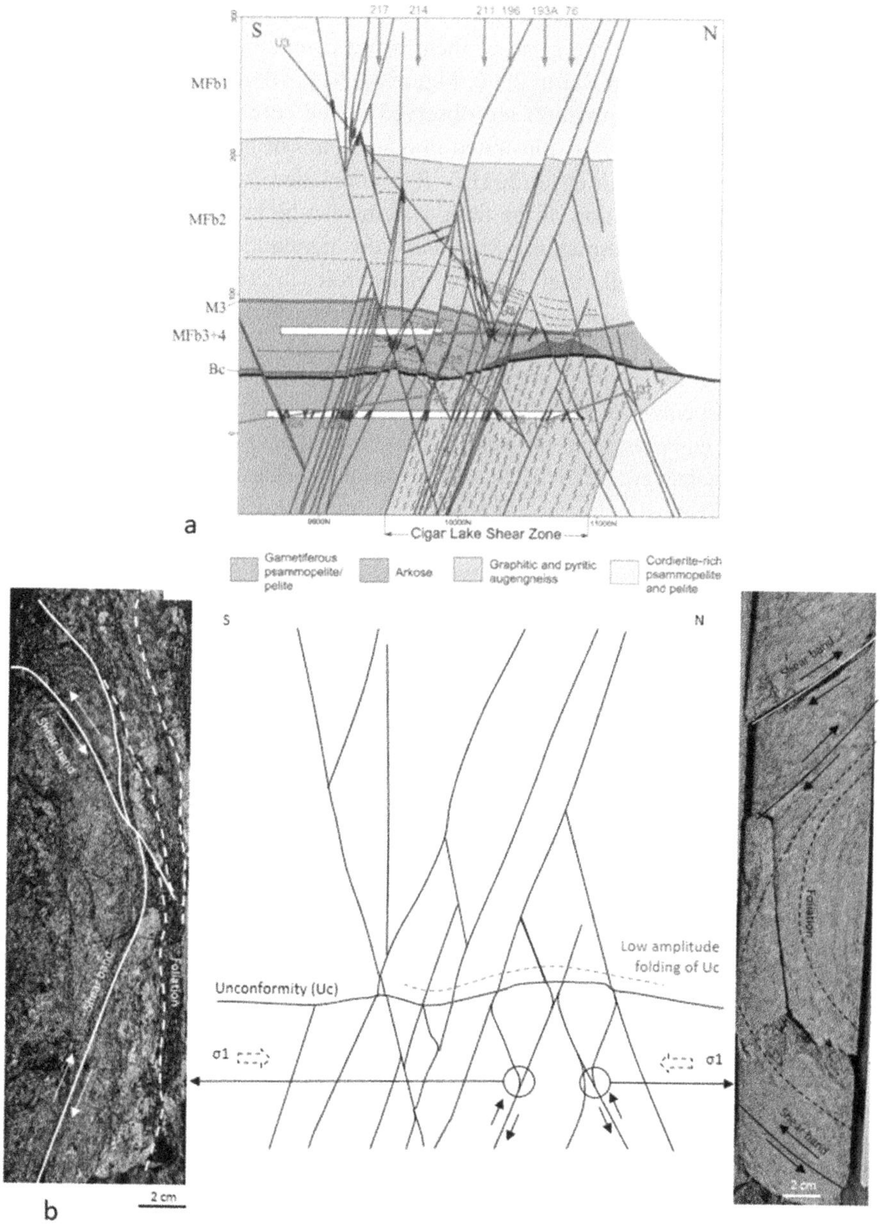

Figure 3.13. *Geological sections across the Cigar Lake deposit. (a) Original cross section from Baudemont and Rafini (2000); (b) re-interpreted cross section considering the conjugate shear zones and associated deformation*

However, drillcore and cross sections reveal the existence of two sets of faults cross cutting the unconformity, one of them being parallel to the graphite-bearing fault zone (Baudemont and Rafini 2000, Figure 3.13(a)). Moreover, conjugate shear zones and associated deformation are observed in the core and damage zones of these faults (Figure 3.13(b)) as illustrated in other zones of the Wollaston-Mudjatik Transition Zone (Benedicto et al. 2021a). It is concluded from these observations that the structural reactivation of the fault systems can be marked by displacement but also by discrete deformation zones and low amplitude folding, up to now considered as a paleohill, that have a major role in the partitioning of the deformation and on the enhancement of the permeability of the whole system.

Understanding the geodynamic trigger of this reactivation is a major stake. Based on the dispersion of the U/Pb ages of the mineralization, many attempts of correlation with compressional far-field tectonic events have been done (Kyser and Cuney 2015; Sheahan et al. 2016). These correlations are based on geochronological data that are too spread out to be conclusive (Figure 3.14, Chi et al. 2018). Several structural observations can be brought forward to constrain this trigger. Because of the unfavorable dip of these faults in the compressive stress field, their reactivation could only occur when fluid pressure exceeded the lithostatic load (Sibson et al. 1988; Sibson 2020). The reactivated faults are mainly the second-order faults and the measured displacement along these faults is limited to a few kilometers along strike (Figure 3.10). As described above for the Cigar North conductor and for almost all deposits (IEA 2018), there is always a concomitance between this tectonic activity and successive paragenesis related to hydrothermal alteration.

Thermo-hydraulic and mechanic (THM) modeling (Chi et al. 2019b) is interesting to constrain thermally-driven free convection (Raffensperger and Garven 1995; Boiron et al. 2010) and to evaluate different scenarios under which the fluid flow is controlled by faulting-related deformation and permeability enhancement. The temperature of the mineralizing fluids (150–250°C), suggested by the alteration mineral assemblages and some fluid inclusion studies (Pagel 1975; Hoeve and Sibbald 1978; Pagel et al. 1980; Hoeve and Quirt 1984; Kotzer and Kyser 1995; Derome et al. 2005; Kister et al. 2005; Richard et al. 2016) can be explained by a normal geothermal gradient, that is, at the base of a 5–7 km thick basin, or by abnormal thermal gradients at much shallower depths of 1–3 km (Boiron et al. 2010; Chi et al. 2018; Martz et al. 2019a). Chi et al. (2018) inferred from the history of the basin, alteration mineral assemblages and fluid inclusion studies that the burial depths of the unconformity surface were likely less than 3 km at the time of mineralization and that local or basin-scale elevation of geothermal gradients and upwelling fluid convection are responsible for this high heat flow.

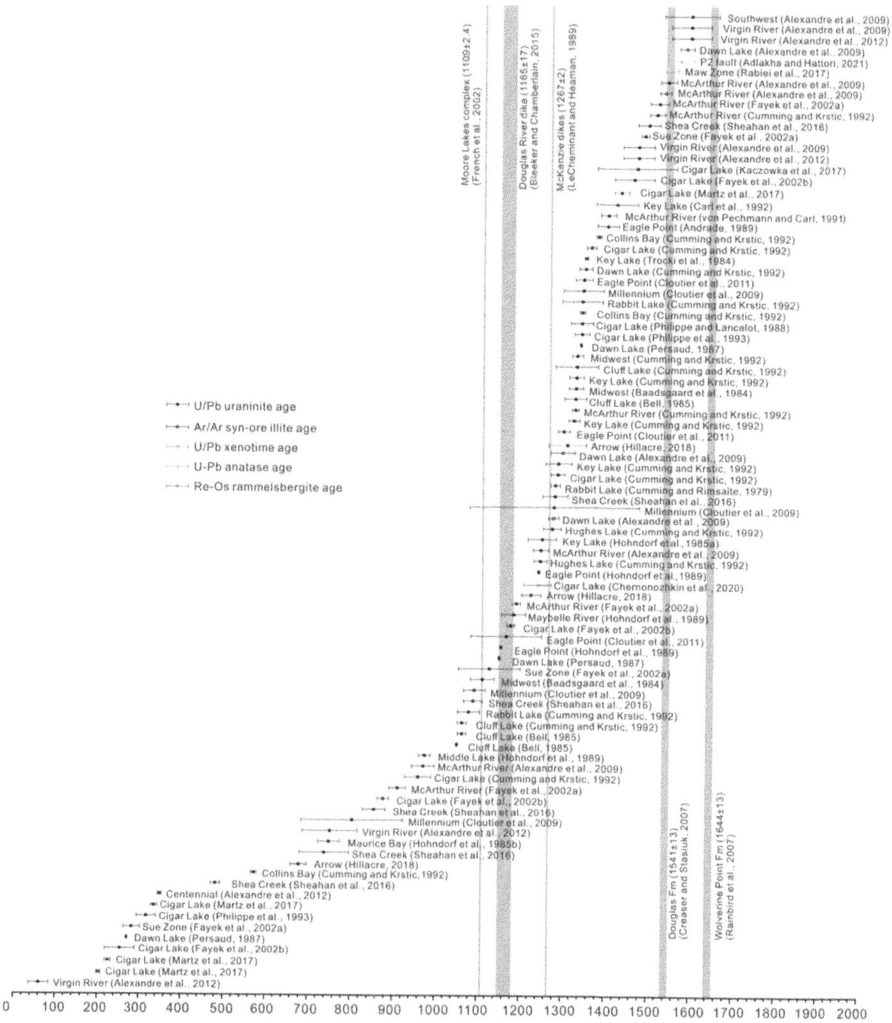

Figure 3.14. *Compilation of isotopic ages associated with U mineralization in the Athabasca Basin, completed from Chi et al. (2018)*

The possible origin of such a high geothermal gradient remains poorly constrained as no intrusion or volcanic activity is synchronous with the main stage of the ore deposit formation. However, a link has already been suggested by Pehrsson et al. (2016) between the location of the main uraniferous trends regarding the Nuna supercontinent breakup and large wavelength anomalies related to mantle upwelling may have developed. A bright band of reflectivity at a depth of about

8,000 m has been interpreted as a series of sub-horizontal sheet like diabase intrusions related to the Mackenzie igneous event at 1,270 Ma (Hajnal et al. 2007) but nothing excludes that they could correspond to an earlier intrusion related to the Nuna supercontinent breakup. Finally, modeling suggests that deeply rooted fault systems inherited from the pre-Athabasca orogens could operate as conduits to bring fluids and heat toward the upper parts of the crust, in particular during episodes of tectonic reactivation (Poh 2019). The discovery of carbonatite dykes within the fault system of the Patterson Lake corridor is another observation that supports such coupling between deep-seated and shallow processes (Johnstone et al. 2021), although this particular event probably took place before the Athabasca Basin.

This high geothermal gradient led Martz et al. (2019a) to propose that, for example, the basement-hosted shear zone below the Cigar Lake deposit could have acted as an upwelling flow zone from deeper zones in the basement to basin between two thermal convection cells perpendicular to the structure as predicted earlier (Hoeve and Quirt 1984) and through 2D numerical modeling (Eldursi et al. 2020; Li et al. 2021). Based on these thermal profiles and taking into account the global compressional tectonics recorded during the reactivation of the basement-hosted fault system, THM numerical modeling of simplified geometries of several deposits (Li et al. 2016, 2017, 2018a, 2018b; Chi et al. 2019b; Eldursi et al. 2020) shows that basement faults generally promote fluid flow. Both ingress flow and egress flow may be generated by either compressional deformation or fluid convection. The actual flow direction for a given basement fault is influenced by a combination of many factors, especially the dip angle and dip direction as well as spacing of neighboring basement faults (Chi et al. 2019b).

The review of these different parameters highlights several critical elements that are compatible with other mineral systems (McCuaig and Hronsky 2014). A new model can be proposed considering that the interplay between compressional fault reactivation and thermal convection is the main driver of this fluid flow system (Figure 3.15; Li et al. 2021). Deformation-driven fluid flow, possibly short-lived, is recorded by the structural and mineralogical evolution of the reactivated basement-hosted fault system. This tectonically active period may be associated with a far-field tectonic event and separated by long periods of quiescence. However, by analogy with enhanced or natural geothermal systems (Cox 2016), the possibility that these episodes are related to transient tectonic activity caused by fluid over-pressure should be also considered, interplay between compressional fault reactivation and thermal convection still being applicable. Thus, the formation of the giant deposits could happen during times when the prevailing geodynamic conditions impose strong threshold barriers to fluid flow, causing fluid-flux systems to become self-organized (Hronsky 2011). In this scenario, the presence of a basement-hosted reservoir is suggested and the fluid expulsion occurred through transient permeable exit pathways, faults acting like valve system ("fluids drive faulting"; Sibson 1981, 2020). In this context, uranium may precipitate in structural traps as jogs and veins within the core and damage zones of the faults, graphite and sulfide being potential reducing agents facilitating the precipitation or reach the unconformity and be trapped in the permeable sediments at the base of the basin if a clay or silicified cap is present. It should be pointed out that given the sandstone-dominated nature of the Athabasca Basin, the fluid pressure within the basin was most likely hydrostatic, as demonstrated by numerical modeling (Chi et al. 2013). Furthermore, due to the fractured nature of the top of the basement, as discussed above, the upper part of the basement immediately underlying the unconformity may be under hydrostatic pressure regime too. The transition between hydrostatic and lithostatic (overpressure), where the fault valve system may be operating, is likely located in the deeper part of the basement (Song et al. In review).

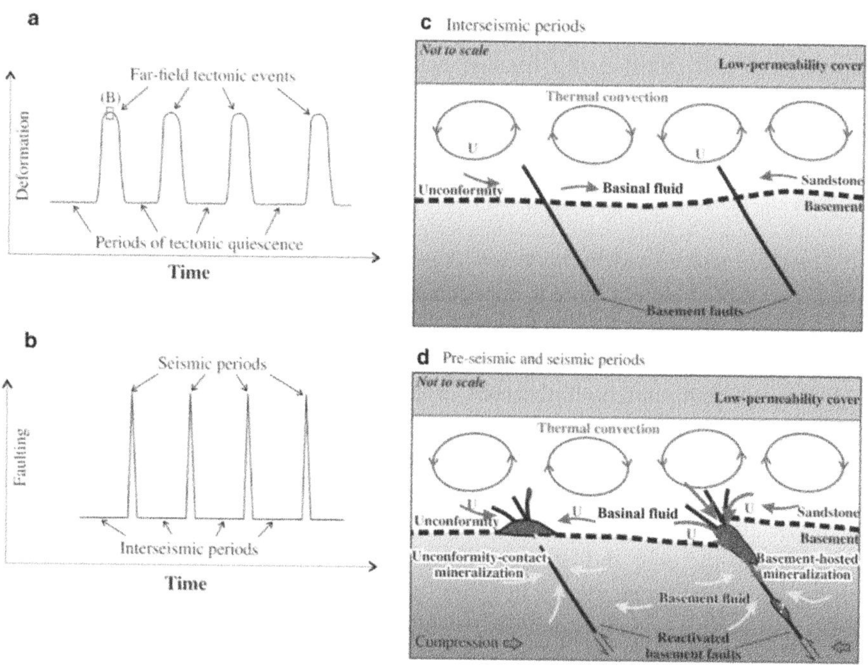

Figure 3.15. *Schematic diagram illustrating an idealized hydrodynamic model (Li et al. 2021)*

COMMENT ON FIGURE 3.15.– *(a) Division of the basin history into periods of tectonic quiescence and tectonic events; (b) division of a period corresponding to a tectonic event into seismic and interseismic periods; (c) dominance of thermal convection during interseismic periods; (d) coexistence of thermal convection and compressive deformation-driven fluid flow during pre-seismic and seismic periods – the convection cell may be temporarily destroyed during seismic events.*

3.2.2.2. *Perspectives for further research to guide exploration*

Regarding the Athabasca uranium mineral system, this model requires further investigation in terms of exploration and the identification of potential traps.

The association of mineralization with a major episode of downward infiltration of brines is recorded by a large range of observations in the basin and the basement and marks a break in the history of the Athabasca Basin. The contemporaneity of fluid circulation and tectonic reactivation of an inherited fault system in a global compressive field is then indicated by faults crosscutting the unconformity, transitional ductile-brittle structures, large dissolution and recrystallization of mainly clays, micas and chlorite in a fracture network. Hydrothermal breccias and veins reveal the polyphased tectonic activity and the evolution of the composition of hydrothermal systems with alternating periods of silicification, desilicification and carbonate infillings that are apparently concomitant with the diagenesis recorded at the base of the basin.

The basement-hosted fault systems have a fundamental role in the mineralizing process. As pointed out above, they constrain the crustal segmentation and are natural links between deep and shallow processes (Chi et al. 2019c). Thus, the connection between basinal fluids with the basement is testified by fluid inclusions and chemical footprints both observed in the basement and the basin. Although the expression of the deformation may be different in both environments, microstructures and mechanisms of deformation illustrate how permeability is enhanced by microfracturing, dissolution and recrystallization processes controlled by a set of fractures and a combination of viscous and brittle deformation, distributed in the core and damage zones of these faults (Abdelrazek et al. 2021). Further investigation is needed in order to evaluate what are the main parameters that condition this level of permeability-enhancement and then the ability of a fault system compared to another to become the locus for maximum fluid-flow. Among these factors that are already identified, one can mention the degree of deformation and associated displacement, the dip angle and orientation of the basement fault in the far- and near-field stress conditions, the number and spacing of neighboring faults, whether the fault extends into the basin or not and the rheological contrasts between lithologies (Chi et al. 2019b). Further structural analysis and modeling should provide additional qualification and quantification in order to evaluate the fertility of these faults and then guide exploration.

The coupling between the structural rejuvenation of the inherited fault system and the hydrothermal system suggests that fluid flow has at least partially controlled the mechanical behavior of active faults and the duration of seismic cycles (Figure 3.15). In such a context where energy gradients are built up and organized fluid flux systems cause fault reactivation, the existence of basement-hosted reservoirs and a lithological or structural barrier is implicit, although they have not been characterized so far as such (Song et al. In review).

The pre-ore hydrothermal alteration is responsible for large dissolution phenomenon creating up to 15–25% void space (Alexandre et al. 2005; Cloutier et al. 2009). It may also have played a role for ground preparation, facilitating fluid flow through enhanced permeability and facilitating basement fault reactivation as a lubricant. Looking for analogs, the development of an enhanced geothermal system (EGS) and the exploration of deep geothermal resources in the Rhine Graben have revealed that naturally fractured reservoirs are found at depth within the granitic basement (Vidal et al. 2017; Vidal and Genter 2018). The main fluid source of the discovered geothermal reservoirs was hosted in the fracture system of granites showing wide desilicification and illitic alteration halo (Sausse et al. 2010; Glaas et al. 2021). In situ experiments have shown that the natural permeability network was able to be stimulated by mechanical and chemical processes, mainly through forceful fluid injection at the base of an injector well (Evans et al. 2005). Structural analysis done in these boreholes has demonstrated that the stimulation of the permeability was mainly related to the development of fracture systems tightly controlled by inherited reactivated structures within these previously altered lithologies. Moreover, induced seismicity following fluid injection illustrates how an increase in fluid pressure can trigger the expression of shear component along a pre-existing fault plane (Dorbath et al. 2009; Cox 2016).

Silicified zones, some of which are pre-Athabasca in the basement, clay alteration halos and the sealed early vein network, prior to the main stage of uranium mineralization, should be seen as potential barriers that may have caused an increase in fluid pressure within a basement-hosted reservoir. This coupling could lead to a new model in which the Ca-rich brines that scavenged metals from the basement (in particular Mn, K, Ba, Sr, Pb, U) could have been accumulated in fractures and zones of dissolution located within the basement and expulsed toward the unconformity when the fluid pressure was sufficient to trigger the expression of a shear component along the inherited fault system. Moreover, although reducing fluids from the basement are described in only few case studies, a contribution of metamorphic fluids cannot be excluded as some of these reactivated fault systems may be extended at a depth below the brittle ductile transition.

In conclusion, the stimulation of the permeability network of the basement is considered as a necessary condition for creating fluid convection that is at the origin of unconformity-related uranium deposit. An analogy between unconformity-related uranium deposits and the EGS is proposed in terms of the role of the basement fault system, heat convection and deformation-driven mechanisms for fluid flow (Ledru 2019). In both cases, a pre-existing permeability network is stimulated, whether by forceful injection of fluids or by tectonic processes. The temperature range to which this stimulation occurs is around 200°C for both environments, and fluid rock interactions are observed, marked by the leaching of a large volume of rocks and progressive enrichment in metals. Thus, it is proposed to consider the unconformity-related uranium deposits of the Athabasca Basin as paleo-enhanced geothermal systems.

3.2.3. *Fertility*

Fertility is a key parameter in terms of the expected prospectivity of the terranes and constitutes the third critical element of the mineral systems (Figure 3.2; McCuaig and Hronsky 2014). Most of the time, it is appreciated according to the matching with the ore deposit model. However, some terranes are considered to be poorly fertile until a discovery is done while "fertile" trends are intensively drilled in the vicinity of known deposits without significant or economic impact. This appreciation is also directly related to the supposed source of metals that for many ore deposits remain highly hypothetical. In the case of the unconformity-related uranium deposits, potential uranium sources exist both in the basement and in the basin and an exhaustive review has recently been published by the IAEA (2018). There have been two schools of thought about the source of uranium – from the basin or from the basement – and this is a continuing debate.

Sources in the basement are found in Paleoproterozoic metasediments, particularly in graphite-rich horizons derived from black shales that generally have a U content above those of Clarke. Paleoproterozoic leucogranites and anatectic pegmatites are also widespread in the vicinity of fertile trends (McKechnie et al. 2013; Mercadier et al. 2013). Thus, magmatic uranium oxides related to intrusions and high-temperature vein-hosted uranium oxides presenting clear evidence of alteration could have provided easily leachable uranium for the metal enrichment of basinal brines during their percolation within the basement (Mercadier et al. 2013).

The hypothesis for a source of uranium in the basin relies on the fact that oxidized fluids have the ability to leach and transport uranium oxide over large distances in permeable detrital rocks, especially at this period during which Earth's atmosphere-hydrosphere-biosphere-lithosphere endured a major change with the change form depleted to enriched oxygen atmosphere (Hoeve and Sibbald 1978; Fayek and Kyser 1997). Recently, Chi et al. (2019a) have discussed the U fertility of the Athabasca Basin by showing that fluid inclusions in quartz overgrowths from the barren part of the basin display U-rich brines more than two orders of magnitude higher than most naturally occurring geologic fluids. By considering that uranium deposits are developed across the whole Athabasca Basin regardless of the difference in basement lithologies between the Rae and Hearne cratons, they conclude that the uranium was mainly sourced from the basin, although the uppermost part of the basement may also be included, as the hydrogeological framework including the sandstone and the basement top facilitated fluid circulation and extraction of U from source rocks.

It should be pointed out that fertility in the context of the minerals systems concept also includes the availability of ligands for metal complexing (McCuaig and Hronsky 2014). Thus, the development of basinal brines in the Athabasca Basin due to seawater evaporation may have been facilitated by its paleogeographic position (Richard et al. 2011) and the right timing of this event with respect to potential geodynamic processes at depth (Chi et al. 2018).

Although knowledge of the source of the mineralization is generally not considered as a priority for guiding the exploration at the prospect scale (McCuaig et al. 2010), a different source model may imply a different exploration strategy for uranium at the district scale. If uranium is sourced from certain fertile basement rocks, these lithologies should be the targets of exploration. Conversely, if uranium is sourced from the basin, then any part of the basement could be mineralized as long as reducing agents are available and structures enabling fluid exchange between the basin and the basement are present. However, as underlined in the analysis of the fluid flow system, as far as the fertility of both basin and basement is established, the interplay between thermal convection and compressional fault reactivation is probably more important in terms of fertility than the presence of specific lithologies.

3.2.4. *Preservation*

The preservation of the main ore deposits is the fourth critical element of mineral systems (Figure 3.2; McCuaig and Hronsky 2014). Remobilization and transportation of the initial uranium deposits is well documented in the basin above many deposits, confirmed by textural observation, isotopic and geochronological data (Kister et al. 2004; Sheahan et al. 2016; Martz et al. 2019b). This is partly related to the ability of uranium to be leached in an oxidized and acidic solution and is also illustrated by the frequent "perched" mineralizations in the Athabasca Basin, found along fault zones. The possibility that extensive trends showing low-grade uranium mineralization represent remnants of originally massive and high-grade orebodies cannot be excluded. The characterization of paleo-transfer zones that could have facilitated the remobilization of the ore deposits is another research topic of interest in order to evaluate if some showings can be considered as pathfinders toward economic deposits.

The generation and role of cap formations like clay-rich early-alteration zones and silicified horizons in the basal formations of the basin remain poorly constrained and require further investigation, as well as the clamped permeability of the faults during the successive stress regimes. Concerning this last element, as some deposits are present within structural traps rooted as deep as 1 km below the unconformity (e.g. Arrow deposit; Johnstone et al. 2020; Abdelrazek et al. 2021; Hillacre et al. 2021) and as mineralizations are found at quite a large distance of the outcropping basins (i.e. the Contact Uranium prospect 12 km south of the Thelon Basin; Grare et al. 2018, 2020), trapping of uranium within basement-hosted fault systems should not be ignored for exploration at the district scale even if the basin strata and the basement have been eroded. Furthermore, for a given fault that hosts known uranium deposits, the possibility of finding additional orebodies at depth should not be excluded from an exploration project, not only because the same fault can be a conduit of ingress or egress fluid flow at different times (Li et al. 2017), but also because the basement-hosted orebodies have a relatively high preservation rate.

3.3. Implication of the mineral system concept applied to the exploration of unconformity-related uranium deposits

In periods of restricted budgets and limited funds available on the stock market, major and junior companies are looking to optimize the drilling campaigns by improving the targeting and by asserting as soon as possible the probability to discover a mineable deposit (Kerr and Wallis 2014; IAEA 2018). Experience from historical discoveries, exploration of fertile trends (i.e. along conductors) and recent discoveries in non-conventional targets provides opportunities to reassess some of the pre-conceived ideas and models. The exploration of unconformity-related uranium deposits mainly focuses on the identification of geophysical, lithological, geochemical and mineralogical footprints with an empirical approach that posits that more footprints equals greater chances of finding a deposit. The collected datasets and processed information and models define pathfinders that are a priori vectors toward mineralization (Jefferson et al 2007; Lesher et al. 2017; IAEA 2018). Thus, common practice for "conventional" drilling programs is to target conductive horizons near the unconformity by trying to quantify alteration halos and geochemical pathfinders. However, in many areas, economic uranium concentration is missing although all footprints considered as positive are present. On the other hand, "non-conventional" deposits are found in zones lacking some of the footprints that are generally associated with mineralization. For example, basin-hosted alteration halos or uranium enrichment are lacking above the basement-hosted Arrow and Gryphon deposits and there is no graphite in the vicinity of the Centennial deposit. The signification of these absences can be related to the origin of the deposits or simply to a lack of preservation of these footprints. Targeting must then evaluate the relative weight of some footprints compared to others. Then, the characterization of the mineral systems, aiming to understand the dynamics of paleo-hydrothermal systems at a regional to camp scale, and the conditions for preservation are likely to improve the delineation of fertile trends. The assessment of the mineral potential at the district and camp scale is crucial and requires a combination of data and knowledge-driven approaches. This review, based on the compilation of recognized scientific results and exploration success stories, proposes the implementation of some practices on the current workflow for project generation by focusing on the discretization of physical properties of terrains and structures on the surface and at depth and the identification of transient events that may correspond to a spatial and temporal coincidence at the origin of large to giant deposits.

At the district scale, the project generation can rely on a knowledge-driven approach based on regional and thematic studies and on the data-driven processing of geophysical surveys. First-order lithological assemblages, tectonic boundaries and structures can be defined, represented and analyzed under different formats using available geographic information systems. The interpretation of the Wollaston-Mudjatik Transition Zone presented in this chapter is an example of such preliminary reassessment of the architecture of the explored district. It could be completed by performing multi-variate analysis, testing different direct and inverse modeling scenarios of single or coupled geophysical datasets and making supervised and un-supervised classification of the physical parameters of the main lithological assemblages and structures. Integration of knowledge and datasets at this scale raises several concerns as, especially in zones that are under cover or poorly surveyed, information and data have most of the time been collected over several decades while standards, analytical means and scientific knowledge have progressed. Thus, harmonization, compilation in a homogenized database and quality control require an operating mode that needs to be formalized and customized according to the geological context in order to guarantee the traceability of the modeling and interpretation. In the case of the Athabasca Basin, one of the difficulties is the lack of a homogenized geological map of the basement below the unconformity synthetizing the knowledge resulting from the exploration performed by junior and major companies. Thus, the change of scale from district to camp requires a careful review and even reprocessing of the results of the district-scale analysis based on the geological maps and cross sections resulting from the drilling campaign and of structural data. This review also makes it possible to identify the second-order structures and subtle changes in the lithologies that may reinforce the interest or lead to the lower ranking of a camp.

At the camp and prospect scale, a geo-referenced 3D geological model is considered to be a pre-requisite to the definition of drilling targets. As the mineral system approach emphasizes the required understanding of the main architecture and evolution of the different structures, the reconciliation between project scale geological information, mainly interpreted from structural and mineralogical datasets from drill hole and field surveys, and geophysical modeling is a real challenge. As far as the collection and compilation of all existing geophysical, geological and geochemical data have been done, a multidisciplinary analysis must be performed to argue and assess the relevance of each dataset and to define additional processing and modeling if needed to fill gaps in data coverage and knowledge. This approach is aimed to converge toward 3D common earth models in which geological and geophysical data can be quantitatively reconciled only through their common representation as physical property models (McGaughey 2007). Such integrated exploration models supported by multiple datasets will improve the characterization of the permeability of the systems that are critical to understand the dynamic of the fluid flow system at the origin of the mineralization. It is also a condition to integrate on the fly results coming out of new drill holes or geophysical surveys and to constrain with a realistic geometry and adequate physical parameters any thermal, hydraulic and mechanical modeling.

As it is more and more illustrated in the multiscale approach of the mineral systems, the space and time dimensions of metallogenic provinces are focused on transient geodynamic events that trigger fluid circulation and tectonic activity. If the formation of giant unconformity-related uranium deposits results from a "self-organized critical" (SOC) system, as defined in Hronsky (2011) and McCuaig and Hronsly (2014), it will then be necessary to look at the scale of a mineralized zone to discover what could be the witness of such a system, that is, the characterization of a paleo-fractured reservoir, threshold barriers and associated tectonic events. The possibility to identify with 3D common earth models the appropriate conditions for building a SOC from observations or modeling represents a major stake for project generation and exploration.

3.4. Conclusion

Multiple decades of exploration of unconformity-related uranium deposits in Saskatchewan and in particular in the Athabasca Basin give access to large datasets available on the Saskatchewan GeoHub (n/a) regarding geology (lithology, mineralogy, alteration, structure), geophysics (airborne, ground and downhole-probing magnetics; gravimetry; electromagnetic (EM); magneto-telluric (MT)) and geochemistry (major and trace elements, isotopic composition and geochronology). Return of experience, data and knowledge-driven approaches regularly lead to the discovery of new deposits and even of new metallogenic provinces, like the Patterson Lake corridor, hosting the Triple R and Arrow deposits and the Spitfire prospect. The review of some of the critical elements defining a mineral system, mainly the architecture of the geological domain and the fluid flow system, as well as recent advances in the understanding of the structural control of hydrothermal alteration and mineralization, provide an opportunity to revisit some of the main concepts that are currently applied for guiding exploration in the Athabasca Basin. Thus, the coupling between the fluid circulation in the vicinity of the unconformity and the reactivation of deeply rooted structures below the unconformity seems to be the main driver for the formation of large to giant deposits hosted within this fault system, both at the unconformity and deep in the basement. Further studies are needed to characterize the markers of suspected transitional episodes corresponding to the development of a self-organized complex, to evaluate more precisely the conditions under which the permeability of the basement has been enhanced and to understand why this coupling may have occurred along some fault zones and not on others.

In parallel, access to these large datasets and the availability of digital platform that enable constrained geometrical 3D modeling, multivariate analysis and the application of new capacities in artificial intelligence and machine learning are also opportunities to update and quantify the footprint of the deposits based on physical and chemical properties of the lithologies and structures. The production of mineral potential models (Bruce et al. 2020) or 3D common earth models, aiming at the reconciliation between geological, geophysical and geochemical data (McGaughey 2007; Lesher et al. 2017), provides innovative workflow for targeting and opens new horizons for a true multidisciplinary approach.

3.5. References

Abdelrazek, M., Benedicto, A., Gerbeaud O., Ledru, P. (2021). Reactivation mechanisms and related-porosity enhancement of shear zones in context of basement-hosted uranium mineralization: Case of the Spitfire discovery in the Patterson Lake corridor, Canada. *Geochemistry: Exploration, Environment, Analysis*, 21(3).

Adlakha, E. and Hattori, K. (2021). Thermotectonic events recorded by U-Pb geochronology and Zr-in-rutile thermometry of Ti oxides in basement rocks along the P2 fault, eastern Athabasca Basin, Saskatchewan, Canada. *GSA Bulletin*, 134(3–4), 567–576.

Alexandre, P., Kyser, K., Polito, P., Thomas, D. (2005). Alteration mineralogy and stable isotope geochemistry of Paleoproterozoic basement-hosted unconformity-type uranium deposits in the Athabasca Basin, Canada. *Econ. Geol.*, 100, 1547–1563.

Alexandre, P., Kyser, K., Thomas, D., Polito, P., Marlat, J. (2009). Geochronology of unconformity-related uranium deposits in the Athabasca Basin, Saskatchewan, Canada and their integration in the evolution of the basin. *Miner. Deposita*, 44, 41–59.

Annesley, I.R., Madore, C., Portella, P. (2005). Geology and thermotectonic evolution of the western margin of the Trans-Hudson Orogen: Evidence from the eastern sub-Athabasca basement, Saskatchewan. *Can. J. Earth Sci.*, 42, 573–597.

Barnicoat, A.C. (2007). Mineral systems and exploration science: Linking fundamental controls on ore deposition with the exploration process. In *Proceedings of the 9th SGA Biennial Meeting*, Dublin.

Baudemont, D. and Pacquet, A. (1996). The Sue D and E uranium deposits, northern Saskatchewan: Evidence for structurally controlled fluid circulation in the Athabasca basin. In *MinExpo'96 Symposium – Advances in Saskatchewan Geology and Mineral Exploration*, Ashton, K.E. and Harper, C.T. (eds). Saskatchewan Geological Survey, Special Publication.

Baudemont, D. and Rafini, S. (2000). Cigar Lake deposit – Eastern Athabasca. Core orientation from underground geotechnical holes. Tentative modelling of post Athabasca faulting. Report 99-CND-29-04, COGEMA Resources Inc.

Benedicto, A., Abdelrazek, M., Ledru, P., Mackay, C. (2021a). Structural controls of uranium mineralization in the Athabasca Basin, Saskatchewan, Canada. *Geofluids*, Article id: 3853468.

Benedicto, A., Harrison, G., Eccles B., Ledru, P. (2021b). Advanced use of borehole acoustic televiewer (ATV) for structural interpretation of unconformity-related uranium deposits. *Econ. Geol.*, 116(6), 1435–1453.

Berman, R.G., Davis, W.J., Pehrsson, S. (2007). Collisional snowbird tectonic zone resurrected: Growth of Laurentia during the 1.9 Ga accretionary phase of the Hudsonian orogen. *Geology*, 35, 911–914.

Boiron, M.C., Cathelineau, M., Richard, A. (2010). Fluid flows and metal deposition near basement/cover unconformity: Lessons and analogies from Pb–Zn–F–Ba systems for the understanding of Proterozoic U deposits. *Geofluids*, 10, 270–292.

Bons, P.D., Fusswinkel, T., Gomez-Rivas, E., Markl, G., Wagner, T., Walter, B. (2014). Fluid mixing from below in unconformity-related hydrothermal ore deposits. *Geology*, 42, 1035–1038.

Bruce, M., Kreuzer, O., Wilde, A., Buckingham, A., Butera, K., Bierlein, F. (2020). Unconformity-type uranium systems: A comparative review and predictive modelling of critical genetic factors. *Minerals*, 10, 738.

Bruneton, P. (1987). Geology of the Cigar Lake uranium deposit (Saskatchewan, Canada). In *Economic Minerals of Saskatchewan*, Gilboy, C.F. and Vigross, L.W. (eds). Saskatchewan Geological Society, Special Publication.

Bruneton, P. (1993). Geological environment of the Cigar Lake uranium deposit. *Can. J. Earth Sci.*, 30, 653–673.

Card, C. (2020). The Patterson Lake corridor of Saskatchewan, Canada: Defining crystalline rocks in a deep-seated structure that hosts a giant, high-grade Proterozoic unconformity uranium system. *Geochem. Explor. Environ. Anal.*, 21 [Online]. Available at: https://doi.org/10.1144/geochem2020-007.

Card, C., Bethune, K.M., Davis, W.J., Rayner, N., Ashton, K.E. (2014). The case for a distinct Taltson orogeny: Evidence from northwest Saskatchewan, Canada. *Precambrian Res.*, 255, 245–265.

Cathelineau, M. and Boiron, M.C. (2010). Downward penetration and mixing of sedimentary brines and dilute hot waters at 5 km depth in the granite basement at Soultz-sous-Forêts (Rhine graben, France). *C. R. Geosci.*, 342, 560–565.

Cernuschi, F., Dilles, J.H., Grocke, S.B., Valley, J.W., Kitajima, K., Tepley, F.J. (2018). Rapid formation of porphyry copper deposits evidenced by diffusion of oxygen and titanium in quartz. *Geology*, 46(7), 611–614.

Chardon, D., Gapais, D., Cagnard, F. (2009). Flow of ultra-hot orogens: A view from the Precambrian, clues for the Phanerozoic. *Tectonophysics*, 477(3–4), 105–118.

Chi, G., Bosman, S., Card, C. (2013). Numerical modeling of fluid pressure regime in the Athabasca basin and implications for fluid flow models related to the unconformity-type uranium mineralization. *J. Geochem. Explor.*, 125, 8–19.

Chi, G., Li, Z., Chu, H., Bethune, K.M., Quirt, D., Ledru, P., Normand, C., Card, C., Bosman, S., Davis, W.J., Potter, E.G. (2018). Shallow-burial mineralization system driven by magmatic heat sources at depth: A new perspective on factors controlling unconformity-related uranium deposits associated with Proterozoic basins. *Econ. Geol.*, 113(5), 1209–1217.

Chi, G., Chu, H., Petts, D., Potter, E.G., Jackson, S., William-Jones, A. (2019a). Uranium-rich diagenetic fluids provide the key to unconformity related uranium mineralization in the Athabasca Basin. *Nature Scientific Reports*, 9, 5530.

Chi, G., Eldursi, K., Li, Z., Bethune, K., Ledru, P., Quirt, D. (2019b). Thermal-hydraulic-mechanical-chemical (THMC) modeling of fluid flow and its control on localization of uranium deposits in the Northeastern Athabasca Basin, Saskatchewan. Technical report, NSERC-CRD and Orano Canada (formerly AREVA) Supported Project.

Chi, G., Li, Z., Xu, D., Ledru, P., Chu, H., Xue, C. (2019c). Coupled control of intra-basinal and deep-seated geologic processes on formation of some super-large uranium and base-metal deposits in sedimentary basins. *Proceedings of the 15th Biennial SGA Meeting*, August 27–30, Glasgow.

Chu, H. and Chi, G. (2016). Thermal profiles inferred from fluid inclusion and illite geothermometry from sandstones of the Athabasca Basin: Implications for fluid flow and unconformity-related uranium mineralization. *Ore Geol. Rev.*, 75, 284–303.

Chu, H., Chi, G., Bosman, S., Card, C. (2015). Diagenetic and geochemical studies of sandstones from drill core DV10-001 in the Athabasca basin, Canada, and implications for uranium mineralization. *J. Geochem. Explor.*, 148, 206–230.

Cloutier, J., Kyser, K., Olivo, G.R., Alexandre, P., Halaburda, J. (2009). The millennium uranium deposit, Athabasca Basin, Saskatchewan, Canada: An atypical basement-hosted unconformity-related uranium deposit. *Econ. Geol.*, 104, 815–840.

Cloutier, J., Kyser, K., Olivo, G.R., Alexandre, P. (2010). Contrasting patterns of alteration at the Wheeler River area, Athabasca Basin, Saskatchewan, Canada: Insights into the apparently uranium-barren zone K alteration system. *Econ. Geol.*, 105, 303–324.

Cloutier, J., Kyser, K., Olivo, G.R., Brisbin, D. (2011). Geochemical, isotopic, and geochronologic constraints on the formation of the Eagle Point basement-hosted uranium deposit, Athabasca Basin, Saskatchewan, Canada and recent remobilization of primary uraninite in secondary structures. *Miner. Deposita*, 46, 35–56.

Condor Consulting, Inc. (2013). Preliminary assessment of aeromagnetic data from the SE Corner of Athabasca Basin. Internal company report, Lakewood, Colorado.

Cox, S.F. (2016). Injection-driven swarm seismicity and permeability enhancement: Implications for the dynamics of hydrothermal ore systems in high fluid-flux, overpressured faulting regimes – An invited paper. *Econ. Geol.*, 111, 559–587.

Creaser, R.A. and Stasiuk, L.D. (2007). Depositional age of the Douglas formation, northern Saskatchewan, determined by Re-Os geochronology. In *EXTECH IV: Geology and Uranium EXploration TECHnology of the Proterozoic Athabasca Basin, Saskatchewan and Alberta*, Jefferson, C.W. and Delaney, G. (eds). Geological Survey of Canada, Ottawa, Ontario.

Dargent, M., Truche, L., Dubessy, J., Bessaque, G., Marmier, H. (2015). Reduction kinetics of aqueous U(VI) in acidic chloride brines to uraninite by methane, hydrogen or C-graphite under hydrothermal conditions: Implications for the genesis of unconformity-related uranium ore deposits. *Geochim. Cosmochim. Acta*, 167, 11–26.

Derome, D., Cathelineau, M., Cuney, M., Fabre, C., Lhomme, T., Banks, D. (2005). Mixing of sodic and calcic brines and uranium deposition at McArthur River, Saskatchewan, Canada: A Raman and laser-induced breakdown spectroscopic study of fluid inclusions. *Econ. Geol.*, 100, 1529–1545.

Dorbath, L., Cuenot, N., Genter, A., Frogneux, M. (2009). Seismic response of the fractured and faulted granite of Soultz-sous-Forêts (France) to 5 km deep massive water injections. *Geophys. J. Int.*, 177, 653–675.

Eglinger, A., Vanderhaeghe, O., André-Mayer, A.S., Goncalves, P., Zeh, A., Durand, C., Deloule, E. (2016). Tectono-metamorphic evolution of the internal zone of the Pan-African Lufilian orogenic belt (Zambia): Implications for crustal reworking and syn-orogenic uranium mineralizations. *Lithos*, 240–243, 167–188.

Eglington, B.M., Pehrsson, S.J., Ansdell, K.M., Lescuyer, J.L., Quirt, D.H., Milesi, J.P., Brown, P. (2013). A domain-based digital summary of the evolution of the Palaeoproterozoic of North America and Greenland and associated unconformity-related uranium mineralization. *Precambrian Res.*, 232, 4–26.

Eldursi K., Chi, G., Bethune, K., Li, Z., Ledru, P., Quirt, D. (2020). New insights from 2- and 3-D numerical modelling on fluid flow mechanisms and geological factors responsible for the formation of the world-class Cigar Lake uranium deposit, eastern Athabasca Basin, Canada. *Mineralium Deposita*, 56, 1365–1388.

Emmermann, R. and Lauterjung, J. (1997). The German continental deep drilling program KTB: Overview and major results. *J. Geophys. Res.*, 102, 18179–18201.

Evans, K.F., Genter, A., Sausse, J. (2005). Permeability creation and damage due to massive fluid injections into granite at 3.5 km at Soultz 1: Borehole observations. *J. Geophys. Res.*, 110, B04203.

Faulkner, D.R., Jackson, C.A.L., Lunn, R.J., Schlische, R.W., Shipton, Z.K., Wibberley, C.A.J., Withjack, M.O. (2010). A review of recent developments concerning the structure, mechanics and fluid flow properties of fault zones. *J. Struct. Geol.*, 32, 1557–1575.

Fayek, M. and Kyser, T.K. (1997). Characterization of multiple fluid-flow events and rare-earth-element mobility associated with formation of unconformity-type uranium deposits in the Athabasca Basin, Saskatchewan. *Can. Mineral.*, 35, 627–658.

Gapais, D. (2018). Tectonics-mineralisation relationships within weak continental lithospheres: A new structural framework for Precambrian cratons. *BSGF – Earth Sci. Bull.*, 189(3), 14.

Gapais, D., Jaguin, J., Cagnard, F., Boulvais, P. (2014). Pop-down tectonics, fluid channelling and ore deposits within ancient hot orogens. *Tectonophysics*, 618, 102–106.

Gibb, R.A., Thomas, M.D., Lapointe, P.L., Mukhopadhyay, M. (1983). Geophysics of proposed proterozoic sutures in Canada. *Precambrian Res.*, 19, 349–384.

Glaas, C., Patrier, P., Vidal, J., Beaufort, D., Genter, A. (2021). Clay mineralogy. A signature of granitic geothermal reservoirs of the Central Upper Rhine Graben. *Minerals*, 11, 479.

Grare, A., Benedicto, A., Lacombe, O., Trave, A., Ledru, P., Blain, M., Robbins, J. (2018). The contact uranium prospect, Kiggavik project, Nunavut (Canada): Tectonic history, structural constraints and timing of mineralization. *Ore Geol. Rev.*, 93, 141–167.

Grare, A., Benedicto, A., Mercadier, J., Lacombe, O., Trave, A., Guilcher, M., Richard, A., Ledru, P., Blain, M., Robbins, J. et al. (2020). Structural controls and metallogenic model of polyphase uranium mineralization in the Kiggavik area (Nunavut, Canada). *Mineralium Deposita*, 56, 1263–1296.

Gratier, J.P., Renard, F., Labaume, P. (1999). How pressure solution creep and fracturing processes interact in the upper crust to make it behave in both a brittle and viscous manner. *J. Struct. Geol.*, 21, 1189–1197.

Hajnal, Z., Takacs, E., White, D.J., Gyorfi, I., Powell, B., Koch, R. (2007). Regional seismic signature of the basement and crust beneath the McArthur River mine district, Athabasca Basin, Saskatchewan. In *EXTECH IV: Geology and Uranium EXploration TECHnology of the Proterozoic Athabasca Basin, Saskatchewan and Alberta*, Jefferson, C.W. and Delaney, G. (eds). Geological Survey of Canada, Ottawa, Ontario.

Hanmer, S., Williams, M., Kopf, C. (1995). Striding-Athabasca mylonite zone: Implications for the Archean and Early Proterozoic tectonics of the western Canadian Shield. *Can. J. Earth Sci.*, 32, 178–196.

Harvey, S. (2018). Structural summary of the Cigar Lake unconformity-type uranium deposit, Athabasca Basin. Report 2018-3, Saskatchewan Geological Survey, December 3 to 5.

Harvey, S. (2020). Structural geology summary of the McArthur River unconformity-type uranium deposit, Athabasca Basin, Saskatchewan. *CIM 4th International Uranium Conference*.

Hillacre, S., Ansdell, K., McEwan, B., McNamara, G. (2017). Structural analysis, paragenesis, and preliminary geochronology of the arrow uranium deposit, Athabasca Basin, northern Saskatchewan, Canada: Implications for controls on mineralization. *Mineral Resources to Discover, 14th SGA Biennial Meeting 2017, Proceedings*, 2, 743–746.

Hillacre, S., Ansdell, K., McEwan, B. (2021). Geology, structural analysis, and paragenesis of the arrow uranium deposit, Western Athabasca Basin, Saskatchewan, Canada: Implications for the development of the Patterson Lake corridor. *Econ. Geol.*, 116(2), 285–321.

Hendry, H.E. and Weathley, K.L. (1985). The Carswell formation, northern Saskatchewan: Stratigraphy, sedimentology, and structure. *Geol. Assoc. Can. Special Paper*, 29, 87–103.

Hoeve, J. and Quirt, D. (1984). Mineralization and host rock alteration in relation to clay mineral diagenesis and evolution of the middle Proterozoic Athabasca basin, northern Saskatchewan, Canada. Research Council Technical Report, 187.

Hoeve, J. and Quirt, D. (1987). A stationary redox front as a critical factor in the formation of high-grade, unconformity-type uranium ores in the Athabasca Basin, Saskatchewan, Canada. *Bull. Mineral.*, 110, 157–171.

Hoeve, J. and Sibbald, T.I.I. (1978). On the genesis of Rabbit Lake and other unconformity type uranium deposits in northern Saskatchewan, Canada. *Econ. Geol.*, 73, 1450–1473.

Hoeve, J., Sibbald, T.I.I., Ramaekers, P., Lewry, J.F. (1980). Athabasca Basin unconformity type uranium deposit: A special class of sandstone type deposits? In *Uranium in the Pine Creek Geosyncline*, Ferguson, J. and Goleby, A.B. (eds). IAEA, Vienna.

Hoffman, P.F. (1988). United plates of America, the birth of a craton: Early Proterozoic assembly and growth of Laurentia. *Annu. Rev. Earth Planet. Sci.*, 16, 543–603.

Hronsky, J.M.A. (2011). Self-organized critical systems and ore formation: The key to spatial targeting? *SEG Newsletter*, 84, 14–16.

Hronsky, J.M.A., Groves, D.I., Loucks, R.R., Begg, G.C. (2012). A unified model for gold mineralisation in accretionary orogens and implications for regional-scale exploration targeting methods. *Miner. Deposita*, 47, 339–358.

IAEA (2018). Unconformity-related uranium deposits. TECDOC, 1857 [Online]. Available at: https://www.iaea.org/publications/13432/.

Jeanneret, P. (2016). Rôle du métamorphisme et de la déformation dans la redistribution et la pré-concentration d'Uranium dans la croûte continentale. Exemple de la zone de transition Wollaston-Mudjatik (Saskatchewan, Canada). PhD Thesis, Université de Besançon.

Jeanneret, P., Goncalves, P., Durand, C., Trap, P., Marquer, D., Quirt, D., Ledru, P., Rivers, T. (2016). Tectono-metamorphic evolution of the pre-Athabasca basement within the Wollaston-Mudjatik Transition Zone, Saskatchewan. *Can. J. Earth Sci.*, 53, 231–259.

Jeanneret, P., Goncalves, P., Durand, C., Poujol, M., Trap, P., Marquer, D., Quirt, D., Ledru, P. (2017). Geochronological constraints on the trans-Hudsonian tectonometamorphic evolution of the pre-Athabasca basement within the Wollaston-Mudjatik Transition Zone, Saskatchewan. *Precambrian Res.*, 301, 152–178.

Jefferson, C.W., Thomas, D.J., Gandhi, S.S., Ramaekers, P., Delaney, G., Brisbin, D., Cutts, C., Portella, P., Olson, R.A. (2007). Unconformity-associated uranium deposits of the Athabasca basin, Saskatchewan and Alberta. In *EXTECH IV: Geology and Uranium EXploration TECHnology of the Proterozoic Athabasca Basin, Saskatchewan and Alberta*, Jefferson, C.W. and Delaney, G. (eds). Geological Survey of Canada, Ottawa, Ontario.

Johnstone, D.D., Bethune, K.M., Card, C.D., Tschirhart, V. (2021). Structural evolution and related implications for uranium mineralization in the Patterson Lake corridor, southwestern Athabasca Basin, Saskatchewan, Canada. *Geochem. Explor. Environ. Anal.*, 21 [Online]. Available at: https://doi.org/10.1144/geochem2020-030.

Jones, A.G., Ledo, J., Ferguson, I.J. (2005). Electromagnetic images of the Trans-Hudson Orogen: The North American Central Plains anomaly revealed. *Can. J. Earth Sci.*, 42, 457–478.

Kaczowka, A.J. (2017). Geometallurgical and geological evaluation of the high-grade polymetallic unconformity-related Cigar Lake uranium deposit. M.Sc. Thesis, Queen's University, Belfast.

Kaczowka, A.J., Kyser, K., Kotzer, T., Leybourne, M., Layton-Matthews, D. (2021). Geometallurgical ore characterization of the high-grade polymetallic unconformity-related uranium deposit. *Can. Mineral.*, 59, 813–845.

Kerr, W. and Wallis, R. (2014). "Real-world" economics of the uranium deposits of the Athabasca basin, Northern Saskatchewan: Why grade is not always king! *SEG Newsletter*, 99, 10–15.

Kister, P., Cuney, M., Golubev, V., Royer, J.J., Le Carlier De Veslud, C., Rippert J.C. (2004). Radiogenic lead mobility in the Shea Creek unconformity-related uranium deposit (Saskatchewan, Canada): Migration pathways and Pb loss quantification. *C. R. Geosci.*, 336, 205–215.

Kister, P., Vieillard, P., Cuney, M., Quirt, D., Laverret, E. (2005). Thermodynamic constraints on the mineralogical and fluid composition evolution in a clastic sedimentary basin: The Athabasca Basin (Saskatchewan, Canada). *Eur. J. Mineral.*, 17, 325–342.

Kotzer, T.G. and Kyser, T.K. (1995). Petrogenesis of the Proterozoic Athabasca Basin, northern Saskatchewan, Canada, and its relation to diagenesis, hydrothermal uranium mineralization and paleohydrogeology. *Chem. Geol.*, 120, 45–89.

Kyser, K. and Cuney, M. (eds) (2015). Basins and uranium deposits. In *Geology and Geochemistry of Uranium and Thorium Deposits*. Mineralogical Association of Canada, Quebec.

Kyser, T.K., Hiatt, E., Renac, C., Durocher, K., Holk, G., Deckart, K. (2000). Diagenetic fluids in Paleo- and Mesoproterozoic sedimentary basins and their implications for long protracted fluid histories. In *Fluids and Basin Evolution*, Kyser, T.K. (ed.). Nevada Mineral & Book Company.

Ledru, P. (2019). The Mineral System concept applied to unconformity-related uranium deposits of the Athabasca Basin (Canada). In *Proceedings of the 15th SGA Biennial Meeting*, Glasgow.

Ledru, P., Benedicto, A., Poh, J., Robbins, J. (2019). Implications of a new tectonic model on the role of pre-Athabasca basement-hosted fault systems in the formation of the unconformity related uranium deposits. Report 2019-3, Saskatchewan Geological Survey.

Lesher, C.M., Hannington, M., Galley, A., Mineral Exploration Research Network (2017). Integrated multi-parameter exploration footprints of the Canadian Malartic Disseminated Au, McArthur river-Millennium unconformity U, and highland valley porphyry Cu deposits: Preliminary results from the NSERC-CMIC footprints research network. In *Proceedings of Exploration 17: Sixth Decennial International Conference on Mineral Exploration*, 325–347.

Lewry, J.F. and Sibbald, T.I. (1980). Thermotectonic evolution of the Churchill province in Northern Saskatchewan. *Tectonophysics*, 68, 45–82.

Li, Z., Chi, G., Bethune, K.M. (2016). The effects of basement faults on fluid convection and implications for the formation of unconformity-related uranium deposits in the Athabasca Basin, Canada. *Geofluids*, 16, 729–751.

Li, Z., Chi, G., Bethune, K.M., Thomas, D., Zaluski, G. (2017). Structural controls on fluid flow during compressional reactivation of basement faults: Insights from numerical modeling for the formation of unconformity-related uranium deposits in the Athabasca Basin, Canada. *Econ. Geol.*, 112, 451–466.

Li, Z., Chi, G., Bethune, K.M., Eldursi, K., Thomas, D., Quirt, D., Ledru, P. (2018a). Synchronous egress and ingress fluid flow related to compressional reactivation of basement faults: The Phoenix and Gryphon uranium deposits, southeastern Athabasca Basin, Saskatchewan, Canada. *Miner. Deposita*, 53, 277–292.

Li, Z., Chi, G., Bethune, K.M., Eldursi, K., Quirt, D., Ledru, P., Gudmundson, G. (2018b). Numerical simulation of strain localization and its relationship to formation of the Sue unconformity-related uranium deposits, eastern Athabasca Basin, Canada. *Ore Geol. Rev.*, 101, 17–31.

Li, Z., Chi, G., Bethune, K.M, Eldursi, K., Thomas, D., Quirt, D., Ledru, P., Thomas D. (2021). Interplay between thermal convection and compressional fault reactivation in the formation of unconformity-related uranium deposits. *Miner. Deposita.*, 56, 1389–1404.

Liskowich, M. (2018). Prefeasibility study report for the wheeler river uranium project Saskatchewan, Canada [Online]. Available at: https://denisonmines.com/site/assets/files/5694/2018-10-30-dml-43-101-pfs-report.pdf.

Manga, M., Beresnev, I., Brodsky, E.E., Elkhoury, J.E., Elsworth, D., Ingebritsen, S.E., Mays, D.C., Wang, C.Y. (2012). Changes in permeability caused by transient stresses: Field observations, experiments, and mechanisms. *Rev. Geophys.*, 50, RG2004.

Martz, P., Cathelineau, M., Mercadier, J., Boiron, M.C., Jaguin, J., Tarantola, A., Demacon, M., Gerbeaud, O., Quirt, D., Doney, A. et al. (2017). C-O-H-N fluids circulations and graphite precipitation in reactivated Hudsonian shear zones during basement uplift of the Wollaston-Mudjatik Transition Zone: Example of the Cigar Lake U deposit. *Lithos*, 294–295, 222–245.

Martz, P., Mercadier, J., Cathelineau, M., Boiron, M.C., Quirt, D., Doney, A., Gerbeaud, O., De Wally, E., Ledru, P. (2019a). Formation of U-rich mineralizing fluids through basinal brine migration within basement-hosted shear zones: A large-scale study of the fluid chemistry around the unconformity-related Cigar Lake U deposit (Saskatchewan, Canada). *Chem. Geol.*, 508, 116–143.

Martz, P., Mercadier, J., Perret, J., Villeneuve, J., Deloule, E., Cathelineau, M., Quirt, D., Doney, A., Ledru, P. (2019b). Post-crystallization alteration of natural uraninites: Implications for dating, tracing, and nuclear forensics. *Geochim. Cosmochim. Acta*, 249, 138–159.

McCuaig, T.C. and Hronsky, J.M.A. (2014). The mineral system concept: The key to exploration targeting. *Soc. Econ. Geol.*, 18, 153–175.

McCuaig, T.C., Beresford, S., Hronsky, J.M.A. (2010). Translating the mineral systems approach into an effective exploration targeting system. *Ore Geol. Rev.*, 38, 128–138.

McGaughey, J. (2007). Geological models, rock properties, and the 3D inversion of geophysical data. Advances in geophysical inversion and modeling. In *Proceedings of Exploration 07: Fifth Decennial International Conference on Mineral Exploration*, Milkereit, B. (ed.). Decennial Mineral Exploration Conferences, Ottawa.

McGill, B.D., Marlatt, J.L., Matthews, R.B., Sopuk, V.J., Homeniuk L.A. (1993). The P2 North uranium deposit, Saskatchewan, Canada. *Explor. Min. Geol.*, 2(4), 321–331.

McKechnie, C., Annesley, I., Ansdell, K. (2012). Medium-to-low pressure politic gneisses of Fraser Lake zone B., Wollaston Domain, Northern Saskatchewan, Canada: Mineral composition, metamorphic P-T-t path, and implications for the genesis of radioactive abyssal granitic pegmatites. *Can. Mineral.*, 50, 1669–1694.

McKechnie, C., Annesley, I., Ansdell, K. (2013). Geologic setting, petrology, and geochemistry of granitic pegmatites and leucogranites hosting U-Th-REE mineralization at Fraser Lakes Zone B, Wollaston domain, Northern Saskatchewan, Canada. *Explor. Min. Geol.*, 21, 1–26.

Mercadier, J., Richard, A., Boiron, M.-C.C., Cathelineau, M., Cuney, M. (2010). Migration of brines in the basement rocks of the Athabasca Basin through microfracture networks (P-patch U deposit, Canada). *Lithos*, 115, 121–136.

Mercadier, J., Cuney, M., Lach, P., Boiron, M.-C., Bonhoure, J., Richard, A., Leisen, L., Kister, P. (2011). Origin of uranium deposits revealed by their rare earth element signature. *Terra Nova*, 23, 264–269.

Mercadier, J., Richard, A., Cathelineau, M. (2012). Boron- and magnesium-rich marine brines at the origin of giant unconformity-related uranium deposits: δ11B evidence from Mg-tourmalines. *Geology*, 40, 231–234.

Mercadier, J., Annesley, I.R., McKechnie, C.L., Bogdan, T.S., Creighton, S. (2013). Magmatic and metamorphic uraninite mineralization in the western margin of the Trans-Hudson Orogen (Saskatchewan, Canada): A uranium source for unconformity-related uranium deposits? *Econ. Geol.*, 108, 1037–1065.

Micklethwaite, S., Sheldon, H.A., Baker, T. (2010). Active fault and shear processes and their implications for mineral deposit formation and discovery. *J. Struct. Geol.*, 32, 151–165.

Moncada, D., Rimstidt, J.D., Bodnar, R.J. (2019). How to form a giant epithermal precious metal deposit: Relationships between fluid flow rate, metal concentration of ore-forming fluids, duration of the ore-forming process, and ore grade and tonnage. *Ore Geol. Rev.*, 113, 103066.

Motta, J.G., Betts, P.G., de Souza Filho, C.R., Thiel, S., Curtis, S., Armit, R.J. (2019). Proxies for basement structure and its implications for Mesoproterozoic metallogenic provinces in the Gawler Craton. *J. Geophys. Res.: Solid Earth*, 124, 3088–3104.

Ng, R., Alexandre, P., Kyser, K. (2013). Mineralogical and geochemical evolution of the unconformity-related McArthur River zone 4 orebody in the Athabasca Basin, Canada: Implications of a silicified zone. *Econ. Geol.*, 108, 1657–1689.

Pagel, M. (1975). Détermination des conditions physico-chimiques de la silicification diagénétique des grès Athabasca (Canada) au moyen des inclusions fluides. *C. R. Acad. Sci., Paris*, 280, 2301–2304.

Pagel, M., Poty, B., Sheppard, S.M.F. (1980). Contribution to some Saskatchewan uranium deposits mainly from fluid inclusion and isotopic data. *Uranium in the Pine Creek Geosyncline*, Ferguson, J. and Goleby, A.B. (eds). IAEA, Vienna.

Pehrsson, S.J., Eglington, B.M., Berman, R.G., Eglington, B.M., Rainbird, R. (2013). Two Neoarchean supercontinents revisited: The case for a Rae family of cratons. *Precambrian Res.*, 232, 27–43.

Pehrsson, S.J., Eglington, B.M., Evans, D.A.D., Huston, D., Reddy, S.M. (2016). Metallogeny and its link to orogenic style during the Nuna supercontinent cycle. In *Supercontinent Cycles through Earth History*, Li, Z.X., Evans, D.A.D., Murphy, J.B. (eds). Geological Society, London.

Poh, J. (2019). Zones de déformation précambriennes dans les régimes tectoniques en compression : une perspective numérique en relation avec le bassin de l'Athabasca et sa minéralisation d'uranium. PhD Thesis, Université de Rennes.

Poh, J., Yamato, P., Duretz, T., Gapais, D., Ledru, P. (2020). Precambrian deformation belts in compressive tectonic regimes: A numerical perspective. *Tectonophysics*, 777, 228350.

Potter, E.G., Tschirhart, V., Powell, J.W., Kelly, C.J., Rabiei, M., Johnstone, D., Craven, J.A., Davis, W.J., Pehrsson, S., Mount, S.M. et al. (2020). *Targeted Geoscience Initiative 5: Integrated Multidisciplinary Studies of Unconformity-Related Uranium Deposits from the Patterson Lake Corridor, Northern Saskatchewan*. Geological Survey of Canada, Ottawa, Ontario.

Quirt, D. (2001). Kaolinite and dickite in the Athabasca Sandstone, northern Saskatchewan, Canada. *Saskatchewan Research Council*, 10400, 1–27.

Rabiei, M., Chi, G., Potter, E.G., Tschirhart, V., MacKay, C., Frostad, S., McElroy, R., Ashley, R., McEwan, B. (2021). Fluid evolution along the Patterson Lake corridor in the southwestern Athabasca Basin: Constraints from fluid inclusions and implications for unconformity-related uranium mineralization. *Geochem. Explor. Environ. Anal.*, 21 [Online]. Available at: https://doi.org/10.1144/geochem2020-029.

Raffensperger, J.P. and Garven, G. (1995). The formation of unconformity-type uranium ore-deposits. Coupled groundwater-flow and heat transport modeling. *Am. J. Sci.*, 295, 581–636.

Rainbird, R.H., Stern, R.A., Rayner, N., Jefferson, C.W. (2007). Age, provenance, and regional correlation of the Athabasca Group, Saskatchewan and Alberta, constrained by igneous and detrital zircon geochronology. In *EXTECH IV: Geology and Uranium EXploration TECHnology of the Proterozoic Athabasca Basin, Saskatchewan and Alberta*, Jefferson, C.W. and Delaney, G. (eds). Geological Survey of Canada, Ottawa, Ontario.

Ramaekers, P., Jefferson, C.W., Yeo, G.M., Collier, B., Long, D.G.F., Drever, G., McHardy, S., Jiricka, D., Cutts, C., Wheatley, K. et al. (2007). Revised geological map and stratigraphy of the Athabasca Group, Saskatchewan and Alberta. In *EXTECH IV: Geology and Uranium EXploration TECHnology of the Proterozoic Athabasca Basin, Saskatchewan and Alberta*, Jefferson, C.W. and Delaney, G. (eds). Geological Survey of Canada, Ottawa, Ontario.

Regan, S.P., Williams, M.L., Chiarenzelli, J.R., Grohn, L., Mahan, K.H., Gallagher, M. (2017). Isotopic evidence for Neoarchean continuity across the Snowbird Tectonic Zone, western Churchill Province, Canada. *Precambrian Res.*, 300, 201–222.

Reid, K.D., Ansdell, K., Jiricka, D., Witt, G., Card, C. (2014). Regional setting, geology, and paragenesis of the Centennial unconformity-related uranium deposit, Athabasca Basin, Saskatchewan, Canada. *Econ. Geol.*, 109, 539–566.

Richard, A., Banks, D.A., Mercadier, J., Boiron, M.C., Cuney, M., Cathelineau, M. (2011). An evaporated seawater origin for the ore-forming brines in unconformity-related uranium deposits (Athabasca Basin, Canada): Cl/Br and δ37Cl analysis of fluid inclusions. *Geochim. Cosmochim. Acta*, 75, 2792–2810.

Richard, A., Rozsypal, C., Mercadier, J., Banks, D.A., Cuney, M., Boiron, M.C., Cathelineau, M. (2012). Giant uranium deposits formed from exceptionally uranium-rich acidic brines. *Nature Geoscience*, 5, 142–146.

Richard, A., Cathelineau, M., Boiron, M.-C., Mercadier, J., Banks, D.A., Cuney, M. (2016). Metal-rich fluid inclusions provide new insights into unconformity-related U deposits (Athabasca Basin and basement, Canada). *Mineral. Deposita*, 51, 249–270.

Saskatchewan GeoHub (n/a). Saskatchewan GeoHub [Online]. Available at: https://geohub.saskatchewan.ca/.

Sausse, J., Dezayes, C., Dorbath, L., Genter, A., Place, J. (2010). 3D model of fracture zones at Soultz-sous-Forêts based on geological data, image logs, induced microseismicity and vertical seismic profiles. *C. R. Géosci.*, 342, 531–545.

Sheahan, C., Fayek, M., Qurit, D., Jefferson, C.W. (2016). A combined ingress-egress model for the Kianna unconformity-related uranium deposit, Shea Creek project, Athabasca Basin, Canada. *Econ. Geol.*, 111, 225–257.

Sibson, R.H. (1981). Fluid flow accompanying faulting: Field evidence and models. In *Earthquake Prediction: An International Review*, Simpson, D.W. and Richards, P.G. (eds). AGU Maurice Ewing Series, Washington, D.C.

Sibson, R.H. (2020). Preparation zones for large crustal earthquakes consequent on fault-valve action. *Earth Planets Space*, 72, 31.

Sibson, R.H., Robert, F., Poulsen, K.H. (1988). High angle reverse faults, fluid pressure cycling, and mesothermal gold-quartz deposits. *Geology*, 16, 551–555.

Skirrow, R.G., Jaireth, S., Huston, D.L., Bastrakov, E.N., Schofield, A., van der Wielen, S.E., Barnicoat, A.C. (2009). Uranium mineral systems: Processes, exploration criteria and a new deposit framework. *Geoscience Australia Record*, 20.

Song, H., Chi, G., Wang, K., Li, Z., Bethune, K.M., Potter, E.G., Liu, Y. (In review). Raman spectroscopy as a tool to decipher the role of graphite in uranium mineralization – A case study of the world-class Phoenix uranium deposit in the Athabasca Basin, Canada. *Am. Mineral.*, 107, 2128–2142.

Sorba, C., Gamelin, G., Liu, Y., Forand, L., Verran, D. (2016). Unravelling and growing the high grade Gryphon uranium deposit, Wheeler River project, Athabasca Basin in Saskatchewan Geological Survey. Miscellaneous report 2016-3, 6, Open House 2016: Abstract Volume, Saskatchewan Geological Survey.

Thiel, S. and Heinson, G. (2010). Crustal imaging of a mobile belt using magnetotellurics: An example of the Fowler Domain in South Australia. *J. Geophys. Res.*, 115, B06102.

Thomas, D.J., Matthews, R.B., Sopuck, V. (2000). Athabasca Basin (Canada) unconformity-type uranium deposits: Exploration model, current mine developments and exploration directions. In *Geology and Ore Deposits 2000: The Great Basin and Beyond*. Geological Society of Nevada Symposium, Reno, Nevada.

Tourigny, G., Quirt, D., Wilson, N.S.F., Wilson, S., Breton, G., Portella, P. (2007). Geological and structural features of the Sue C uranium deposit, McClean Lake area, Saskatchewan. In *EXTECH IV: Geology and Uranium EXploration TECHnology of the Proterozoic Athabasca Basin, Saskatchewan and Alberta*, Jefferson, C.W. and Delaney, G. (eds). Bulletin 588, Geological Survey of Canada, Ottawa, Ontario.

Tran, H.T. (2001). Tectonic evolution of the Paleoproterozoic Wollaston Group in the Cree Lake Zone, Northern Saskatchewan, Canada. PhD Thesis, University of Regina.

Tran, H.T., Lewry, J.F., Bethune, K., Ashton, K., Yeo, G., Ansdell, K. (2000). Crustal thrusting and tectonic imbrication in the Cree Lake Zone, northern Saskatchewan: Implications for the tectonic evolution of western foreland of the Trans-Hudson Orogen. Annual Meeting, Calgary, Alta., 25, CD 579.

Tran, H.T., Ansdell, K., Bethune, K.M., Ashton, K., Hamilton, M.A. (2008). Provenance and tectonic setting of Paleoproterozoic metasedimentary rocks along the eastern margin of Hearne craton: Constraints from SHRIMP geochronology, Wollaston Group, Saskatchewan, Canada. *Precambrian Res.*, 167, 171–185.

Tschirhart, V., Pehrsson, S., Card, C., Potter, E.G., Powell, J., Pană D. (2020). Interpretation of buried basement in the southwestern Athabasca Basin, Canada, from integrated geophysical and geological datasets. *Geochem. Explor. Environ. Anal.*, 21(1): geochem2019–061. doi: https://doi.org/10.1144/geochem2019-061.

Vidal, J. and Genter, A. (2018). Overview of naturally permeable fractured reservoirs in the central and southern Upper Rhine Graben: Insights from geothermal wells. *Geothermics*, 74, 57–73.

Vidal, J., Genter, A., Chopin, F. (2017). Permeable fracture zones in the hard rocks of the geothermal reservoir at Rittershoffen, France. *J. Geophys. Res., Solid Earth*, 122, 4864–4887.

Wang, K., Chi, G., Bethune, K.M., Li, Z., Blamey, N., Card, C., Potter, E.G., Liu, Y. (2018). Fluid P-T-X characteristics and evidence for boiling in the formation of the Phoenix uranium deposit (Athabasca Basin, Canada): Implications for unconformity-related uranium mineralization mechanisms. *Ore Geol. Rev.*, 101, 122–142.

Wilson, M.R. and Kyser, T.K. (1987). Stable isotope geochemistry of alteration associated with the Key Lake uranium deposit. *Econ. Geol.*, 82, 1450–1557.

Witherly, K. (2014). Geophysical expressions of ore systems – Our current understanding. *Soc. Econ. Geol.*, Special Publication, 18, 177–208.

Wyborn, L.A.I., Heinrich, C.A., Jaques, A.L. (1994). Australian Proterozoic mineral systems: Essential ingredients and mappable criteria. *Proceedings of the Australasian Institute of Mining and Metallurgy Annual Conference*, Darwin, Northern Territory.

4

North African Mississippi Valley-Type Deposit and Its Link with the Alpine Chain Evolution

Mohammed Bouabdellah
Laboratoire des gîtes minéraux, University Mohammed Premier, Oujda, Morocco

4.1. Introduction

Mississippi Valley-type deposits, which we will refer to from now on as MVTs, refer to a class of epigenetic carbonate-hosted Pb-Zn ± Cu ± Ba ± F deposits whose genesis is related to the geodynamic evolution of their host basins, without direct connection with synchronous magmatic activity. Etymologically, MVTs owe their nomenclature to the fact that several classic districts such as Upper Mississippi Valley, Southeast Missouri (Old Lead Belt, Viburnum Trend, Indian Creek), Central Missouri, Tri-State and Northern Arkansas are located within the drainage basin of the Mississippi River in the Central United States. The MVTs account for 24% of the global resources for Pb and Zn, and 38% of the global tonnage for sediment-hosted deposits (Leach et al. 2010a, 2010b). The most valuable attribute of MVTs relates to their reputation as nonpolluting, owing to their host carbonates, which act as a buffer, tending to neutralize the released acids, following sulfide precipitation.

For a color version of all figures in this chapter, see www.iste.co.uk/decree/metallic2.zip.

Metallic Resources 2,
coordinated by Sophie Decrée.

In addition to lead and zinc, MVT deposits produce subeconomic proportions of silver, copper, fluorite and baryte locally accompanied by elements such as In, Ge, Ga, Cd and Co as by-products. The Ag grades, though highly variable, are on average of the order of 32.5 g/t, and can locally reach 120 g/t. This is the case of the Touissit-Bou Beker district in Morocco where the Ag average grade is of about 120 g/t (Bouabdellah et al. 2012) and Viburnum Trend in the central part of the United States where the annual average quantities extracted during 1960–1990 exceeded 46 tons (Leach et al. 2010a, 2010b). It is noteworthy that the majority of MVT deposits are predominantly Zn-dominant with a Zn/Pb mass ratio of 10:1, though a minority of other deposits are Pb-dominant such as those of Viburnum Trend in United States and Beddiane, and Mekta in Morocco. Stratabound sulfide-bearing mineralization occurs either as moderate to massive replacements of calcareous-dolomitic host rocks giving rise to stratiform mineralization, or open spaces fillings (karsts, veins, veinlets and vugs), and in collapse breccia cement. The mineral paragenesis is simple, being reduced to the occurrence of sphalerite and galena as main sulfides with subordinate pyrite, marcassite and chalcopyrite ± gray copper of the terahedrite-tennantite series. Gangue phases consist predominantly of hydrothermal dolomite (*saddle dolomite*) and calcite, and to a lesser extent siderite, baryte, fluorite and quartz.

From a genetic point of view, MVT deposits originate from hydrothermal circulation of warm, dense, basinal brines that, during their long-distance migration through appropriate aquifers, become enriched in metals in response to extensive fluid-rock interactions, and the resulting Pb-Zn ore is precipitated along the edges of the host basins within platform carbonate successions (mainly dolostone and limestone). Having been for a long time considered a North American curiosity, MVT deposits are currently being recognized as a distinct type of base metal-bearing deposits with mineralogical characteristics and geochemical attributes, which set them apart from other Pb-Zn deposits. In French literature including that emanating from North African countries (Morocco, Algeria, Tunisia), this type of ore deposit has been referred to as an "ore deposit associated with carbonated rocks" or "ore deposits associated with unconformities". An exhaustive listing describing the main attributes of MVTs was compiled by Leach and Sangster (1993) and the reader should refer to it for further details.

In North Africa, the intracontinental Atlas system that stretches over a distance of 2,000 km along the Moroccan passive Atlantic continental margin to the Tunisian Mediterranean coast (Figure 4.1) hosts more than 100 MVT deposits, and prospects with an estimated total Pb + Zn production in excess of 20 Mt (Bouabdellah and Sangster 2016). The main deposits are distributed along two distinct geographical areas referred therein as the Pb-dominant "Atlas province" to the west and the Zn-dominant "Diapir Zone province" to the east (Figure 4.1).

There is, however, no economic Pb-Zn mineralization between these two metallogenic provinces. It is noteworthy that, regardless of the considered province, stratabound sulfide-bearing orebodies are hosted in a succession of unmetamorphosed, flat-lying Lower Jurassic to Upper Cretaceous shallow-water carbonate platform strata, mainly limestones and dolostones. The common feature to all of these carbonate strata is how they have experienced the combined effects of multiple overlapping episodes of dolomitization, dissolution (i.e. karstification) and mineralization during the Mesozoic breakup of the Pangean continental shelf of West Africa. (Bouabdellah and Sangster 2016).

Stratigraphically, the main aquifers include the Lower–Middle Jurassic carbonates of the "Atlas" province and those of the Cretaceous from the "Diapir Zone" province. The Triassic carbonate strata are known to host subeconomic Pb-Zn prospects within the dolomitized limestone ridge of the Rif belt (Bouabdellah and Sangster 2016). It should be stressed, however, that there is no known MVT deposit of economic interest in the Precambrian or Paleozoic terrains.

The mineralogical and geochemical investigations performed over the last two decades using a combination of fluid inclusions studies, REE + Y geochemistry, stable (C, O, H, S) and radiogenic (Sr, Pb) along with noble gases isotope analyses have significantly improved our knowledge on the nature, source(s) and origin(s) of the mineralizing fluid(s), the source(s) of sulfur and metals, as well as the main mechanism(s) and ore depositional processes at the origin of North African MVTs. Nevertheless, we should keep in mind that despite these remarkable advances, and in common with MVT deposits worldwide, the age(s) of mineralization remain(s) the subject of ongoing debates. This temporal controversy hampers the establishment of robust genetic models based on which exploration should focus on for further targeting new Pb-Zn resources.

The aim of this contribution is to:

– update our bibliographic database relative to the geological settings of the main MVT deposits and districts in North Africa by providing a synthesis on new mineralogical, petrographic and geochemical data;

– constrain the origin(s) and evolutionary history of the mineralizing plumbing system(s), sources of sulfur and metals, mechanism(s) of metal transport and ore depositional processes, as well as the main ore controls and timing of mineralization;

– propose reliable genetic models for exploration purposes.

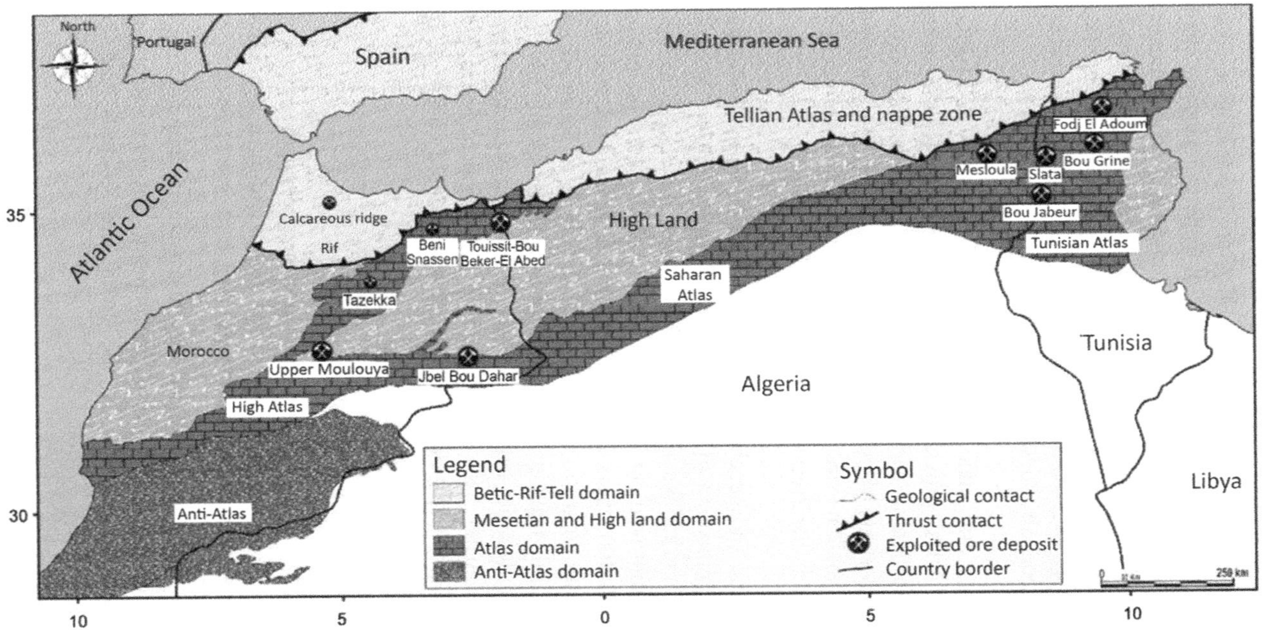

Figure 4.1. *Geologic map of northwest Africa showing the main structural domains and location of major Mississippi Valley-type districts from the Atlas system (source: Bouabdellah and Sangster 2016)*

Nonetheless, it should be noted that the listing of North African carbonate-hosted Pb-Zn ± Cu ± F ± Ba deposits described therein does not include mineral deposits associated with sandstone and feldspar-bearing sandstone (i.e. arkoses, case of the Moroccan Zeida deposit considered as of syndiagenetic origin with mineral resources estimated at 16 Mt of ore at 3% Pb (Bouabdellah and Sangster 2016)), the structurally controlled fluorite-baryte deposits, and those vein-type Pb-Zn ± Cu occurrences enriched in Hg-Cu-As-Bi-Sb such those of "zone des nappes" in Tunisia and Tell domain in Algeria as such enrichments may reveal a magmatic epithermal setting.

On the other hand, in agreement with the recommendations of Leach et al. (2010), our synthesis includes the Pb-Zn Cu ± F ± Ba deposits associated with salt domes (i.e. peridiapiric deposits), which are considered as a variant of MVTs because they share many similarities with the large MVT family (petrography of the host rocks, epigenetic origin of mineralization, sulfide-bearing mineral parageneses, structural and paleogeographical control on ore deposition, nature and origin of the mineralizing fluids).

4.2. Geological settings of MVT deposits of the Atlasic system: main ore deposits and districts

As indicated above, the North African Atlas system is remarkably endowed with the largest and the most productive MVT deposits and prospects in Africa, with the highest grade orebodies being concentrated in two distinct metal-bearing provinces: the "Atlas" province to the west and the "Diapir Zone" province to the east (Figure 4.1).

In Morocco, the most significant MVT deposits and prospects occur in the Early to Middle Jurassic carbonate successions with the most productive being Pb-dominant. The largest districts are those of Touissit-Bou Beker (>100 Mt averaging 4% Pb, 3.5% Zn, <1% Cu and 120 g/t Ag) (Bouabdellah et al. 2012, 2015), the Upper Moulouya (>30 Mt at 4% Pb) and Jbel Bou Dahar (>30 Mt at ~4% Pb, ~4% Zn) (Bouabdellah and Sangster 2016).

In Algeria, the main occurrences are distributed either along the northern edge of the "Hauts Plateaux" (High Land) or in the "Diapir Zone" where over 2,000 Pb-Zn ± Cu ± Ba ± F deposits, prospects and showings have been documented (Vydrine 1972; Touahri 1991) (Figure 4.1). Stratigraphically, the orebodies are hosted in two major carbonate aquifers of Jurassic (case of deposits of the "Hauts Plateaux") and Lower Cretaceous ages (case of deposits of the "Diapir Zone"). Based on the available economic data, the Algerian MVTs produced a total (production +

reserves) of over 5 Mt Pb-Zn with 68% of the metal estimates being derived from peridiapiric deposits (Touahri 1991). Of these, the largest and most productive deposits are those of El Abed in the Touissit-Bou Beker district with total production and reserves estimated at over 30 Mt of ores averaging 4.5% Zn and 1% Pb (Bouabdellah et al. 2012) and Mesloula whose total production has been estimated at 300,000 tons of Pb concentrates (François 1970; Laouar et al. 2016).

In Tunisia, the Mesozoic–Cenozoic calcareous-dolomitic sequences of the three major structural domains that compose the country (zone des nappes, Diapir Zone and Saharan Atlas), host numerous Pb-Zn ± Ba ± F ± Sr deposits, prospects and showings. However, it is in the Diapir Zone that the largest and most productive MVT deposits are reported occurring as high-grade stratabound orebodies intimately related to the Triassic diapiric structures and associated carbonate formations (peridiapiric deposits) (Figure 4.2).

The descriptions referred to in this contribution are from some previous studies (Bouladon 1949; Claveau et al. 1952; Voirin 1965; Rajlich 1983; Dupuy and Touray 1986; Adil et al. 2004; Bouabdellah et al. 2012, 2015; Bouabdellah and Sangster 2016; Rddad et al. 2016) for Moroccan MVTs (Boutaleb 2001; Haddouche et al. 2004, 2014, 2016; Salmi-Laouar 2004; Salmi-Laouar et al. 2004, 2007; Prochaska et al. 2007; Sami 2011; Ysbaa et al. 2019) for the Algerian deposits, and finally for the Tunisian deposits (Perthuisot 1981; Rouvier et al. 1985; Charef and Sheppard 1987; Perthuisot 1992; Shepard et al. 1996; Perthuisot et al. 1998; Bejaoui et al. 2013; Bouhlel et al. 2016; Rddad et al. 2019).

4.2.1. *Morocco*

During the last two centuries, Morocco ranked among the top 10 world leading countries in the production of Pb-Zn ± baryte ± fluorite. This ranking reflects the key role of the mining industry in the economy of the country. In this respect, economic data show that approximately 70% of the Moroccan production of Pb-Zn derived from MVT deposits. The exploited orebodies occur within the Atlas system, including the High and Middle Atlas and the "Hauts Plateaux" (Bouabdellah and Sangster 2016). Indeed, it is exclusively in the Lower to Middle Jurassic carbonate strata of the High and Middle Atlas along with their lateral extensions to the NE, namely "la chaine-des-Horsts" and the "Hauts Plateaux" that the majority of the Moroccan MVT deposits and prospects are located, among which the most productive are those of Touissit-Bou Beker, Upper Moulouya and Jbel Bou Dahar (Bouabdellah and Sangster 2016) (Figure 4.1).

In addition to these world-class districts, there are several subeconomic Pb-Zn deposits and prospects such as Ali Ou Daoud, Aguerd N'Tazoult and Bou Arhous in Central High Atlas (Ibouh and Mouguina 2012; Mouguina et al. 2012; Rddad et al. 2018), Izilane, Adeldal and Cadenar prospects in the calcareous ridge of the Rif belt (Benaissa 1996), Sidi Bou Houria, Bellaïne, Zegzel, Isarhdaouen, Foughal and Sidi Amar Acherki in "la chaîne-des-Beni Snassen" (Bouabdellah et al. 2004), and Khalifa and Aïn Hallouf in the Tazekka massif (Auajjar and Boulègue 1999, 2002; Noronha et al. 2007) (Figure 4.1).

4.2.1.1. *Touissit-Bou Beker district*

The Touissit-Bou Beker district is located in the northeastern part of "la chaîne-des-Horsts" adjacent to the Algeria–Morocco border (Figure 4.3). The district comprises five main deposits, four of which (Bou Beker, Touissit, Mekta and Beddiane) are located in the Moroccan territory and the fifth, El Abed, in the Algerian counterpart (Figure 4.3). It is noteworthy that all the deposits of the district are hosted in the Missouine graben within the Aalenien–Bajocian dolomitic formation, which is stratigraphically assigned to the so-called "Dalle des Hauts Plateaux". This mineralized horizon can therefore be delineated as a lithological and stratigraphic metallotect at the scale of "la chaîne-des-Horsts" and its northeastern extension in Algeria (Rhar Rhouban mounts) (Figure 4.3).

The host rocks consist predominantly of a Paleozoic basement complex unconformably overlain by a succession of Middle Jurassic (i.e. Aalenian–Bajocian) unmetamorphosed flat-lying carbonate strata (Figure 4.4). The basement comprises Lower to Upper Paleozoic volcaniclastic metasediments locally cut by Visean rhyodacitic plugs. The unconformably overlying Jurassic cover starts with a thin (~5 m) sequence of Lower Jurassic (Infra-Lias) conglomerates and limestones, followed by subhorizontal Aalenian–Bajocian dolomitic strata i with thin and clay-marly interbeds, indicating a low-energy intertidal to supratidal platform deposit medium.

The yellowish succession of sandstones, marls and carbonates of the Late Bajocian–Lower Bathonian age, locally known as "Toit Jaune", approximately 20-m thick, constitutes the hanging wall of sulfide mineralization. The Late Bajocian sequence grades upward into a red ferruginous Bathonian oolitic regional marker unit (~2 m in thickness), which is in turn overlain by a thick Callovian–Oxfordian section consisting of thick (600 m) impermeable shales and carbonates forming a "*cap rock*" (Figure 4.3). As for the "Toit Jaune", the reddish bathonian facies constitute a stratigraphic marker at the scale of "la chaîne-des-Horsts" and serve as an exploration tool for local miners during the drilling operations. The appearance of the first reddish sludge during drilling indicates the proximity of mineralization. The

Kimmeridgian, a succession of up to 100-m thick of dolostones and limestones, caps the lithostratigraphic column.

Structurally, the host rocks are truncated by a network of ENE–WSW, E–W and NNW–SSE-trending faults (Figures 4.3 and 4.4). Paleogeographic reconstructions indicate that the Pb-Zn deposits of the Touissit Bou-Beker district are located above a Paleozoic high structure known as "Touissit shelf" against which the Triassic and Lower Jurassic formations thin and pinch out (Figure 4.5).

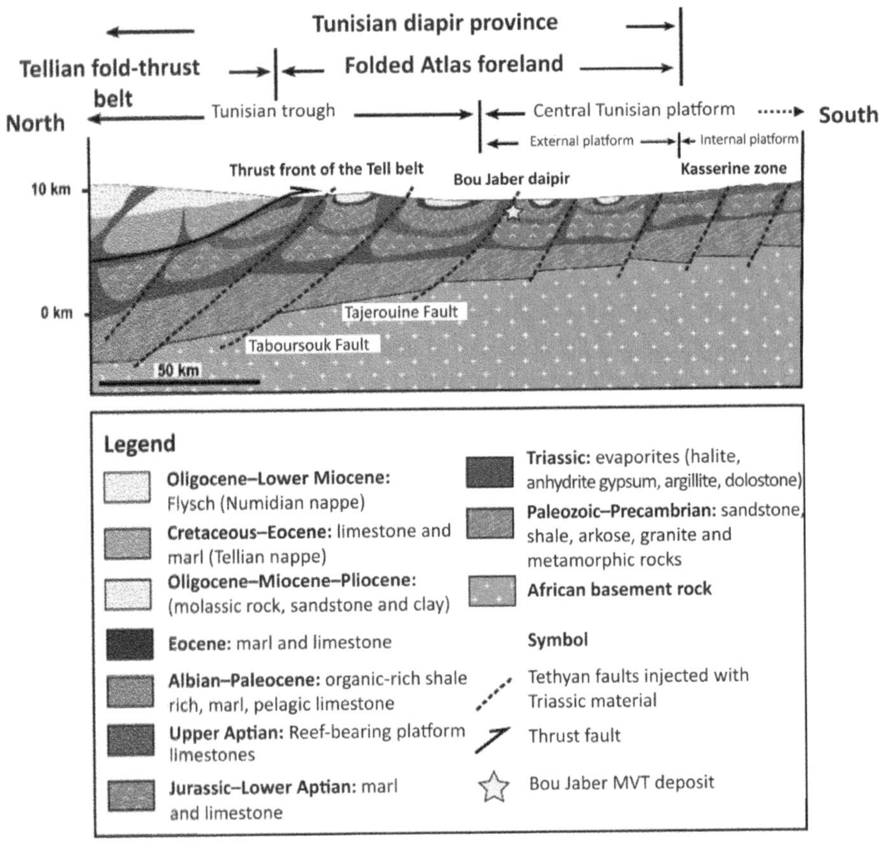

Figure 4.2. *Present-day interpretative N–S cross-section along the Tunisian salt diapir province and its relationship to the Tell and the foreland Atlassic domains from the Early Cretaceous to the Quaternary. Modified from Perthuisot (1978) and Bouhlel et al. (2016)*

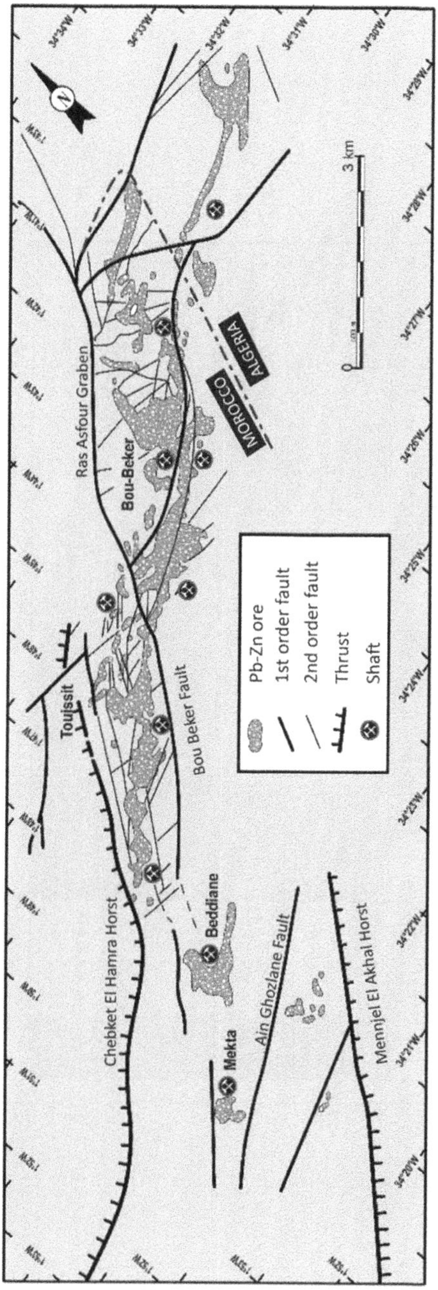

Figure 4.3. *Structural setting of Touissit-Bou Beker Mississippi Valley-Type orebodies projected to surface (source: Bouabdellah et al. 2012). Note the parallelism between the elongation of the orebodies and the regional ENE–WSW- and E–W-trending faults*

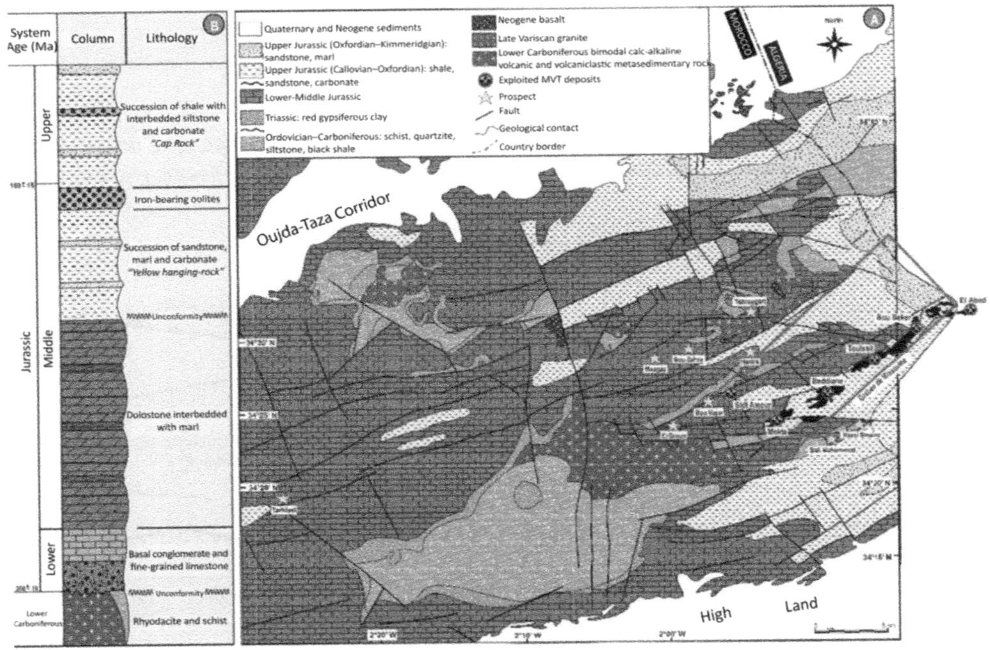

Figure 4.4. *(a) Geological map of the northeastern part of "a chaîne-des-Horsts" showing the location of the main Mississippi Valley-type deposits and prospects (projected to surface) of the Touissit-Bou Beker district and their relationship with the main lithostratigraphic units, igneous rocks and the major fault systems; (b) idealized lithostratigraphic column for the Touissit-Bou Beker district (source: Bouabdellah et al. 2012)*

Dolomitization is by far the most widespread form of wall-rock alteration at the scale of the deposits in the district. Two major chronologically and genetically distinct phases of dolomitization are distinguished, including (1) an early regionally extensive dolomitization in relation with the diagenetic evolution of the host basin, and (2) a later epigenetic ore-stage dolomite cement superimposed on the early diagenetic dolostones referred to as hydrothermal dolomite (Bouabdellah et al. 2012, 2015). The occurrence of hydrothermal dolomite (i.e. "baroque" or "saddle" dolomite) is considered as an important exploration guide in the search for new hidden Pb-Zn deposits at the scale of "la chaine-des-Horsts".

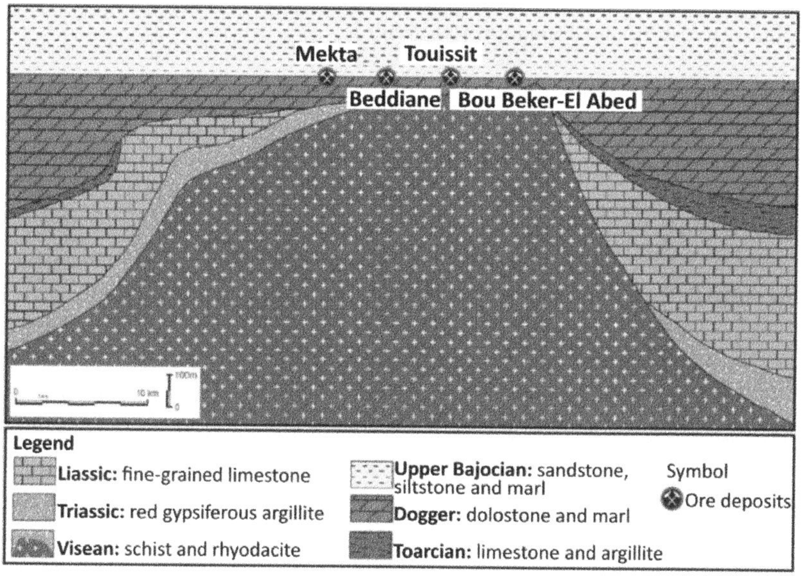

Figure 4.5. *Longitudinal ENE–WSW cross-section through the "Touissit shelf" showing the location of the main Mississippi Valley-type deposits of the Touissit-Bou Beker district. It is important to note the pinch-out of the Triassic and overlying Liassic and Toarcian strata against the "Touissit shelf". Modified from Samson (1973) and Bouabdellah et al. (2015)*

Typologically, hydrothermal dolomite occurs as massive open space fillings (i.e. karsts, veins, veinlets, geodes and breccia cements) or metasomatic replacement of the diagenetic dolostones giving rise to the so-called recrystallized dolostones (Samson 1973). The spatial coexistence of these two types (diagenetic and hydrothermal) of dolomite is at the origin of the development of typical textures and structures such as the zebra structure, lobe-shaped and cruciform dolostones, collapse breccia, and more importantly the *snow-on-roof* texture considered as the most diagnostic feature of MVTs (Leach and Sangster 1993). Stratabound sulfide mineralization whose spatial distribution is restricted to the uppermost 10 m of the coarse-grained host ore-bearing dolomite (D_2 and D_3 dolomites) (Bouabdellah et al. 2012, 2015) occurs as massive infills of interconnected karstic cavities and collapse breccia cement. Hence, sulfide mineralization is described as being epigenetic and stratabound. Sulfide paragenesis consists predominantly of galena and sphalerite with subordinate chalcopyrite, tetrahedrite and pyrite-marcasite in addition to hydrothermal dolomite as the main gangue phase. Conversely, the supergene mineral assemblages are composed of lead-zinc-copper-bearing carbonates and sulfates (cerussite, anglesite, smithsonite, hydrozincite, malachite, azurite), Ca-sulfates (gypsum) and vanadates (wulfenite).

The spatial distribution of Pb/Zn mass ratios through the deposits delineates a typical metal zoning pattern typified by the predominance of Zn-rich ore to the east (El Abed deposit) and Pb-rich ore farther to the west (Beddiane and Mekta deposits) (Figure 4.3). In addition to this regional lateral zonation, there is also a vertical zoning pattern of Pb, Zn and Cu displayed by the predominance of Zn-rich ore at the lowermost part of the Aalenian–Bajocian dolomitic strata followed by Pb-rich ore at the upper part of the host dolostones close to the "Toit Jaune" hanging wall rocks, though occasionally economic mineralization extends into the overlying "Toit Jaune" and even into the Callovian–Oxfordian sequences. Cu-rich ore, though restricted to the Beddiane and Bou Beker deposits, is located stratigraphically higher than the Pb-rich ore.

4.2.1.2. *Upper Moulouya district*

The Upper Moulouya district is the second producer of lead-zinc ± baryte ± fluorite in Morocco with a total production + reserves exceeding 31 Mt of Pb-Zn ore at an average grade of ~4.5% Pb and <1% Zn (Annich and Rahhali 2002). The main exploited mines include the Aouli vein system (9.6 Mt of ore at 5% Pb), the stratiform syndiagenetic Zeida deposit (16 Mt of ore at 3% Pb) and the Mibladen stratabound orebodies (6.5 Mt of ore at 5% Pb) (Figure 4.6). Among these, only the Mibladen deposit is classified as a typical MVT (Bouabdellah and Sangster 2016).

The Mibladen mining district extends in ENE direction, approximately 10 km long and 5 km wide (Figure 4.7). The host rocks consist of a Paleozoic basement complex mainly composed of Cambro–Ordovician schists with quartzite interbeds and amphibolite (Vauchez 1976; Filali et al. 1999) unconformably overlain by a thick (up to 4–4.5 km) succession of rift-related Permian–Triassic series. The latter consist of a succession of basal conglomerates that grade upward into red evaporite-bearing argillites with interbedded dolerite-diabase sills dated from ~210 ± 2.1 to 196 ± 1.2 Ma ($^{40}Ar/^{39}Ar$) (Fiechtner et al. 1992). The post-rift Liassic to Dogger successions consist of marine carbonates deposited in a strongly subsiding basin (Fedan et al. 1989).

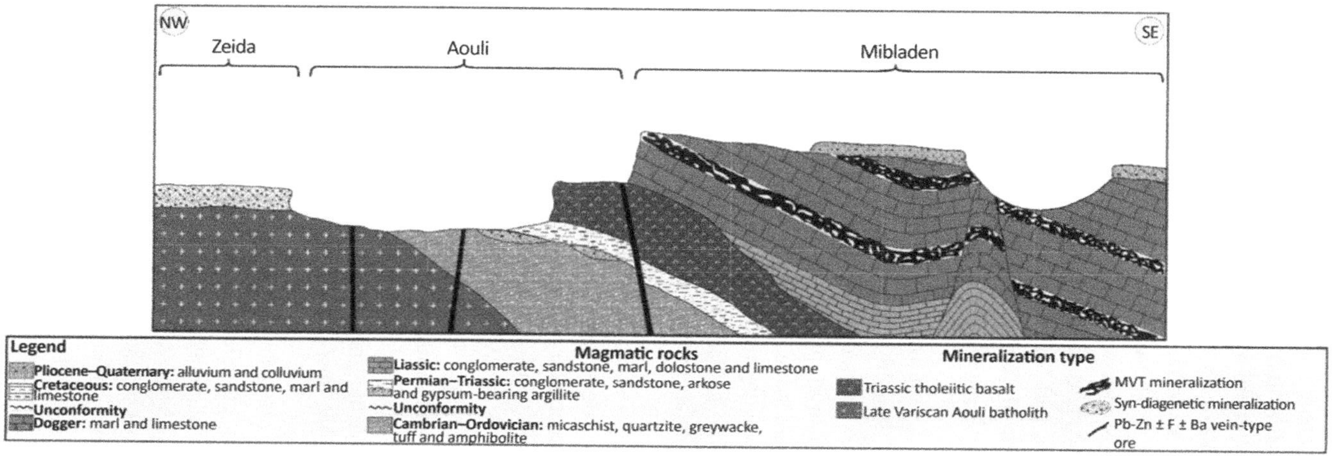

Figure 4.6. *Longitudinal NW–SE cross-section through the Upper Moulouya district showing spatial distribution of Aouli, Zeida and Mibladen base-metal ± fluorite ± barite deposits and their relationship to stratigraphy and fault systems time. Modified after Emberger (1965) and Bouabdellah and Sangster (2016)*

The Liassic and Dogger carbonate strata are unconformably overlain by Lower Cretaceous and Cenomanian successions essentially composed of shallow carbonates, sandstones, marls and shales with evaporite-bearing interbeds. The Cretaceous sequences are overlain by up to 1,500-m-thick Mio–Pliocene continental detrital sediments locally cut by ca 14.6–0.5 Ma alkaline basalt lavas (Harmand and Cantagrel 1984; Duggen et al. 2005, 2009; Wittig et al. 2010) (Figure 4.7). It should be stressed that the Domerian (Lower Jurassic) dolomitized carbonates which host the main exploited orebodies underwent the interplay of several dissolution episodes (karstification) related to the short-term exposure of the carbonate platform under an extending regime (Dagallier and Macaudière 1987). Structurally, the country rocks are truncated by a network of four pluri-km E–W-, ENE–WSW-, NW–SE- and WNW–ESE-trending faults (Figure 4.7).

The pre-ore diagenetic dolomitization event is by far the most widespread form of wall-rock alteration in the Mibladen deposit. The effects of silicification are restricted to rare occurrences located in the vicinity of the sulfide orebodies. Similar to the Touissit-Bou Beker district, pre-ore replacement dolomitization is generalized to all the Domerian carbonate strata, giving rise to layered successions of dolostones and/or dolomitic limestones with strong intercrystalline porosity. It should however be stressed that, unlike the Touissit-Bou Beker district, hydrothermal dolomite genetically related to the emplacement of sulfide mineralization is lacking.

Lead-baryte-bearing mineralization is hosted by the uppermost part of the Domerian limestone-dolomitic strata and is spatially distributed along two mineralized lobes that geometrically delineate two subparallel corridors referred to as ("Faisceau supérieur" and "Faisceau inférieur") (Emberger 1965). The latter are limited on both sides by the ENE–WSW transcrustal faults of Aouli to the north and Amourou to the south (Figure 4.7). In summary, lead-baryte mineralization, which is clearly epigenetic and stratabound, occurs as massive open-space fillings of horizontal and vertical karsts, veins, tension gashes and vugs, and of collapse breccia cement. Conversely, the uneconomic replacement mineralization occurs as fine disseminations dusting the host dolomitic limestones.

Sulfide paragenesis is one of the simplest, ever described in MVTs, being reduced to the occurrence of galena with subordinate pyrite and chalcopyrite. Sphalerite is virtually lacking. Gangue phases consist exclusively of baryte with no dolomite nor calcite. Hence, Mibladen deposit is characterized by the predominance of Pb over Zn in a 10:1 mass ratio and the occurrence of baryte as the exclusive gangue mineral. Supergence mineralization, which derived from oxidation of primary sulfides, consists of a wide range of exotic minerals, such as vanadinite, wulfenite and cerussite, whose gem quality is challenged only by the beauty of crystals displayed in the world's largest museums.

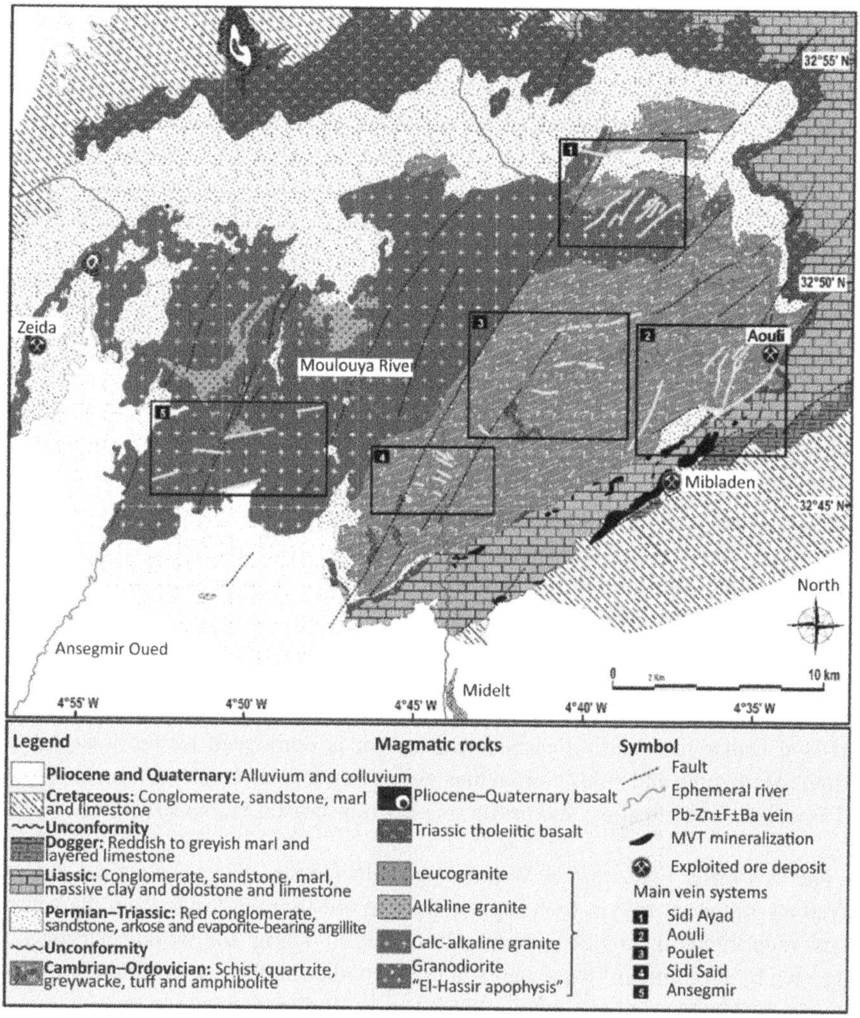

Figure 4.7. *Geological map of the Upper Moulouya district showing regional geology along with the main igneous rocks and major faults. The location of historically mined Aouli, Mibladen and Zeida base-metal ± fluorite ± barite deposits is also indicated (modified from Emberger 1965; Bouabdellah and Sangster 2016)*

4.2.1.3. *Jbel Bou Dahar district*

The Jbel Bou Dahar plateau is located in the eastern part of the High Atlas mountains, and it is host for over 100 of carbonate-hosted and fault-controlled Pb-Zn-Ba deposits and prospects occurring along its northern and southern edges.

The core of the reef-bearing carbonate platform, which is exposed in the northern part of the reefal structure at Sebbab-Kébir inlier, consists of a succession of greenschist-facies Cambro–Ordovician to Silurian metasedimentary and volcaniclastic rocks made dominantly of quartzites and schists (Figure 4.8).

Unconformably overlying the Paleozoic basement rocks are the syn-rift ~400- to 500-m-tick Triassic red bed series with intercalations of salt-bearing argillites and CAMP basalts. During Early and Middle Liassic times, the Lower Paleozoic formed a topographic high on which the Jbel Bou Dahar reef-bearing carbonate platform and its related lagoon were built, incompletely separated from the open sea by a barrier reef that developed to the south (Figure 4.9).

The reef deposits consist of massive, coral-algal- and bivalve-bearing Pleinsbachian, flat-lying, thickly-bedded carbonate mudstone, skeletal wackestone, skeletal pelletal packstone, laminated carbonate mudstone and calcareous clay. Toarcian shales and Aalenian lime mudstones progressively onlap the edges of the Domerian platform and conformably overlie the platform roof, recording an abrupt cessation of platform evolution at the Domerian–Toarcian transition (Crevello 1990; Campbell and Stafleu 1992).

It is noteworthy that dolomitization is restricted to the Sinemurian limestones and does not overprint the overlying Domerian strata. The structural history of the Jbel Bou Dahar reef-bearing carbonate platform is dominated by the interplay of both the Hercynian and Atlasic orogenies giving rise to a succession of ductile (folds and associated schistosities) and brittle (mainly transcrustal faults) structures.

The first-order ENE- and E–W-trending faults (Figure 4.8) are assumed to have played an important role in focusing the flow of ore-forming fluids from the source zones (depocenters) to depositional sites located along the basin margins, as suggested by proximity of these faults to the orebodies and by the general elongation of mineralized veins and ore zones in ENE and E–W directions

The economic orebodies are distributed within the Lower to Middle Jurassic carbonate strata along three main lithostratigraphic units, the Lotharingian oolitic reef-bearing limestones and associated sub-reefal brecciated carbonates (i.e. Sebbab deposit), the Sinemurian–Pliensbachian reef-build-up lithotypes and the Aalenian limestones (Bouabdellah and Sangster 2016) (Figure 4.9).

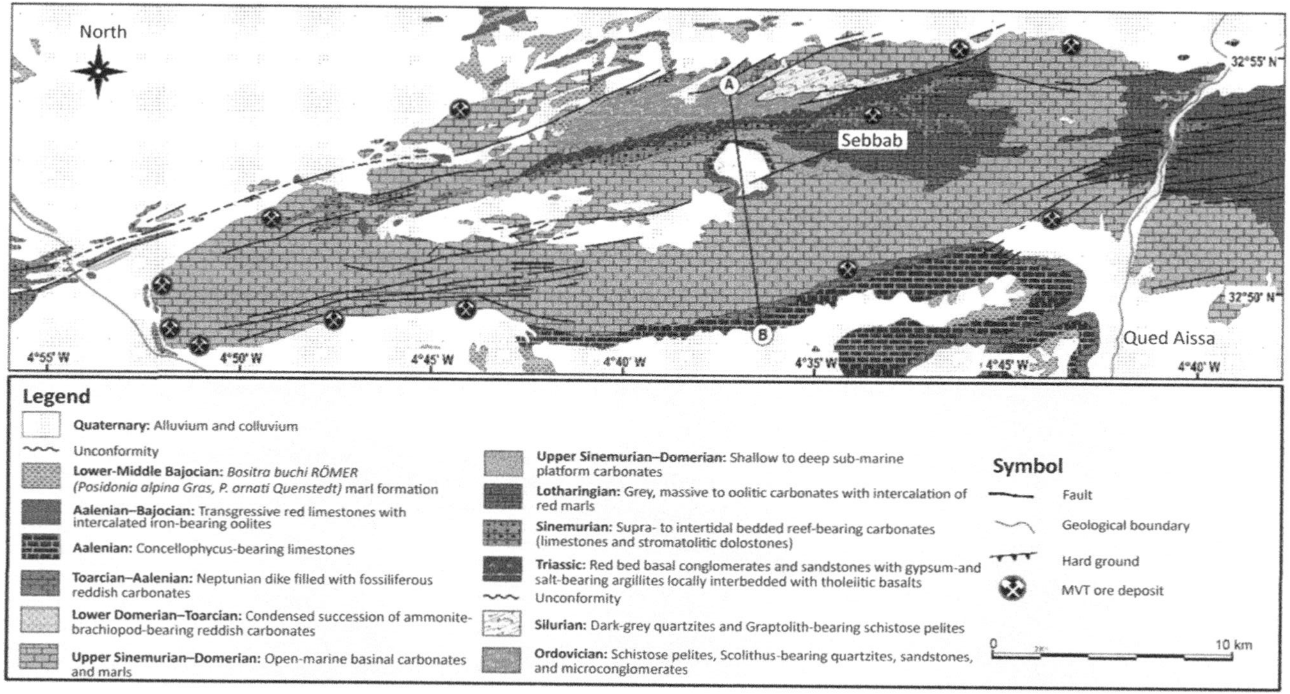

Figure 4.8. *Detailed geological map of Jbel Bou Dahar reef-bearing complex showing regional geology, main tectonic structures and stratigraphic distribution of major Mississippi Valley-type deposits and prospects. Modified after Agard and Du Dresnay (1965)*

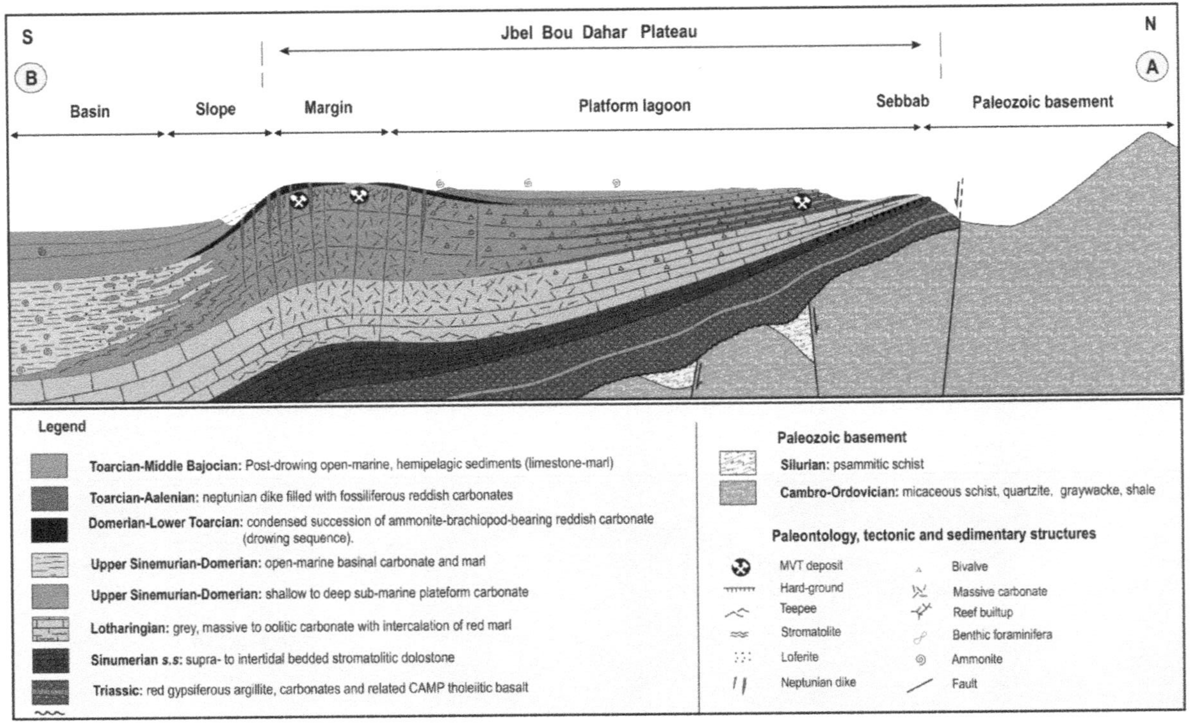

Figure 4.9. *Longitudinal N–S cross-section through the Liassic reef-bearing carbonate platform of Jbel Bou Dahar showing the the different paleogeographical domains along with the location of major Mississippi Valley-type deposits and prospects. Modified after Agard and Du Dresnay (1965)*

Sulfide mineralization occurring as massive karsts, veins and veinlets infills and in breccia cement, is clearly epigenetic and commonly stratabound, though stratiform mineralization predominates in the Sebbab deposit (Figure 4.8). The mineral paragenesis is quite simple, being reduced to galena and sphalerite as the predominant sulfides with subordinate pyrite-marcassite and chalcopyrite. Major gangue phases consist predominantly of calcite and baryte and to a lesser extent celestine. Overall, five mineralizing stages and substages of various economic interest have been recognized (Bouabdellah and Sangster 2016).

4.2.2. *Algeria*

Algerian MVTs occur along two contrasting geographical areas located on both sides of the Algeria–Morocco and Algeria–Tunisia borders (Figure 4.1). These two opposite geographical areas encompass the MVT deposits of the "Atlas" province to the west and those of the "Diapir Zone" province to the east. It is noteworthy that MVT deposits and prospects of the "Diapir Zone" are spatially and genetically related to Triassic salt-bearing diapirs, and as such these are referred to as peridiapiric deposits (Laatar 1980; Rouvier et al. 1985; Bouzenoune et al. 1995, 2006; Orgeval 1995; Sheppard et al. 1996; Salmi-Laouar et al. 2004; Laouar et al. 2016). In contrast, MVT deposits of the Atlas province are hosted in the Triassic to Jurassic carbonate plateform of "la chaine-des-Horsts" and "Rhar Rouban" belt and the "Hauts Plateaux" domain.

In addition to the aforementioned provinces, some MVT deposits and prospects are documented in the autochtonous infratellian domain, the most significant of which are those of Ouarsenis or Bou Caïd-Teniet El Haad districts with a cumulated Pb-Zn production totaling 300,000 tons including 110,000 tons of non-sulfide zinc (Touahri 1991). From a mineral resource point of view, the Algerian MVTs are of medium size except of El Abed and Mesloula deposits whose Pb-Zn resources are estimated to have exceeded 2 Mt of Pb+Zn and 300,000 tons of Pb concentrates, respectively (Touahri 1991; Bouabdellah et al. 2012). The cumulated production of the remaining deposits and prospects is estimated to be low (<100,000 tons) (Touahri 1991).

4.2.2.1. *Ore deposits of the Atlas province*

The Lower to Middle Jurassic dolomitic strata of the northwestern part of the Algerian territory host many MVT deposits, prospects and showings, the most significant of which are those of El Abed, Deglen, Oued Zebboudj, Takhemert, Aïn Melaz and Fillaoucène (Touahri 1991). Among all of these, only the El Abed deposit was mined and exploited over a time period exceeding 20 years.

The El Abed deposit is located in the eastern termination of the Missouine graben in stratigraphic, structural and mineralogical continuum of the stratabound sulfide orebodies of the Moroccan Bou Beker deposit (Figure 4.3). Moreover, both deposits share a common mining history as they were exploited together by the "Société de Zellidja". In 1966, following up the decision of the Algerian authorities to nationalize the mining industry, a feasibility study for the reopening of the El Abed mine led to mineral reserve estimates of 0.9 Mt Pb + Zn (Touahri 1991). Abandoned since its closure in 2002–2003, the El Abed mine was taken over by the Chinese consortium Cecomines based on a joint venture signed with the Sonatrach, but there was no practical follow-up. The mine is presently closed due to the depletion of reserves and/or low Pb-Zn grades.

Similar to other deposits of the Touissit-Bou Beker district, the host rocks in the El Abed(-Bou Beker) deposit consist of a Paleozoic basement complex unconformably overlain by a thick succession of Triassic to Upper Jurassic sediments. Unlike deposits of the Western part of the Touissit-Bou Beker district, where the rhyodacitic volcanites predominate (Mekta and Beddiane deposits), the Paleozoic basement rocks of the El Abed(-Bou Beker) deposit consist predominantly of Ordovician to Silurian variably deformed and metamorphosed lower greenschist facies metasediments and metavolcanic rocks.

The unconformably overlying Mesozoic sedimentary rocks consist, from bottom to top, of a non-fossiliferous limestone level attributed to the Lower Liassic (Infra-Liassic) overlain by the flat-lying Aalenian–Bajocian dolomitic strata, truncated by an unconformity, by the Bajocian yellowish sandstones and marls that grade upward into a ferruginous Bathonian oolitic regional marker unit (Figures 4.4 and 4.10). These latter are overlain by a Callovian–Oxfordian succession of thick impermeable shales and carbonates referred to in Algeria as "Saïda clays" (Touahri 1991). The Oxfordian Boumeddine sandstones or their Bou Beker sandstones Moroccan counterpart, the Zarifet limestones, along with the Kimmeridgian Tlemcen dolostones cap the lithostratigraphic column.

Similar to other deposits of the Touissit-Bou Beker district, the Aalenian–Bajocian ore-bearing dolomite which hosts the main stratabound sulfide-bearing experienced the effects of two chronologically and genetically distinct dolomitization events: an early regional-scale premineralization diagenetic dolomitization overprinted by a later epigenetic hydrothermal dolomitization whose spatial distribution delineates halos around the deposits and laterally for several kilometers beyond the deposits. The tectonic history of the El Abed deposit is in line with the overall setting of "la chaîne-des-Horsts" and Ghar Rouban mounts (Elmi 1973).

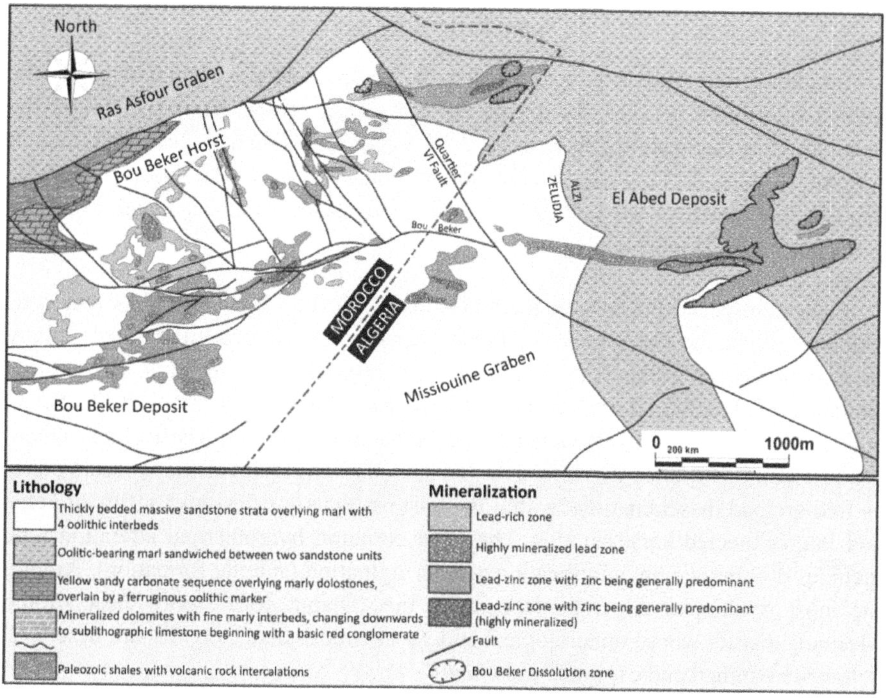

Figure 4.10. *Geologic map of Bou Beker-El Abed deposits (Morocco–Algeria border) showing regional geology, major faults, distribution of ore-bearing dolomite and related orebodies (projected to surface)*

Sulfide mineralization occurs either as moderate to massive replacement of the dolomitic host strata giving rise to the stratiform ore style or open space fillings of interconnected karst cavities, veins and veinlets and cementing breccias. Economically, the open space filling mineralization accounts for most of the higher grade orebodies (Figure 4.10). Sulfide paragenesis consists predominantly of various proportion of sphalerite and galena with subordinate pyrite, chalcopyrite and tetrahedrite. Gangue phases are exclusively composed of various generations of white hydrothermal dolomite "saddle dolomite". It should be noted that the emplacement of mineralization is structurally controlled as evidenced by the geometry of the orebodies that parallels the major WNW–ESE to N–S, NE–SW- and E–W-trending faults (Figures 4.3 and 4.10).

4.2.2.2. *Ore deposits of the Diapir Zone province*

Straddling the Algeria/Tunisia border, the south of Medjerda mountains of the Tellian fold-thrust belt along with the eastern margin of the "Diapir Zone" of the Saharan Atlas delineate together the second largest MVT metallogenic province in Algeria (Figure 4.11). From an economic point of view, the most productive deposits are those of El Ouasta, Ouenza, Bou Jaber, M'Khiriga and Mesloula described below (Figure 4.11).

Collectively, the Pb-Zn orebodies are hosted by the Lower–Middle Cretaceous (Albian–Aptian and Campanian) carbonate successions that deposited in the vicinity to or at the margin of the Triassic diapiric structures (Bouzenoune et al. 1995, 2006; Boutaleb et al. 2000; Boutaleb 2001; Haddouche 2004, 2014; Sami 2011; Ysbaa et al. 2019). Typologically, the stratabound sulfide mineralization whose emplacement is structurally controlled by major NE–SW and NW–SE fault corridors occurs either as fine-grained disseminations within the host carbonates or massive infills of veins and interconnected karst cavities. The most common hydrothermal alteration type includes dolomitization, silicification and ferrugination (siderite formation). Among the most productive Algerian deposits of the "Diapir Zone", only those of the Mesloula district whose mineralogical and geochemical characteristics are described below were mined and exploited industrially.

The Mesloula district, which extends in a NE–SW direction over 10 km in length and 1–2 km in width, contains numerous Pb-Zn(-Ba) deposits, of which the most productive are those of Kef Bouguern, Bardo, main deposit, Kef Rakhma, Petit Jabeur and Kef M'Khiriga (Figure 4.12). Available production data indicate that the whole mines of the district produced approximately 300,000 tons of Pb concentrates, most of which being mined from the main deposit (François 1970; Laouar et al. 2016).

Geologically, the Mesloula massif appears as an antiform structure whose core consists of Triassic salt-bearing domes. Lithologically, the host rocks consist of a chaotic succession of multicolored gypsum-bearing marls, dolostones, limestones, sandstones and Triassic rock fragments, all of which being overlain by a sedimentary cover of Cretaceous to Miocene age (Figure 4.12). Post-Triassic formations whose cumulated thickness reaches 9,000 m are composed of limestones, sandstones and marls (Burollet and Stainfeld 1956; Perthuisot 1992, 1994). Since their deposition, the host rocks were subjected to the combined effects of two major Alpine phases dating from the Middle–Upper Eocene–Oligocene and the Upper Miocene–Pliocene (Vila et al. 1994, 1996; Frizon de Lamotte et al. 2000).

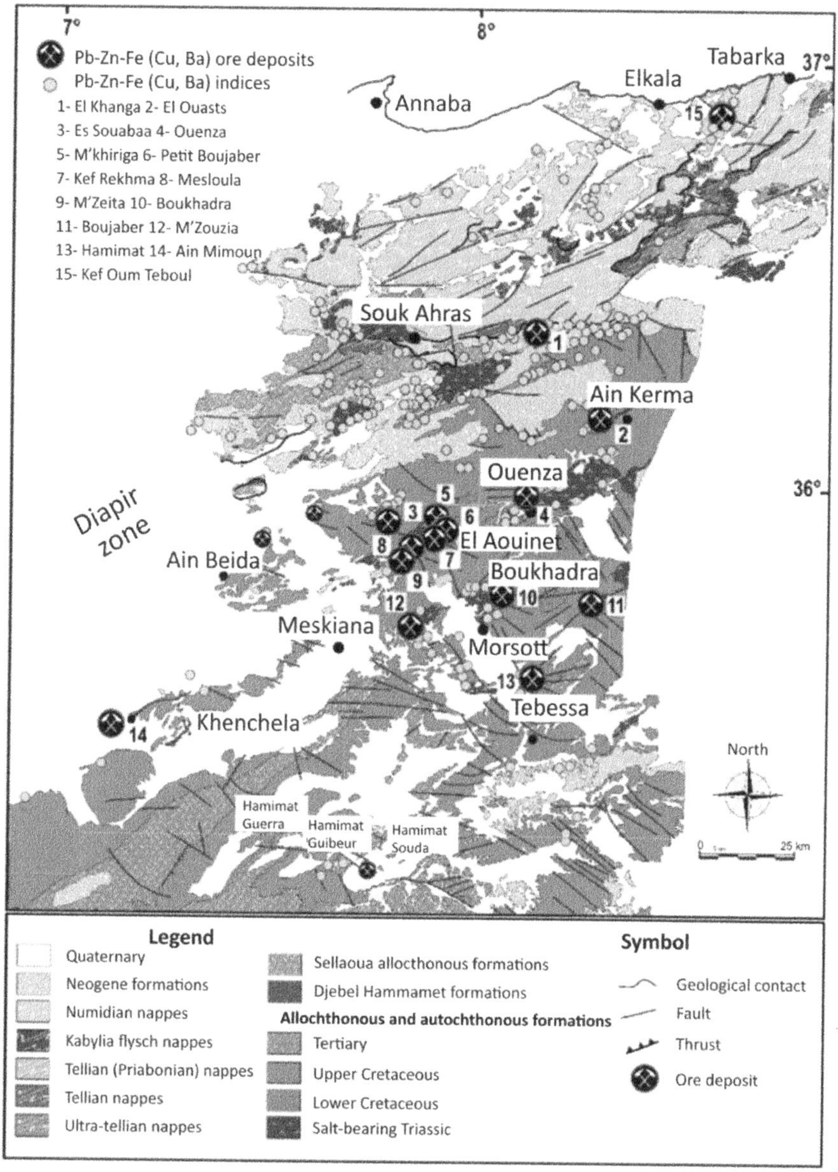

Figure 4.11. *Geological setting of the northeastern termination of the Algerian territory showing regional geology, extent of structural styles and distribution of ore-bearing carbonates and associated Pb-Zn-Fe-Ba ± Cu ± F ± Sr ± Au ± Ag deposits and prospects. Modified after Vila (1980), Haddouche et al. (2014) and Ysbaa et al. (2019)*

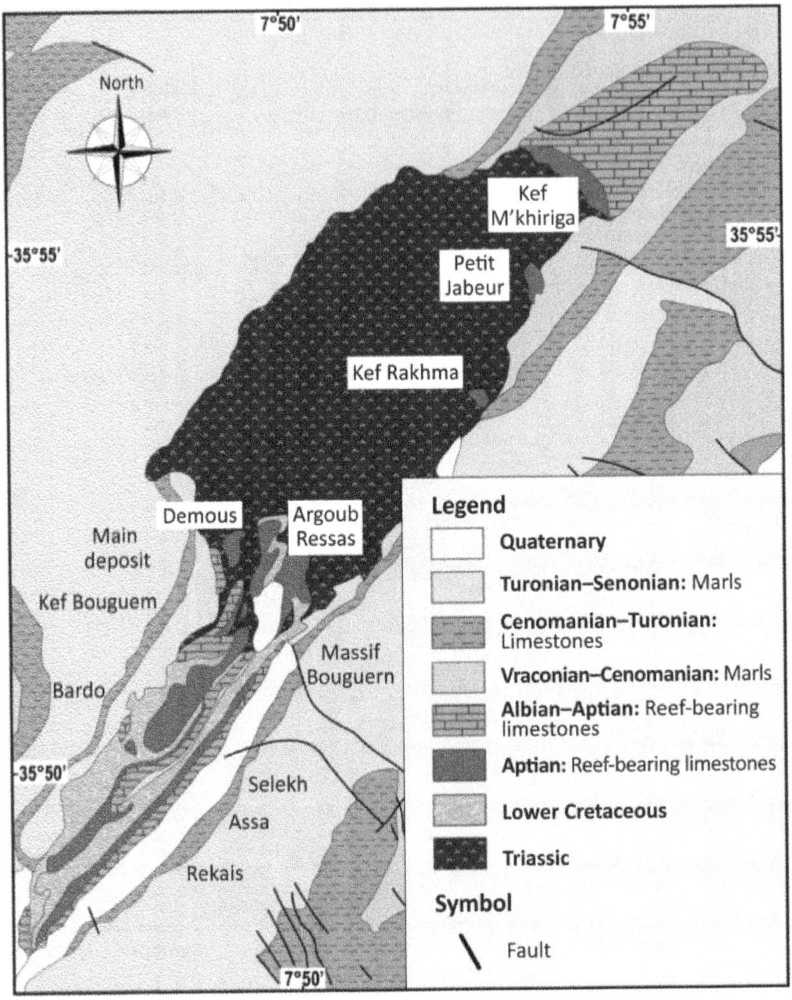

Figure 4.12. *Geological map of the Mesloula district showing the main geological features along with the location of major exploited orebodies. Modified from Dubourdieu (1959) and Laouar et al. (2016)*

Sulfide mineralization occurs close to or at the margin of salt domes as moderate to massive replacements of the Aptian reef-bearing limestones and reworked carbonate blocks contained within the Triassic chaotic aggregate or massive fillings of interconnected karst cavities, veins and veinlets, and as breccia cements. The resulting structures vary widely and include striped, dissolution-reprecipitation and colloform and vuggy structures. The sulfide paragenesis consists predominantly of galena and sphalerite with subordinate pyrite and tetrahedrite. Gangue minerals include hydrothermal dolomite, calcite, baryte, fluorite and quartz. Non-sulfide Pb-Zn ± Cu phases resulting from supergene oxidation of the primary sulfides consist of cerussite, smithsonite, malachite and azurite. The most extensive hydrothermal alteration whose effect increases toward the orebodies include the diagenetic and hydrothermal dolomitization, silicification and carbonatation (Laouar et al. 2016).

4.2.3. *Tunisia*

As for the Algerian counterpart, the Tunisian "Diapir Zone" or "Dome Zone" is host to numerous MVT deposits, prospects and showings, of which the most productive are those of Bou Grine, Fedj-El-Adoum, Bou Jaber, Slata and Oued Jebs (Figure 4.13). This structural domain, which lies between the thrust front of the Tellian belt to the northwest and the Saharan Atlas to the south, is characterized by the development of a large number of Triassic diapiric structures whose lithostratigraphic formations have intruded through the overlying thick Jurassic to Cretaceous sedimentary marine cover during the Aptian–Albian rifting phase and the intra-Miocene tectonic tightening episodes.

Lithostratigraphically, the Triassic salt diapirs consist of a chaotic aggregate of gypsum, anhydrite, sandstones and carbonates overlain by post-Triassic formations that comprise up to 5,000-m-thick succession of Cretaceous limestones, sandstones and marls, confomably overlain by over 4,000 m of Eocene sediments (Burollet and Stainfeld 1956; Perthuisot 1978). The uplift of salt-bearing bodies (halokinesis) is assumed to have started during the Upper Aptian–Lower Albian in a submarine setting (Perthuisot 1981) and continued after deposition of Tertiary sediments to reach its paroxysmal uprise during the Upper Tertiary. The extensional tectonic regimes, which were active in the Upper Cretaceous and Oligo–Miocene resulted in the reorganization of basins in a series of tilted blocks and strongly subsident troughs (Gharbi et al. 2005).

Typologically, the stratabound sulfide-bearing mineral systems occur in various deposit types including (Sainfeld 1952; Monthel et al. 1986; Orgeval 1995):

– stratabound orebodies that straddle the Triassic–Cretaceous contact giving rise to medium size ore deposits (100–300,000 t metal), the most illustrative of which is exemplified by the Fedj-El-Adoum deposit (Charef and Sheppard 1987; Sheppard et al. 1996);

– veins and tabular to columnar orebodies located within the Cretaceous peridiapiric carbonate strata with the most illustrative example being represented by Bou Grine deposit (Orgeval et al. 1985; Bouhlel et al. 2016);

– Aptian–Vraconian reef-bearing carbonate-hosted faults, fractures and karsts infills along with moderate to massive wall-rock replacement giving rise to stratabound orebodies, of which the most significant are those of Bou Jaber and Slata deposits (Bouhlel et al. 2016; Rddad et al. 2019);

– late vein-type deposits, of lesser economic importance, cutting across the Cretaceous carbonate host strata.

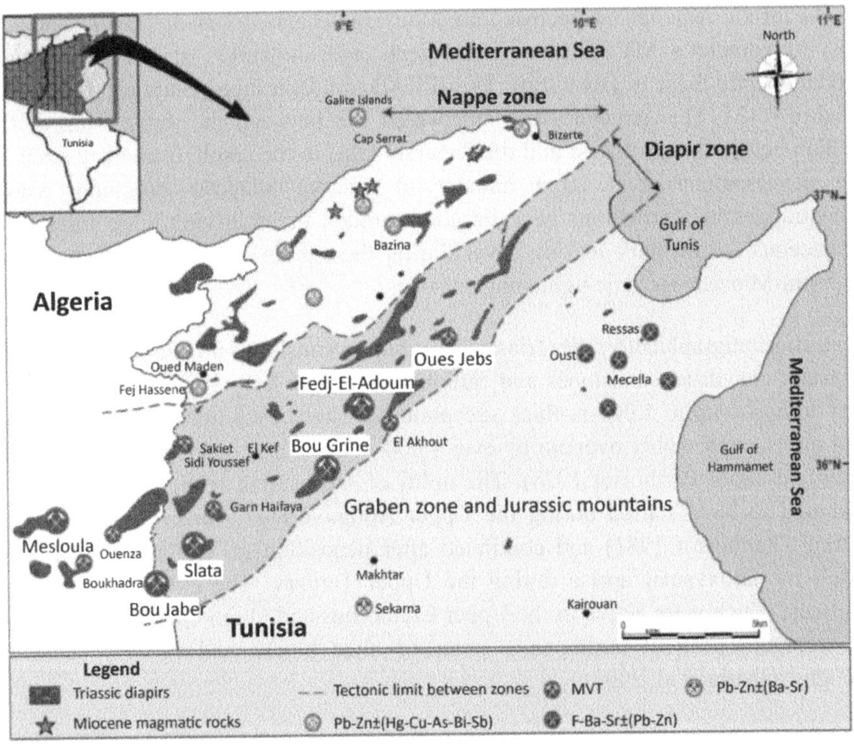

Figure 4.13. *Geologic and metallogenic map of Northern Tunisia showing the main structural domains, regional geology and location of major Pb-Zn-Fe-Ba-Sr-F deposits including the historically mined Mississippi Valley-type deposits in relation with the Triassic diapiric halokinesis and the Miocene igneous intrusions. Modified from Rouvier et al. (1985) and Bouhlel et al. (2016)*

4.2.3.1. *Bou Grine deposit*

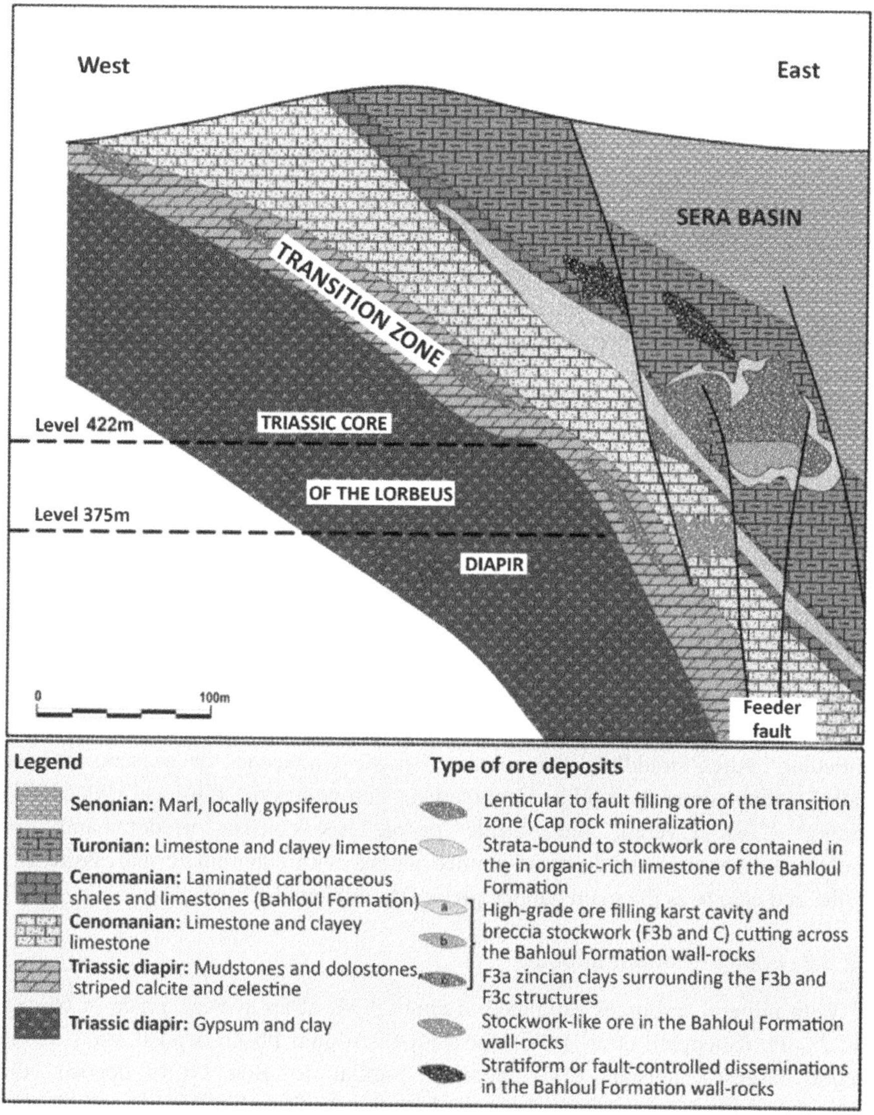

Figure 4.14. *Longitudinal E–W cross-section through the Bou Grine deposit showing the main lithostratigraphic units along with the location of major orebodies and their relationship to the Triassic diapiric structure and the associated fault system. Modified from Schmidt (1999) and Bouhlel et al. (2016)*

Discovered in 1979, the Bou Grine Pb-Zn deposit is by far the most productive mine in the "Diapir Zone" with an overall production + reserves estimated at 5.5 Mt of ore at an average grade of 12% Zn and 2.5% Pb (Orgeval 1994, 1995; Bouhlel et al. 2016). Exploited between 1993 and 1996 by the Bou Grine mining company, the mine closed in August 2005 while its production reached approximately 0.6 Mt of Zn and 0.1 Mt of Pb (Orgeval 1994).

Geologically, the Bou Grine deposit is located along the northern margin of the Jbel Lorbeus diapiric structure, whose core consists of Triassic salt-bearing formations overlain by a succession of Cretaceous to Tertiary sediments (Figure 4.14). In the mining area, the host rocks comprise a succession of Triassic to Paleocene sedimentary strata (Monciardini and Orgeval 1987).

Lithostratigraphically, the monoclinal Triassic series consist of a succession of alternating gypse-bearing argillites, dolostones, marls and clayey sandstones conformably overlain by Cenomanian marls, nodular limestones and clayey limestones. The conformably overlying Lower–Middle Cretaceous strata comprise a succession of Cenomanian organic-rich layered limestones stratigraphically assigned to the Bahloul Formation. In this respect, it should be stressed that most of the higher grade orebodies are confined within the carbonate lithologies of the aforementioned formation.

Sulfide mineralization occurs either as moderate to massive replacement of the layered organic-rich limestones or open space fillings giving rise to stratabound lenticular bodies straddling the contact Triassic–Cretaceous or as semi-massive bodies cutting across the Middle Cenomanian–Turonian series (Orgeval 1994, 1995) (Figure 4.14). Overall, the sulfide paragenesis consists predominantly of fine-grained sphalerite and coarse-grained galena with subordinate marcassite and pyrite, and calcite as the main gangue phase (Bouhlel et al. 2016).

4.2.3.2. *Fedj-El-Adoum deposit*

With mineral resources estimated at about 3 Mt at an average grade of 6% Zn and 2% Pb (Sheppard et al. 1996), the Fedj-El-Adoum Pb-Zn deposit ranks as the second largest Tunisian MVT deposit. Similar to Bou Grine deposit, the Fedj-El-Adoum mineral system is located in the vicinity of a Triassic salt-bearing diapiric structure intruding the Lower Cretaceous to Neogene sediments (Laatar 1980; Perthuisot 1981) (Figure 4.15).

The host rocks consist of a succession of Triassic diapiric evaporites, dolostones and compacted clay breccias of the Transition Zone Formation (TZF) overlain by the Albian to Cenomano–Turonian marly limestones (Laatar 1980). It is noteworthy that most of the economic orebodies are confined within the organic-rich dolomitic clasts of the TZF. In a such setting, stratabound sulfide mineralization occurs either as moderate to massive replacement of the dolomitic blocks giving rise to massive ore or open space fillings (faults, fractures, dissolution vugs) (Laatar 1980; Rouvier et al. 1985; Charef and Sheppard 1987) (Figure 4.15).

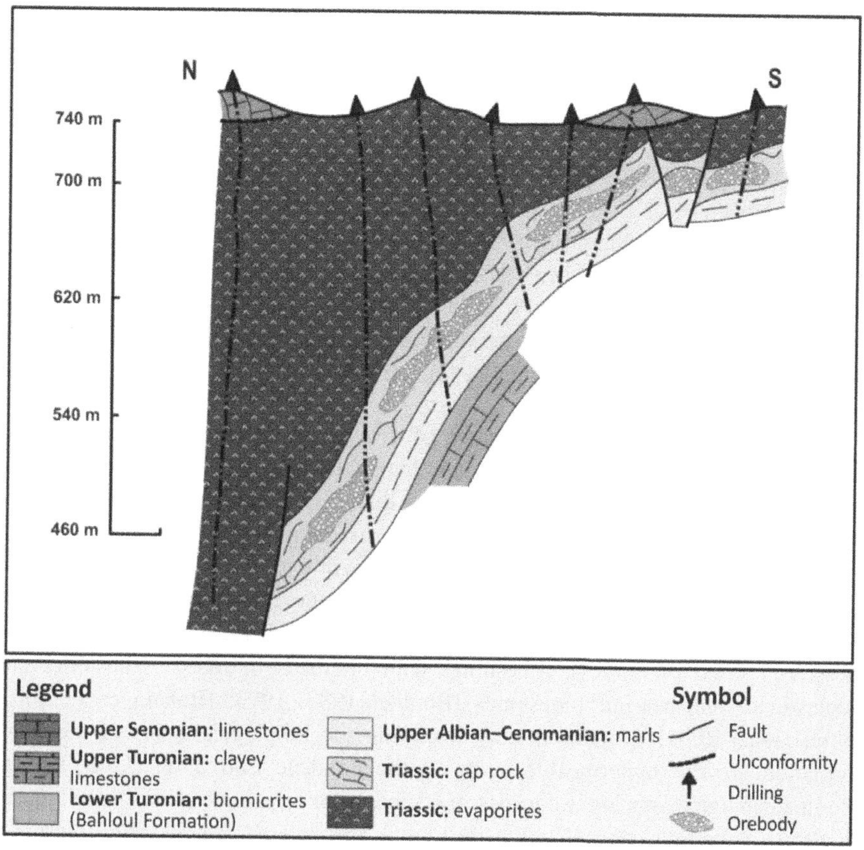

Figure 4.15. *Longitudinal N–S cross-section through the Fedj-El-Adoum deposit depicting the various lithostratigraphic units and the stratigraphic distribution of the main orebodies and their relationship to the Triassic diapiric structure. Drilling shafts are also shown. Modified from Laatar (1980) and Rouvier et al. (1985)*

Based on Pb/Zn mass ratios, Bouhlel et al. (2007) distinguished four mineralization styles: a Zn-dominant ore accounting for up to 80% of the estimated mineral resources of the deposit, and a Pb-dominant ore occurring as massive orebodies or vein and *stockworks* infills. The two remaining styles have been referred to as zebra calcite and native sulfur (Bouhlel et al. 2007). Nonetheless, it should be noted that, irrespective of the considered style of mineralization, the sulfide paragenesis consists predominantly of variable proportions of sphalerite and galena, with subordinate pyrite, all of which being embedded in a calcite and quartz matrix.

4.2.3.3. *Bou Jaber deposit*

Similar to other Pb-Zn deposits in the "Diapir Zone", the underground Bou Jaber deposit is hosted in a NE–SW anticlinal structure whose core consists of a 10 × 2 km Triassic salt dome (Amouri 1989) (Figure 4.16(a)). The deposit has been intermittently exploited between 1972 and 2004 by state-owned mining companies producing a total of 50,000 t of Pb + Zn and 0.5 Mt of baryte. Upon the closure of the mine in 2004, the available reserves were estimated at 8 Mt of ore averaging 1.5% Pb, 3.5% Zn, 33% $BaSO_4$ and 9% CaF_2 (Bouhlel et al. 2016).

The host rocks consist of a complex of Triassic rocks forming the core of the Bou Jaber salt diapir, which are overlain in fault contact by Uppar Aptian carbonates (Figures 4.16(a) and (b)). Lithologically, the diapiric structure consists of a chaotic complex of mega-breccias with hectometric to metric-sized blocks of blackish dolostones embedded in a matrix of evaporites (mainly gypsum resulting from halite transformation) and clays. The Upper Aptian strata consist of a succession of platform carbonates stratigraphically assigned to the Serdj Formation (Burollet 1956) composed of massive limestones with interbeds of marls, sandstones and bioclastic to reef-bearing limestones (Bouhlel 1987, 1993; Braham et al. 2002; Salmi-Laouar 2004). Unconformably overlying the lithological units of the Serdj Formation are the impermeable strata of the Fahdène Formation (Late Albian–Cenomanian age). This latter consists of a succession of pelagic shales, black marls, siltstones, glauconite, phosphate-bearing rocks and layered organic-rich limestones (Figure 4.16(c)). The Fahdène Formation sequences are capped by the finely layered organic-rich limestones of the Bahoul Formation, which in turn are overlain by a series of ~2,000-m-thick succession of Turonian to Paleocene limestones and marls. A series of NE–SW, NW–SE to N–S-trending faults segmented and downdropped the Upper Aptian peridiapiric carbonates to produce the "piano key"-like configuration of local (mine district-scale) horsts and grabens.

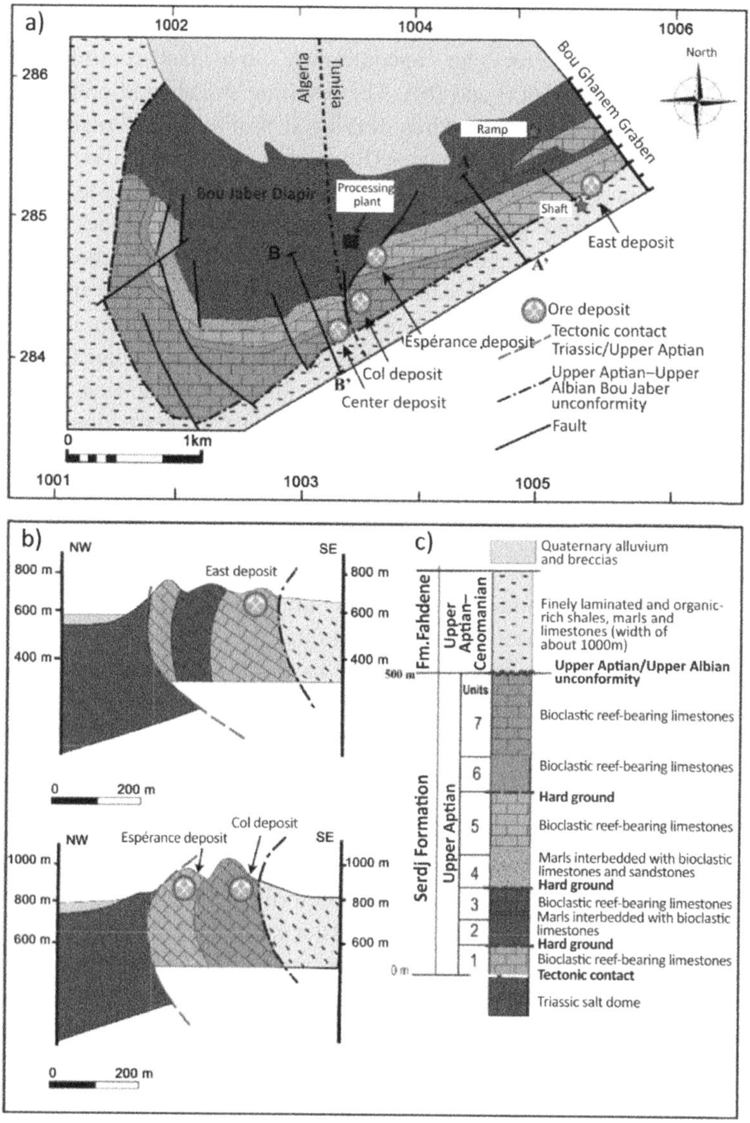

Figure 4.16. *(a) Simplified geological map of the Bou Jaber district showing the location of the main exploited F-Ba-Pb-Zn orebodies and their relationships to the Triassic salt diapiric structure and fault systems; (b) representative longitudinal NW–SE cross-sections showing the location of the Eastern, Col and Esperance orebodies within the carbonate succession of the Serdj Formation; (c) generalized stratigraphic column for the Bou Jaber deposit showing main lithostratigraphic units. Modified from Bouhlel et al. (2016)*

The Upper Aptian carbonates located on the SE margin of the Bou Jaber diapir are the principal host for the unconformity and fault-controlled sulfide mineralization (Figures 4.16(a) and (b)). The resulting stratabound orebodies occur as subvertical column-shaped structures developed in dissolution-collapse breccias and in cavities within the host carbonates (Bouhlel et al. 2016). Sulfide paragenesis consists predominantly of various proportions of sphalerite and galena, with baryte and fluorite as the main gangue phases. The distribution of the ore mineral assemblages including the gangue phases delineate four distinct orebodies referred to as Col Pb + Zn + Ba, Center Pb-calcite, Espérance Pb-Zn + calcite and East Ba-F ± Pb-Zn (Bouhlel et al. 2016) (Figure 4.16(b)). In this respect, it should be highlighted that the economic orebodies are elongated along the major fault systems with the higher grade ores being located in faulted organic-rich dolomitized carbonates and in the "noeuds tectoniques" formed at the intersection of NE–SW and NW–SE-trending faults (Bouhlel et al. 2016).

4.2.3.4. *Slata deposit*

With a total production exceeding 160,000 tons of ore at an average grade of 2% Pb and ~0.5% Zn in addition to subeconomic proportions of baryte reaching a production rate of 2,000 tons per year during the 1960s (Smati 1986), the peridiapiric Slata Ba-Pb-(± Zn) deposit ranks as the fourth largest Tunisian MVT deposit. The host rocks consist of a Triassic salt dome unconformably overlain by Cretaceous to Miocene sediments (Smati 1986; Perthuisot et al. 1998; Chihaoui et al. 2010; Jaillard et al. 2013) (Figure 4.17).

The Triassic succession consists of a chaotic mixture of fragments of black dolostones and lenticular sandstones embedded in a matrix of multicolored clays (illite-kaolinite-chlorite) and gypsum. Unconformably overlying the Triassic lithologies are the clayey limestones, sandstones, dolostones and reef-bearing limestones of the Upper Aptian Sedj Formation, sandstones, limestones and marls of the Lower–Middle Albian Hameima Formation, and the organic-rich limestones of the Upper Albian Fahdène Formation (Smati 1986).

The overlying Tertiary rocks consist of a succession of Lower Oligocene–Miocene shales and sandy limestones overlain by Upper Miocene sandstones and clays of the molassic complex (Figure 4.17). Similar to other deposits of the "Diapir Zone", the host rocks are truncated by a dense network of fault systems, among which the Aptian NW–SE-trending faults are the most prominent.

Mineralogically, the main Pb-Ba-(± Zn) orebodies are those of Sidi Amor Ben Salem and Slata Fer, both of which are located along the WSW margin of the diapiric salt intrusion (Figure 4.17). Sulfide mineralization occurs as moderate to massive replacement of the host carbonates giving rise to stratiform mineralization or open-space fillings of interconnected karst cavities, veins and fractures within the carbonates of the Serdj Formation. Locally, the wall-rocks display the effects of intensive silicification (Smati 1986; Rddad et al. 2019). In this respect, it should be noted that the higher grade orebodies are located in the "neouds tectoniques" displayed by the intersection of major fault systems. Sulfide paragenesis consists of variable proportions of sphalerite and galena accompanied by hydrothermal carbonates (calcite, siderite, ankerite), baryte and quartz as the main gangue phases.

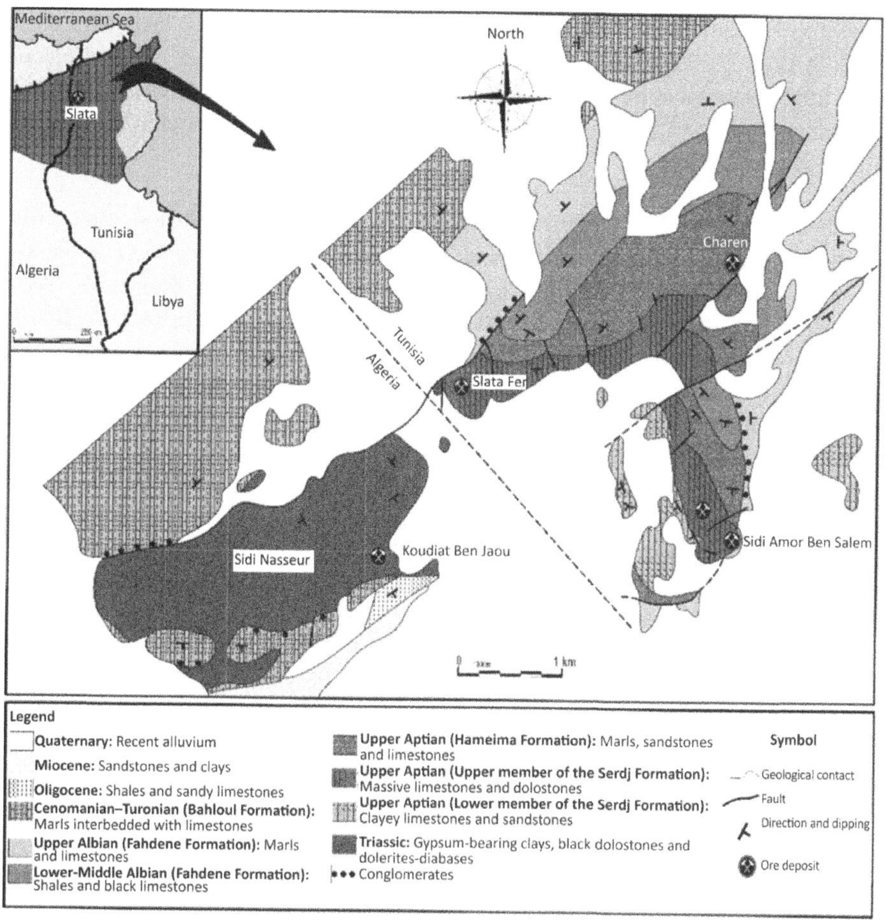

Figure 4.17. *Geological map of the Slata Pb-Ba-(± Zn) district and its southwestern extension in the Algerian territory showing major lithological units, structures and location of significant peridiapiric Mississippi Valley-type deposits. Modified from Smati (1986)*

4.3. Discussion

4.3.1. *Physicochemical characteristics of the mineralizing fluids*

Microthermometric data on fluid inclusions in ore phases (sphalerite exclusively) and associated gangue minerals (dolomite, baryte, fluorite and quartz) reveal relatively homogeneous intervals of temperature and salinity for the studied deposits, with homogenization temperatures (T_h) ranging on average from 80 to 160°C and salinities from ca. 10 to 30 wt.% equiv. NaCl (Dupy and Touray 1986; Charef and Sheppard 1987; Sheppard et al. 1996; Aboutahir 1999; Makhoukhi et al. 2003; Adil et al. 2004; Haddouche et al. 2004; Bouabdellah et al. 2012; Bejaoui et al. 2013; Bouhlel et al. 2016; Laouar et al. 2016; Rddad et al. 2019). Calculated entrapment pressures in the range of 150–200 bars indicate depth emplacement of sulfide mineralization below 1 km.

Nonetheless, while T_h values on sphalerite-hosted fluid inclusions are fairly homogeneous for all the studied deposits, measured T_h values on gangue minerals are extremely variable, reaching up to >200°C (Charef and Sheppard 1987; Sheppard et al. 1996; Bejaoui et al. 2013; Bouhlel et al. 2016; Laouar et al. 2016). These temperatures are significantly higher than the average T_h values considered as typical for MVTs (Leach and Sangster 1993). In this respect, it should be stressed that the aforementioned gangue phases, besides their late appearance in the paragenetic sequence, are known for their ability to undergo post-entrapment processes, such as *stretching, necking down and leaking* (Roedder 1984; Ulrich and Bodnar 1988) which could explain the anomalously high T_h values. Besides the *stretching* phenomenon, the recorded upward increasing values may also reflect entrapment of an organic component characterized by the presence of immiscible gases such as methane (Hanor 1980).

These assumptions are confirmed by Raman data that indicate the occurrence, in nearly all the studied deposits, of hydrocarbon-bearing inclusions (Makhoukhi et al. 2003; Bouabdellah et al. 2012; Bouhlel et al. 2016; Laouar et al. 2016; Rddad et al. 2016) whose compositions consist of a mixture of aliphatic (CH_4 and ethane, among others) and aromatic components (Makhoukhi et al. 2003; Bouabdellah et al. 2012; Bouhlel et al. 2016). Consistent with variations of T_h values, the large variations of salinities of the mineralizing fluids, ranging from 10 to 30% wt% equiv. NaCl, are consistent with the involvement of several types of fluids, including those of meteoric origin. In this respect, it is noteworthy that the lowest salinities (<15% wt% equiv. NaCl) have been reported in gangue minerals mainly baryte, fluorite, quartz and calcite from the paragenetically later stages. These observations support incursion of higher amounts of meteoric fluids into the mineralizing system during the waning stage of hydrothermal activity.

Furthermore, the low first-melting temperatures (eutectic; Te) whose values range from –70°C to –37°C (Bouabdellah et al. 2012; Bouhlel et al. 2016) are far lower than that of the eutectic characterizing the system NaCl-H_2O (Te = –20.2°C). These low eutectic temperatures suggest the presence of divalent cations, other than NaCl, such as magnesium chloride (Te($MgCl_2$) = –35°C) (Dubois and Marignac 1997) and calcium chloride (Te($CaCl_2$) = –52°C) (Davis et al. 1990). Moreover, the fact that hydrohalite melts before the ice at temperatures ranging from –28°C to –22°C indicates NaCl/(NaCl + $CaCl_2$) ratios from 0.4 to 0.5 and Na/Ca ratios from 4 to 5 (Bouabdellah et al. 2012; Bouhlel et al. 2016). Based on these observations, the compositions of the mineralizing brines at the origin of stratabound sulfide ores of the North African MVTs can be described in the H_2O-NaCl-$CaCl_2$-$MgCl_2$ ± KCl ± LiCl ± CH_4 ± CO_2 ± ethane system. These compositions, common to all MVTs worldwide (Leach and Sangster 1993), are similar to those characterizing the oil field brines. These similarities raise the question of whether the observed association between sulfide mineralization and oil fields is spatial and/or genetic as discussed below. It is notably the case of ore deposits in the "Diapir Zone" such as Bou Grine in Tunisia where there is a close association between sulfide-bearing mineralization and *oil seeps* (Belayouni 1992).

4.3.2. *Source and origin of fluid salinity*

The high salinities that characterize the mineralizing fluids of MVT deposits, including those from North Africa, have been attributed to the evaporation of seawater or dissolution of ancient evaporites (mainly halite) or mixtures of both sources (Hanor 1994; Kesler et al. 1996; Viets et al. 1996; Heijlen et al. 2001; Bouabdellah et al. 2012, 2013). Crush-leach fluid inclusion compositions make it possible to distinguish between the above-cited processes.

In this respect, Na-Cl-Br compositions of fluids related to sphalerite and the associated gangue minerals (Adil et al. 2004; Prochaska et al. 2007; Bouabdellah et al. 2012; Bouhlel et al. 2016) delineate two *brine provinces* with extreme compositions, including (1) the "Atlas" province to the west and (2) the "Diapir Zone" province to the east. Notably, the electrolyte compositions of the brines from the "Atlas" province show Na/Br and Cl/Br ratios consistent with evaporation of seawater while those ratios characterizing ore deposits of the "Diapir Zone" province suggest dissolution of halite (Prochaska et al. 2007; Bouhlel et al. 2016). Nonetheless, the high K/Na and Ca/Mg ratios of the mineralizing fluids of the two provinces reflect the importance of fluid-rock interactions and suggest that the mineralizing brines acquired these compositions through various phases of dolomitization and extensive interactions with the basement siliciclastic rocks.

In summary, solute data for fluid inclusions in sphalerite and associated gangue minerals from North African MVTs display two distinct brine provinces with contrasting origins. While the Na-Cl-Br compositions of the mineralizing fluids from the "Atlas" brine province are consistent with seawater evaporation, those from the diapir province are more compatible with halite dissolution.

4.3.3. *Origin(s) and source(s) of mineralizing fluids*

As discussed above, fluid inclusion data reflect the involvement of multiple fluids (Sheppard et al. 1996; Adil et al. 2004; Bouabdellah et al. 2012; Bejaoui et al. 2013; Bouabdellah et al. 2016; Bouhlel et al. 2016; Laaouar et al. 2016; Rddad et al. 2019) whose sources and origins are to be determined. In this respect, the $\delta^{18}O_{V\text{-}SMOW}$ and $\delta^{13}C_{PDB}$ values obtained on various generations of hydrothermal carbonates (dolomite, calcite, siderite) show relatively homogeneous compositions for a given deposit, but vary widely from one deposit to another, and from a given metallogenic province to another (Sharef and Sheppard 1987; Orgeval 1994; Sheppard et al. 1996; Aboutahir 1999; Bouabdellah et al. 2012; Bouhlel et al. 2016; Laouar et al. 2016; Rddad et al. 2016, 2019).

The MVT deposits of the "Diapir Zone" province show the heaviest isotopic compositions, with $\delta^{18}O_{V\text{-}SMOW}$ and $\delta^{13}C_{PDB}$ of 29.4 ± 2.2‰ and –9.5 ± 9.3‰ (1σ) (n = 24) for Fedj-El-Adoum deposit (Charef and Sheppard 1987), 25.8 ± 2.6‰ and 2.6 ± 1.5‰ (1σ) (n = 38) for Bou Jaber deposit (Bouhlel et al. 2016), 25.4 ± 2.6‰ and 0.8 ± 3.2‰ (1σ) (n = 20) for Slata deposit (Rddad et al. 2019) and 25.5 ± 4.1‰ and 3.7 ± 1.3‰ (1σ) (n = 7) for Mesloula deposit (Laouar et al. 2016) (Figure 4.18). Conversely, MVT deposits of the "Atlas" province display depleted isotopic compositions, with $\delta^{18}O_{V\text{-}SMOW}$ and $\delta^{13}C_{PDB}$ ratios of 22.2 ± 3.2‰ and 1.9 ± 0.8‰ (1σ) (n = 7) for Bou Dahar deposit (Rddad et al. 2016), 21.0 ± 1.4‰ and 0.0 ± 1.9‰ (1σ) (n = 115) for deposits of the Touissit-Bou Beker district (Bouabdellah et al. 2012), and 22.7 ± 2.5‰ and 0.1 ± 2.1‰ (1σ) (n = 10) for El Abed deposit (Touahri 1991).

It is noteworthy that the majority of $\delta^{18}O_{V\text{-}SMOW}$ and $\delta^{13}C_{PDB}$ ratios of the host carbonates (dolostones and limestones) reflect the isotopic compositions of contemporaneous seawater (Jurassic versus Cretaceous) (Figure 4.18). Nonetheless, $\delta^{13}C$ values clustering around 0 ± 2‰ are consistent with an origin of carbon by replacement or dissolution of the host carbonates with a slight contribution of organic carbon.

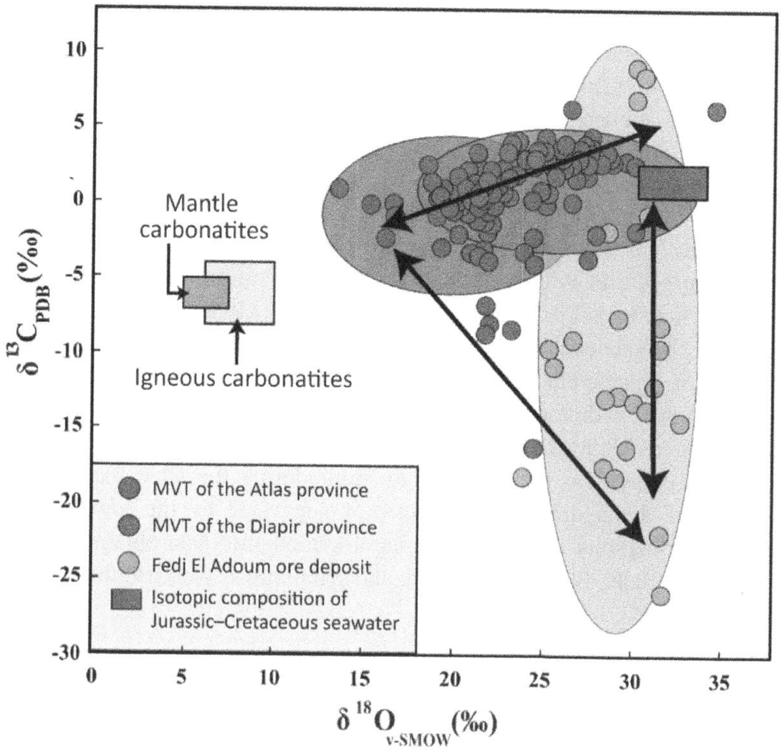

Figure 4.18. *Distribution of $\delta^{18}O_{V\text{-}SMOW}$ and $\delta^{13}C_{PDB}$ values for the main generations of ore- and post-ore related gangue carbonates from the main paragenetic stages depicted in the North African Mississippi Valley-type deposits. Source data are from Charef and Sheppard (1987); Orgeval (1994); Sheppard et al. (1996); Aboutahir (1999); Bouabdellah et al. (2012); Bouhlel et al. (2016); Laouar et al. (2016); Rddad et al. (2016, 2019). The light gray and dark gray compositional fields refer to compositions of igneous carbonatites and mantle-derived carbonatites as defined by Taylor et al. (1967) and Keller and Hoefs (1995). The Jurassic to Cretaceous seawater field is from Veizer et al. (1999) and Jenkyns et al. (2002)*

Moreover, $\delta^{18}O$ compositions of the mineralizing fluids, calculated from the calcite-H_2O and dolomite-H_2O isotopic fractionation factors of Zheng (1999), assuming an average temperature of 120°C as inferred from microthermometric data, show calculated $\delta^{18}O_{H2O}$ values ranging from 5.0 ± 1.4‰ to 13.4 ± 2.2‰. These compositions indicate the involvement and subsequent fluid-rock interactions, at high fluid-rock ratios, among three main types of fluids, including basinal brines, organic waters and meteoric waters (Figure 4.17). The input of magmatic waters is excluded as the recorded $\delta^{18}O$ and $\delta^{13}C$ ratios do not fall within the ranges of values expected to be diagnostic of magmatic waters ($\delta^{18}O_{V\text{-}SMOW}$ = 5–10‰ and $\delta^{13}C_{PDB}$ = –5.0‰ to –8.0‰) (Taylor et al. 1967; Keller and Hoefs 1995). Similarly, the involvement of metamorphic fluids is unlikely due to the lack of regional metamorphic thermal overprint that affected the various domains of the Atlas system during the Alpine cycle.

Remarkably, organic waters (Sheppard 1986) played a crucial role in the mineralizing process as evidenced by the high concentrations of total organic carbon (TOC) of the host rocks (0.23–35.6%) (Montacer et al. 1986, 1988; Montacer 1989; Touahri 1991; Bouabdellah et al. 2001; Affouri et al. 2012). Further supports for the involvement of organic components are provided by the abundance of N_2 and H_2S, the ^{13}C depletion of hydrothermal carbonates (average = –9.5 ± 9.3‰, n = 24), and the strongly negative δD ratios (–110‰) (Charef and Sheppard 1987). Nevertheless, the involvement of meteoric fluids was significant during the later stages related to the collapse of the hydrothermal system, as suggested by the lowest values of temperature and salinity displayed by microthermometric data obtained on gangue phases such as calcite, fluorite and baryte and sphalerite as well (Bouhlel et al. 2016; Laouar et al. 2016).

Moreover, the metal-bearing NaCl-$CaCl_2$-rich basinal brines are identified as the dominant ore-forming fluids responsible for ore formation of North African MVTs (Charef and Sheppard 1987; Sheppard et al. 1996; Bouhlel et al. 2016; Laouar et al. 2016; Rddad et al. 2019). These brines percolated through the traversed faulted aquifers, including the Mesozoic host carbonates and underlying basement rocks. Intense fluid-rock interactions between the brines and the traversed aquifers explain the apparent homogeneity of the obtained $\delta^{18}O_{V\text{-}SMOW}$ and $\delta^{13}C_{PDB}$ values throughout the North African MVTs, as well as the lack of correlation between trapping temperatures and corresponding salinities of fluid inclusions.

4.3.4. *Origin of sulfur and ore depositional processes*

Sulfur isotope compositions of sulfides (sphalerite, galena, pyrite, tetrahedrite) and the associated sulfate phases (baryte, celestine and gypsum) display a wide range of $\delta^{34}S$ values from –24‰ to +14‰ and +12‰ to +33‰, respectively (Touahri 1991; Adil et al. 2004; Bouabdellah et al. 2012, 2015; Bouabdellah and Sangster 2016; Bouhlel et al. 2016; Laouar et al. 2016; Rddad et al. 2019). The $\delta^{34}S_{sulfates}$ ratios are consistent with values for Triassic to Cretaceous seawater sulfate. Remarkably, galena separate fractions from Mibladen deposit exhibit the lightest compositions, with $\delta^{34}S_{galena}$ values ranging from –26‰ to –20‰ (Bouabdellah and Sangster 2016).

In the Touissit-Bou Beker district, Bouabdellah et al. (2015) have shown that the $\delta^{34}S_{sulfides}$ ratios decrease laterally and vertically (1) with advancing paragenetic sequence, and upward stratigraphic position; (2) toward distal prospects; and (3) within single grains from core to rim. Collectively, the $\delta^{34}S_{sulfides}$ ratios are lower than the isotopic signatures that characterize the Jurassic to Cretaceous seawater-derived sulfate sources ($\delta^{34}S_{sulfates}$ = 13–20‰ (Claypool et al. 1980; Kampschulte and Strauss 2004)).

The large inter- and intra-deposit $\delta^{34}S_{sulfide}$ variations suggests a multisource origin of sulfur although many other factors, such as chemical kinetics, the transmissivity of fluids entrapped in pore sediments, the availability of sulfates (open vs. closed reservoirs) and the fractionation between the various sulfide-bearing or sulfate-bearing phases, could be invoked (Ohmoto 1972; Machel et al. 1995; Butler et al. 2004; Amrani et al. 2006; Bouabdellah et al. 2015). Temperature fluctuations are unlikely as isotopic fractionation between the co-precipitating sulfides (sphalerite and galena) and H_2S is low and widely independent of temperature (≤±2‰) (Ohmoto and Rye 1979). Similarly, pH variations and oxygen fugacity (f_{O2}) cannot be invoked as at temperatures below 200°C such as those that prevailed in the studied deposits, fluctuations in the values of the latter parameters would produce insignificant variations of $\delta^{34}S$ ratios.

Moreover, it is noteworthy that in sedimentary environment, sulfur is reduced either bacterially (bacterial sulfate reduction [BSR]) or inorganically (thermochemical sulfate reduction (TSR)). The BSR commonly produces large isotopic fractionation while TSR-derived H_2S involves lower fractionation degrees, leading ultimately to $\delta^{34}S$ values near zero (Ohmoto and Rye 1979). In North African MVTs, the large variations of $\delta^{34}S_{sulfides}$, oscillating between negative and positive values, are explained in terms of inorganic and/or bacteriogenic reduction, at different rates, of seawater sulfate in seawater or pore fluids, in reservoirs both open or closed with respect to sulfate. Mixture of various sulfur sources showing different $\delta^{34}S_{\Sigma S}$ ratios may also be considered. In this respect, three possible sulfur-bearing reservoirs are suggested: magmatic, seawater or evaporitic, or a combination of these.

A magmatic origin of sulfur is unlikely since fluid inclusion data and C-O isotopic constraints are inconsistent with such possibility, as already shown. Hence, the two remaining options, namely evaporitic and seawater sulfate sources, are considered as the most likely sulfur sources, consistent with microthermometric data and Na-Cl-Br leachate (Adil et al. 2004; Prochaska et al. 2007; Bouabdellah et al. 2012; Bouhlel et al. 2016). In regard of MVT deposits of the "Diapir Zone" province, the close association of sulfides with evaporite-bearing successions (apex of salt domes), the occurrence of native sulfur and ^{13}C depletion of hydrothermal carbonates ($\delta^{13}C << 0$) combined with the high degradation level of organic matter are all consistent with TSR of sulfates of the salt domes or thiosulfates as the main mechanism at the origin of H_2S production in the fluid.

4.3.5. *Sulfide ore controls*

Constraining the main ore controls responsible for the emplacement of sulfide-bearing mineralization, or metallotects as defined in French literature by Laffite et al. (1965), is crucial for the explorationists searching for new deposits. This is particularly true when dealing with MVT deposits for which conventional exploration methods such as geophysics (seismic reflection, gravimetry and induced polarization) and stream sediments geochemistry fall to predict the potential locations of buried deposits.

Concerning North African MVTs, field mapping combined with mineralogical, sedimentological, structural and geochemical investigations reveal that the main ore controls include stratigraphy and lithology (lateral sedimentary facies variations, unconformities, reef and barrier complex), carbonate dissolution, sedimentology (dolomitization and the presence of organic matter), paleogeography (basement highs, salt diapirs and shale edges), structure (structural lineaments, network of tectonic faults and related fractures) and geochemistry (availability of sulfur and metals). It is noteworthy that the synchronous interplay of all these factors, over a defined period of time, leads to the formation of an ore deposit of economic interest.

The lateral sedimentary facies variations exemplified by lithology transitions (i.e. limestone-dolostone) exert a strong control on the location of sulfide-bearing orebodies because of the enhanced secondary porosity and permeability along with transmissivity contrasts induced by these lithological variations. This environment provides ideal conditions for fluid mixing and subsequent sulfide precipitation. Moreover, the higher porosity and permeability displayed by the host dolomitic rocks explains why coarse-grained dolostones are the most important host for North African MVT ores. The most illustrative example is provided by the Touissit-Bou Beker district whose higher grade orebodies are hosted in dolomitic strata rather than their limestone counterparts. Altogether, the five deposits of the district have produced over twice as much ore as all the other deposits (Bouabdellah et al. 2012). In those MVT deposits associated with reef-bearing carbonates and barrier complexes such as those of the Jbel Bou Dahar district (Bouabdellah and Sangster 2016), the sulfide-bearing orebodies are not associated with the reef itself but instead with the carbonate rocks of the reefal talus and the associated brecciated horizons (bioclastic facies (biostroms-biohermes); Figure 4.9).

Paralleling the increase of permeability and transmissivity related to lateral sedimentary facies variations, faults and related fractures acted as major fluid conduits for MVT mineralization by channelizing ascent of metalliferous fluids into contrasting aquifers. Evidence includes the positive correlation between deposits and areas of maximum and close-spaced faulting and fracturing, location of the ore-bearing structures along the edges and hinges of anticlines and, more importantly, the parallelism between the elongation of the sulfide orebodies and the major ENE–WSW, E–W and NNW–SSE-trending fault systems (Adil et al. 2004; Haddouche et al. 2016; Ysbaa et al. 2019).

Pre-ore dissolution collapse processes related to emersion of carbonate platforms are major ore controls in most North African MVTs as they provided the needed open spaces (karst cavities and vugs) that served as receptacles for sulfide-bearing mineralization, in addition to providing extreme contrasting fluid transmissivity in carbonate platform successions. Indeed, location of the sulfide-bearing orebodies in the upper part of the ore-bearing carbonates close to major unconformities reveal the occurrence of emersion episodes during which the carbonate platform strata were subjected to meteoric dissolution (karstification), thus preparing future traps for mineralization. In this respect, the most illustrative example is given by the higher grade MVT deposits of the Touissit-Bou Beker district, where over 80% of the Pb-Zn ore is hosted by a dense interconnected network of horizontal and vertical karsts and *runs* (Bouabdellah et al. 2012, 2015).

The overlying impermeable cap rocks (shales, marls and argilites), which seal the uppermost part of the sulfide-bearing host carbonates, served as aquitards preventing upward hydrothermal flow out of the Jurassic (case of MVT deposits of the "Atlas" province) to Cretaceous (case of MVT deposits of the "Diapir Zone" province) ore-bearing carbonates are equally as important.

In addition to the above-discussed factors, paleogeographic reconstructions reveal that all MVT deposits of North Africa are located above or marginal to positive paleogeographic structures (basement highs and/or salt domes) against which the host strata *pinch-out*. The presence of these paleogeographic structures gives rise to lateral sedimentary facies variations, stratigraphic pinch-outs, faults and brecciations, which are all favorable factors for channeling mineralizing fluids from the source zones to the deposit sites. In these environments, fluid-rock interactions induce pH changes and the initiation of redox reactions which, together, trigger sulfide precipitation.

Equally important is the degradation of organic matter, which, through its catalyst role in the kinetics and thermodynamics of redox reactions, is strongly involved in sulfide precipitation. The most noticeable example is provided by the Tunisian Bou Grine deposit which was discovered based on the close association between organic matter and sulfide-bearing orebodies (Orgeval et al. 1986). Similarly, Bouabdellah et al. (1996, 2001) have used organic matter and clay mineral assemblages as unconventional prospection tools in the search for new MVT in anticipation of the closing of the mines of the Touissit-Bou Beker district. Practically, a typical zoning pattern of measured vitrinite reflectances and clay mineral assemblages around the exploited MVT deposits of the Touissit-Bou Beker district has been delineated (Bouabdellah et al. 1996, 2001).

4.3.6. *Role of organic matter*

Type I, II and/or III organic matter (Espitalié et al. 1977, 1985) is present in almost all host carbonate sequences of North African MVTs, either as mappable interbeds within the Jurassic and Cretaceous limestone-dolomitic strata or organic compounds closely associated with sulfide and related gangue phases (Montacer et al. 1986, 1988; Orgeval 1994; Sheppard et al. 1996; Bouabdellah et al. 1996, 2001; Bouhlel et al. 2016; Rddad et al. 2019). The organic matter has also been detected as hydrocarbon droplets trapped along growth zones of sphalerite and associated gangue minerals (dolomite, calcite, fluorite and baryte).

Moreover, organic geochemistry data show COT values ranging from 0.23% to 35.6% with the highest concentrations being documented in the thinly laminated carbonates with marly interbeds of Bahloul Formation (Montacer et al. 1986, 1988; Montacer 1989; Affouri et al. 2012). In this respect, it is noteworthy that the Bahloul Formation, which is well known to be a source rock for petroleum in Upper Cretaceous reservoirs and seepage oil in Northern Tunisia (Bishop 1988; Montacer et al. 1988; Belayouni 1992; Layeb and Belayouni 2000; Lüning et al. 2004; Yukler et al. 1994), also hosts significant sulfide Pb-Zn-hosted deposits, including the studied Bou Grine deposit.

In Touissit-Bou Beker-El Abed district, organic matter occurs as interbedded coal-bearing horizons within the Middle Jurassic dolomitic strata. Petrographic and geochemical studies reveal that the organic component consists of macerals, amorphous kerogen and solid bitumen of type III origin (Touahri 1991; Bouabdellah et al. 1996; Bouabdellah et al. 2001). Measured vitrinite reflectances ($PR_{vitrinite}$) range on average from 0.5% to 1.3%. The $PR_{vitrinite}$ values of ~1.3% would correspond to burial temperatures of about 100–130°C, consistent with microthermometric fluid inclusion data (Dupy and Touray 1986; Aboutahir 1999; Makhoukhi et al. 2003; Bouabdellah et al. 2012). Nonetheless, TOC values range from 0.15% to 0.96% with T_{max} of 400°C (Touahri 1991). Elsewhere in other North African MVTs, the occurrence of organic matter has been reported by many authors including Adil et al. (2004), Laouar et al. (2016), and Rddad et al. (2019).

The depletion in ^{13}C of the ore-related hydrothermal carbonates ($\delta^{13}C$ ratios << 0), along with the low δD values of about −110‰ (Charef and Sheppard 1987), the occurrence of biodegraded organic compounds and hydrocarbon-bearing fluid inclusions all suggest oxidation of low molecular weight hydrocarbons. Indeed, the degradation of organic matter by TSR produces native sulfur, H_2S and HCO_3, as indicated in reactions [4.1] and [4.2] (Orr 1974; Charef and Sheppard 1987):

$$CaSO_4 + CH_4 \rightarrow CaCO_3 + H_2S + H_2O \quad [4.1]$$

$$15(C_{12}H_{26}) + 8CaSO_4 \rightarrow 8CaCO_3 + 8\,H_2S + 8\,H_2O + 10\,(CH_{0,7}) \quad [4.2]$$

where CH_4, $C_{12}H_{26}$ and $CH_{0.7}$ are methane, petroleum undergoing oxidation and residual bitumen, respectively. In this respect, Anderson (1991) showed that at higher temperatures than those typical of the oil window (>120 °C), methane derived from thermal decomposition of kerogene diffuses through the mineralized horizons and reduces sulfates by producing hydrogen sulfide (H_2S) which by complexing with a metal-bearing fluid (Pb and Zn in our case) triggers precipitation of galena and sphalerite (Anderson 2015). If the amount of Mg^{2+} ions in the ore-forming fluid is sufficient, as in Touissit-Bou Beker (Bouabdellah et al. 2012), dolomite precipitates according to the following reaction:

$$2CaSO_4 + 2CH_4 + Mg^{2+} \rightarrow CaMg(CO_3)_2 + 2H_2S + Ca^{2+} + 2H_2O$$

In the “Diapir Zone” province, the intense biological activity that would have been developed in the deposit sites at the expense of the organic matter dispersed within the Bahloul Formation and sulfates from the salt domes would have produced the amounts of reduced sulfur (H_2S) required for sulfide precipitation.

In summary, organic matter dispersed within the host carbonate rocks of North African MVTs would have constituted the main catalyst for the redox reactions at the origin of sulfide precipitation. In addition to its role as a reducing agent, organic matter also would have acted as both source and transport agent of metals during the early diagenesis phases (Saintilan et al. 2019).

4.3.7. *Metal source(s)*

The source of metal(s) remains one of the outstanding controversies to be solved in MVT deposit genesis. In this ore deposit type, Pb isotopic compositions of sulfides and potential metal rock sources provide robust constrains on the provenance and the source(s) of the ore metals owing to similarities of the geochemical behavior of Pb, Cu and Zn (Tosdal et al. 1999) and the occurrence of Pb-bearing phases in the same paragenetic stages as those containing Zn, Cu, Ag and Cd (Wood et al. 1987). Moreover, the Pb isotopic ratios are not subjected to fractionation in measurable quantities during various redox reactions and fluid-rock interactions, and therefore the isotopic compositions of sulfides accurately reflect those of possible crustal rock sources (Tosdal et al. 1999).

Overall, the Pb isotopic data obtained both on galena separates and a wide range of potential rock sources show $^{206}Pb/^{204}Pb$, $^{207}Pb/^{204}Pb$ and $^{208}Pb/^{204}Pb$ ratios ranging from 18.12 to 18.95, 15.60 to 15.75 and 37.39 to 39.10, respectively, for galena separate fractions, and from 17.95 to 18.76, 15.51 to 15.99, 37.78 to 39.13, respectively, for whole rock powders (Bouabdellah et al. 2012, 2015; Decrée et al. 2014; Bouhlel et al. 2016; Rddad et al. 2016, 2019). These isotopic compositions contrast significantly from those characterizing North American MVT deposits, but are similar to the Pb isotopic compositions of the European Alpine MVTs including Upper Silesia in Poland, Cévennes in France and those occurrences in Austria, Slovenia and Italy, Appalachian and South American (mainly Peru) (Figure 4.19). It is noteworthy that all the Pb isotopic compositions of North African sulfide-bearing ores plot well above the average crustal Pb growth curves of Stacey and Kramers (1975) and Cumming and Richards (1975), but between the upper crustal and the orogene growth curves of Zartman and Doe (1981). These compositions suggest that Pb, and by inference accompanying metals, were derived mostly from a radiogenic reservoir with a significant upper crustal component (Figure 4.19). While the Pb isotopic compositions of galena separates are extremely homogeneous in a given deposit leading to clustering of Pb values (Figure 4.19), there are, however, significant inter-deposit and especially inter-province variations. Similar to $\delta^{13}C$ and $\delta^{18}O$ isotopic ratios (Figure 4.18), the Pb-Pb systematics display the existence of two distinct metallogenic provinces whose $^{206}Pb/^{204}Pb$, $^{207}Pb/^{204}Pb$ and $^{208}Pb/^{204}Pb$ ratios are correlated with the ages of the carbonate host rocks.

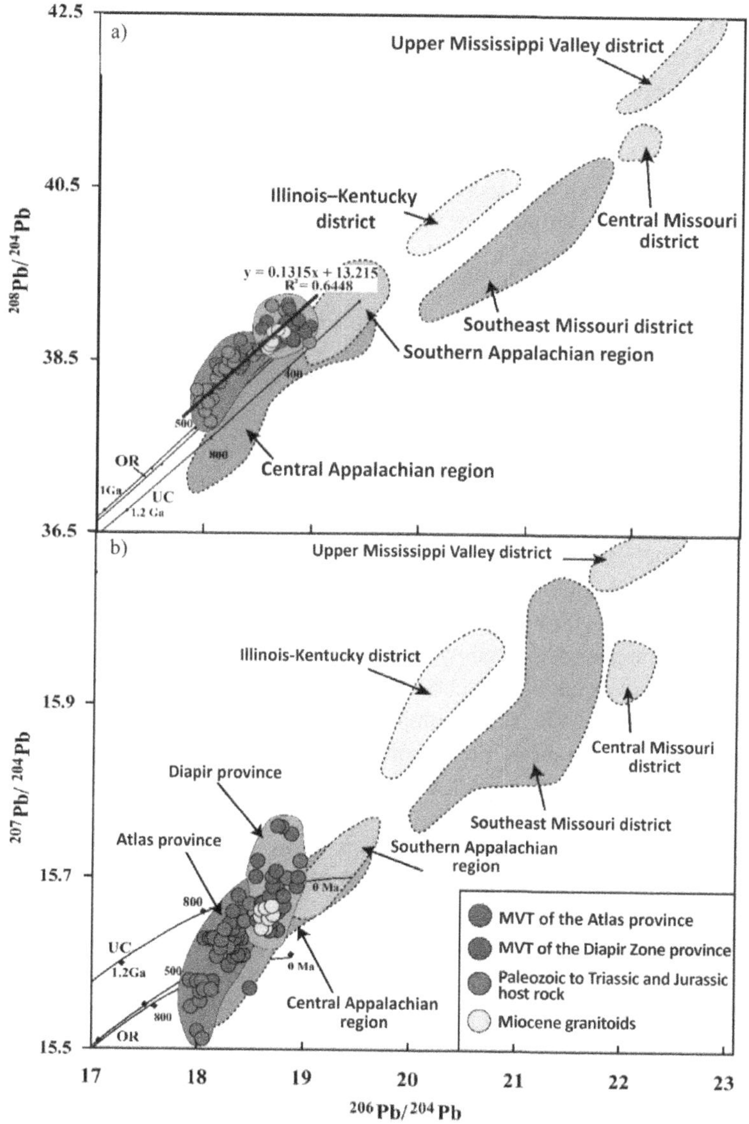

Figure 4.19. *Lead isotopic compositions of sulfide separate fractions along with the whole-rock potential rock sources from the North African MVT deposits. (a) $^{208}Pb/^{204}Pb$ versus $^{206}Pb/^{204}Pb$. (b) $^{207}Pb/^{204}Pb$ versus $^{206}Pb/^{204}Pb$. Compositional fields for North American- and Alpine-type MVTs are adapted from Potra et al. (2018) and are shown for comparison. Mantle, Orogen and Upper Crustal evolution curves are from Zartman and Doe (1981)*

Remarkably, Jurassic carbonate-hosted MVT deposits of the "Atlas" province are far less radiogenic than their counterparts from the "Diapir Zone" province, whose sulfide-bearing orebodies are hosted in Cretaceous carbonate successions. These Pb isotopic variations, which correlate with the age of the host carbonates and the geographical location of the studied MVT deposits ("Atlas" vs. "Diapir Zone" provinces), can be explained either by age differences in the emplacement of mineralization (heterochronism) or alternatively by the introduction and mixing of Pb from various sources during the leaching and transport of metals during migration of the ore-forming fluids. In this respect, uranogenic and thorogenic plots (Figures 4.19(a) and (b)) show that the spread of the Pb isotopic ratios of sulfide separate fractions from the two MVT provinces along a linear array is interpreted either as an *errorchron* or a mixing trend. Moreover, the slope of the reference line ($y = 0.1315x + 13.215$ with $R^2 = 0.6448$) displayed by the distribution of Pb isotopic ratios indicates a Pb model age of 2.2 Ga, which is significantly older than the ages of the surrounding host rocks. Hence, the calculated model age of 2.2 Ga is interpreted as representing the age of Pb extraction from an *Orogen Lead Reservoir*, rather than the emplacement age of mineralization. Accordingly, the linear clustering of the Pb isotopic data for galena ores across the paragenetic sequence (Figure 4.19) reflects mixing of at least two contrasting Pb reservoirs. The radiogenic endmember displayed by MVT deposits from the "Diapir Zone" province suggests the involvement of a U-enriched but Th-depleted reservoir. The lithologies that are consistent with such compositional constraints are typified by the organic-rich shales of Bahloul Formation, the Triassic evaporite-bearing successions and/or the Cretaceous phosphate-rich carbonates.

In this regard, organic geochemical data show that the Bahloul Formation is enriched in metallic elements, particularly Pb-Zn (Montacer et al. 1986, 1988) with the latter elements likely being mobilized and transported by hydrocarbon-bearing components released during maturation of buried organic matter. Except for the oil-bearing shales, the overlap between the Pb isotopic compositions of Miocene granitoids and those of galena ores (Figure 4.19) suggests that the destruction of K-feldspars from the Miocene granitoids was the main metal source in the "Diapir Zone" province. Conversely, the less radiogenic endmember shown by galena ores from the "Atlas" province reveals the involvement of Variscan basement rocks as a potential source of metals, as suggested by the similarities of Pb isotope compositions of K-feldspars of late Variscan granitoids and galena separates (Jébrak et al. 1998; Bouabdellah et al. 2012, 2015; Bouabdellah and Sangster 2016). Moreover, the overall tendency for the galena separate data to disperse along the Paleozoic, and more importantly the Mesozoic, sedimentary host-rock trends, argues for a quantitatively significant proportion of the Pb having been derived from the country rocks (Bouabdellah et al. 2015).

In summary, the Pb isotopic constraints reveal the local provenance of metals from both the sedimentary and magmatic reservoirs containing radiogenic and common Pb, respectively. It also seems that each deposit possesses its local potential metal rock sources independently from the other deposits depending on the petrographic composition of the host rocks. Mixing of compositions from all these rock sources explains the apparent homogeneity of the isotopic compositions documented at the scale of each of the two North African MVT provinces.

4.3.8. *Timing of sulfide ore formation*

Owing to the lack of radiometric ages for the North African MVTs, combined geological field observations along with paleogeographic reconstructions and geodynamic evolution of host basins were used to estimate the relative timing of mineralization.

During 1960–1980s, many of the North African MVT deposits have been interpreted as of syngenetic origin based on the laminated and stratiform features of the orebodies along with the occurrence of reef-bearing structures such as zebra and "lobate" dolomite interpreted at that time as a result of the recrystallization of stromatolitic beds (Samson 1973; Amouri 1989). Syngenetic models have rapidly been transcended by epigenetic models which, in contrast, suggest a late and structurally controlled emplacement of mineralization either as late-diagenetic replacement of the host carbonates or dissolution cavities, veins and/or veinlets infills or as cement of collapse breccia (Rouvier et al. 1985; Bouabdellah et al. 2012, 2015; Bouabdellah and Sangster 2016; Bouhlel et al. 2016; Laouar et al. 2016; Rddad et al. 2019). Hybrid models involving syngenetic plus overprinting epigenetic mineralization have also been proposed (e.g. Touahri 1991). In North African MVTs, structural, sedimentological and textural constraints show that sulfide-bearing mineralization is clearly epigenetic postdating the lithification of the host rocks. Moreover, $PR_{vitrinite}$ of about 1.3% (Touahri 1993; Bouabdellah et al. 1996, 2001) coupled with the occurrence of petroleum-bearing (liquid or gas) fluid inclusions (Makhoukhi et al. 2003; Bouabdellah et al. 2012; Bouhlel et al. 2016; Laouar et al. 2016; Rddad et al. 2019) indicate thermal maturity from the early oil window to the dry gas zone. Such degrees of thermal maturity indicate burial depths in excess of 5 km. Remarkably, paleogeographic reconstructions indicate that such depths would have been reached during the Upper Cretaceous–Lower Eocene and the peak of hydrocarbon migration during the Eocene–Miocene synchronous with the major phase of Atlas orogeny (Belayouni 1992; Bouabdellah et al. 2012, 2013). Furthermore, field mapping constraints reveal that the emplacement of sulfide-bearing orebodies is structurally controlled by NE–SW, E–W and WNW–ESE fault systems whose interplay coincide with the major phase of the Atlas orogeny during the Oligocene–Miocene time (Frizon de Lamotte et al. 2000, 2006, 2009). The combination of all these observations suggests an age span for the emplacement of sulfide-bearing ore between the Eocene and the Miocene.

From the rarely-available chronological data that constrain the age emplacement of sulfide-bearing mineralization of the Tunisian "Diapir Zone" province, three hydrothermal mineralization events peaking at 93, 68 and 24 Ma were recognized (Sheppard et al. 1996). As discussed earlier, the two peaks at 93 and 68 Ma may be a priori discarded as combined field mapping and structural constraints bracket the emplacement of sulfide-bearing mineralization between Upper Cretaceous and Miocene time window. Conversely, the significance of the age peaking at 24 Ma could be of interest in the light of the new Pb isotopic data provided by Decrée et al. (2014). Interestingly, the overlap between the Pb isotopic compositions of the Miocene magmatic rocks dated from 13 to 7 Ma (<15 Ma) (Decrée et al. 2014) and sulfide separates (Bouhlel et al. 2016; Rddad et al. 2019) suggest either direct input of a magmatic component and/or leaching of metals and progressive extraction of Pb, and by inference other metals, from the Miocene igneous rocks. Regardless of the considered hypothesis, the age of MVT mineralization of the "Diapir Zone" province could be therefore constrained as early as 13 Ma (i.e. Upper Miocene). Similarly, Bouabdellah et al. (2012) suggested similar age emplacement at <15 Ma for MVT deposits of the "Atlas" province.

In summary, it is noteworthy that most MVT deposits of North Africa formed during the same period of time (<15 Ma), consistent with time spans proposed by Bouladon (1949), Claveau et al. (1952), Voirin (1965), Rajlich (1983), Dupuy and Touray (1986), Touahri (1991), Sheppard et al. (1996), Boutaleb (2001), Torbi and Gélard (2000), Haddouche et al. (2004, 2014, 2016), Sami (2011), Bouabdellah et al. (2012, 2015), Bejaoui et al. (2013), Jemmali et al. (2013), Bouhlel et al. (2016), Laouar et al. (2016), Rddad et al. (2019) and Ysbaa et al. (2019).

4.3.9. *Genetic model(s) – conclusions*

Recent genetic models for MVT deposit invoke long-distance migration (>100 km) of crustally-derived metal-bearing, warm, dense basinal brines in response to multiple phases of deformation and uplift of foreland basins within a collisional setting (Leach and Rowan 1986). In this respect, three largely accepted models are proposed: (1) sediment compaction of the sedimentary host rock (i.e. episodic overpressure dewatering) in a host basin subjected to high subsidence rates (Jackson and Beales 1967); (2) orogenic squeezing (Oliver 1986); and (3) gravity-driven flow (Bethke and Marshak 1990; Garven et al. 1993; Appold and Garven 1999; Appold and Nunn 2005). In addition to these conceptual models, gas-driven fluids (Eisenlohr et al. 1994; Tompkins 1994), thermal and density brine reflux (Jones et al. 2004), and thermal convection due to disintegration of radioactive elements in the basement rocks (Spirakis and Heyl 1996) have all been suggested as alternative mechanisms. Nonetheless, review of the applicability of all of these models reveals that none of the above-mentioned models meet the needed mineralogical and hydrodynamic requirements for their generalization to all MVTs worldwide. In this regard, numerical simulations of mass and heat transfer and hydrodynamic flow studies show that the gravity-driven groundwater flow model offer the best mechanism for maintaining temperatures and moving fluids over long distances, during which heat and ore-fluid components are scavenged from the traversed aquifers (Garven et al. 1993; Appold and Garven 1999, 2000). As a result, the gravity-driven model is proposed as the most robust in providing optimum conditions for MVT ore genesis (Leach et al. 2010a, 2010b).

Based on this conceptual model, calculated flow velocity rates are high enough to advect heat from the deeper parts of the basin to raise temperatures to the levels recorded by ore-stage fluid inclusion T_h (75–200°C (Bethke and Marshak 1990; Garven et al. 1993; Appold and Garven 1999; Appold and Nunn 2005)). It should, however, be stressed that the applicability of the assumptions related to this model, initially designed and developed for North American MVTs and then systematically transposed to MVTs worldwide, still remains subject of ongoing debates (Bouabdellah et al. 2015).

Notably, the gravity-driven flow model relies on a set of paleogeographic, paleoclimatic and geochemical requirements that are not necessarily encountered in a large number of MVTs worldwide, including the so-called Alpine type which belongs to the North African MVTs. Indeed, the gravity-driven flow model assumes, even though it is not always clearly expressed (Bouabdellah et al. 2015), (1) the involvement of large basins characterized by the development of megaplatforms (several hundred kilometers); (2) the presence of continuous and interconnected aquifers over long distances (>100 km) as a requisite for lateral fluid conduits; (3) the existence of a recharge zone that is permanently fed by high rates of rainfall and rapidly uplifted mountains for the production of an efficient hydraulic head to create the high fluid flow velocities required to generate a fluid flow system efficient enough to form an economic ore deposit (Ge and Garven 1992; Appold and Garven 1999); and (4) the existence of thermal and/or geochemical gradients from depocenters to the margins of the host basins. Unfortunately, most of the above-mentioned requirements are not met by the majority of Alpine MVT deposits, including those of North Africa where the host basins have limited lateral extension with the host carbonate platforms being fragmented into a succession of horsts and grabens and isolated from each other by a series of trans-crustal, deep-rooted, subvertical faults playing the role of impermeable seals for the passage of mineralizing fluids (Bouabdellah et al. 2015). In addition to these paleogeographic and structural constraining factors, the paleoclimatic arid conditions that prevailed during ore deposition and the low uplift rates (<1,500 m) (Frizon de Lamotte et al. 2006, 2009) weaken the gravity-driven fluid hypothesis as an efficient driving force for fluid flow in North African MVT deposits.

Moreover, the occurrence of petroleum-bearing fluid inclusions (Makhoukhi et al. 2003; Bouabdellah et al. 2012; Bouhlel et al. 2016; Laouar et al. 2016) along with the high thermal maturity of organic matter ($PR_{vitrinite}$ 1.3%; T_{max} = 439–453°C) (Touahri 1991; Bouabdellah et al. 1996, 2001; Rddad et al. 2019) reflect the mature, and even ultra-mature state of the organic compounds dispersed within the host carbonates. The resulting high thermal maturation rates reflect either catagenetic transformation of organic matter in response to increased burial depth (>2 km), or to the prevalence of an abnormal geothermal gradient directly related to hydrothermal flow. In this respect, paleogeographic reconstructions reveal that the overburden sediments reached their maximum thickness of less than 2 km during the Upper Cretaceous (Campanian)–Lower Miocene. Such depths are largely insufficient to be consistent with T_h values displayed by microthermometric measurements (up to 220°C). Indeed, assuming a normal geothermal gradient of about 30°C/km, the resulting 2 km depth-related temperatures would not exceed 60°C.

Equally as important are the Pb isotope data constrained the emplacement of MVT mineralization at <15 Ma (Upper Miocene). Regionally, this time span coincides with the Neogene to Quaternary late- to post-orogenic collapse and relaxation phases related to the Alpine orogeny. Such geodynamic environment is characterized by (1) intense bimodal magmatic activity exemplified by the emanation of a large volume of post collisional high K, calc-alkaline to alkaline anorogenic basaltic rocks (Juteau et al. 1986; Coulon et al. 2002; Duggen et al. 2005; Lustrino and Wilson 2007; Decrée et al. 2014; Chazot et al. 2017); (2) emplacement of mineralization in an extensional setting during the Middle Miocene–Lower Tortonian time despite convergence between the African and the Eurasian plates acting as the primary tectonic processes (Frizon de Lamotte et al. 2000, 2006, 2009; Bouaziz et al. 2002); (3) second phase of halokinesis (Perthuisot 1992; Perthuisot et al. 1998; Kowalski and Hamimed 2000); (4) migration of hydrocarbons (Bouabdellah et al. 2012, 2015); and (5) the occurrence of a major paleoclimatic event at 5.96–5.33 Ma referred to as the Messinian salinity crisis (Hsü et al. 1973; Krijgsman et al. 1999).

Based on the preceding observations, we propose an alternative genetic model for the North African MVTs that reflects a complex interplay among paleoclimate (i.e. Messinian salinity crisis), geodynamics (i.e. extensional regime), magmatism (high heat flow), brine and hydrocarbon migration, and fluid mixing. Nevertheless, this model does not pretend to be exhaustive and applicable in its integrality to all Alpine MVT deposits, but variants of this model can be proposed for a given MVT deposit based on the local and regional geological specificities of each deposit and/or district.

Indeed, the contemporaneity of the emplacement of sulfide-bearing mineralization and the Neogene–Quaternary late- to post-collisional magmatic events may explain the mature, and even ultra-mature features of the organic matter dispersed within the host carbonates. Moreover, geophysical data indicate that the lithosphere under the Atlas is abnormally thin (Teixell et al. 2003, 2005), being isostatically accommodated by the diapiric upwelling of the asthenospheric mantle. This geodynamic setting would result in high heat flow and consequently an increased geothermal gradient and subsequent buoyancy-driven fluid convection (Facenna and Becker 2020). In support of this hypothesis, noble gas data including the $^{3}He/^{4}He$ ratios of 0.15 to 0.51 R_A suggest the involvement of a mantle-derived (^{4}He) component in the range of ~3–8% (<10%) (Bouabdellah et al. 2015). The interplay among all of these factors therefore would have constituted a favorable environment for producing a long-lived convective hydrothermal system, consistent with the homogeneity of the physicochemical conditions (homogenization temperatures and salinities) and isotopic fluid compositions (Bouabdellah et al. 2012, 2015). Furthermore, the large volumes of the dolomitized host rocks would reflect the involvement of large-size convective cells (Gomez-Rivas et al. 2014). In MVT deposits of the "Atlas" province, the role of Neogene to Quaternary post-collisional magmatism was probably limited to providing the heat flow required for triggering an efficient convective system allowing the circulation of mineralizing brines from the deepest parts to the margins of the host basins (Bouabdellah et al. 2015).

The ore-forming brines expulsed from overpressured compartments (i.e. graben zones) are inferred to have rapidly migrated along the transcrustal, deep-rooted, ENE–WSW and E–W-trending transcrustal faults adjacent to positive paleogeographic structures (basement highs and salt domes), and then spread laterally along the Jurassic (i.e. MVTs of the "Atlas" province) and Cretaceous (i.e. MVTs of the "Diapir Zone" province) carbonate host rocks. The prevalence of an extensional regime with high heat flow centered on positive paleogeographic structures such as basement highs and salt domes apexes, coupled with the availability of sulfur in sufficient amounts seem to have been the main factors at the origin of the concentration of MVT mineralization at high tonnages and ore grades. Incursion of meteoric waters during the waning stage of hydrothermal activity coincident with the main uplift stage is assumed to have been essential in the dilution of mineralizing brines and the extinction (collapse) of the hydrothermal system.

4.4. Supergene non-sulfide Pb-Zn mineralization associated with MVT deposits

4.4.1. *Introduction*

Long ignored by the mining industry owing to problems with the mineral processing and recovery of the non-sulfide ore, supergene Pb-Zn ± Cu deposits are becoming attractive exploration targets worldwide due to the new developments in hydrometallurgy (Boni and Large 2003; Bouabdellah et al. 2021). In addition to these metallurgical improvements, the renewed interest in supergene non-sulfide ore deposits resides in their higher Pb-Zn grades averaging up to 20 wt.% Zn and 18.1 wt.% Pb (Johnson and Skinner 2003; Bouabdellah et al. 2012), and less environmental impact than sulfide ore (e.g. no sulfur emissions or wastes, little to no acid mine drainage).

Just as important is how non-sulfide Zn ore significantly improves melting of sulfide-rich Zn and Pb ore. In addition to conventional base metals, supergene non-sulfide deposits provide, as by-products, a large range of strategic metals such as Ge, Ga and In. The soaring demand for these high-tech commodities reflects the increasing interest for supergene non-sulfide deposits, which are considered as viable alternative sources complementary to carbonatites and alkaline magmatic rocks.

The current state of our knowledge reveals that supergene ore deposits formed in a near-surface environment by low-temperature oxidation of a sulfide-rich protolith by the action of meteoric fluids. Ore-forming processes involve a series of reactions between downward-percolating meteoric waters, metal solubility and reactive host rocks contemporaneously with tectonic uplift and/or water table depression related to alternating dry and humid seasons and to the biologic activity (Hitzman et al. 2003; Tornos et al. 2013; Bouabdellah et al. 2021). The released metals could be trapped locally, directly above the sulfide protolith, or could be transported and precipitated far away from the sulfide protore (e.g. Hitzman et al. 2003).

Concerning MVT deposits of the North African Atlas system, all the above studied deposits and districts show evidence of significant supergene alteration typified by development of non-sulfide gossan and laterite bodies overlying the sulfide-bearing protore. In this respect, it should be stressed that the largest and higher grade non-sulfide orebodies are related to Moroccan MVTs, including those of Touissit-Bou Beker, Upper Moulouya and Jbel Bou Dahar districts (Figure 4.1). In Tunisia, subeconomic non-sulfide mineralization superimposed on primary sulfide Pb-Zn orebodies has been documented in the "Diapir Zone" (Garnit et al. 2018). For Algerian MVTs, mineralogical and geochemical data on supergene Pb-Zn ores are missing or at the least unavailable. The following section focuses on the description of the largest and economically the most attractive supergene non-sulfide deposits in the Atlas system. Nevertheless, it should be stressed that while significant advances have been made in understanding the genesis and depositional mechanisms of sulfide ores, the current state of our knowledge on mineralogy and genesis of supergene non-sulfide mineralization is still in its early stage (Choulet et al. 2014, 2016, 2017; Garnit et al. 2018; Bouabdellah et al. 2021).

4.4.2. *Supergene non-sulfide mineralization associated with MVT deposits of the Touissit-Bou Beker and Upper Moulouya districts*

The MVT deposits of the Touissit-Bou Beker and Upper Moulouya districts (Figure 4.1) have a common specificity, which is related to the fact that 60–70% of the extracted ore derived from oxidized non-sulfide phases (cerussitic ore) (Bouabdellah et al. 2012). In surface outcrops as well as in underground exposures, supergene orebodies could be recognized by the reddening of the host carbonates, the crumbling of the host dolostones giving rise to sandy dolomite and development of vertical karsts and related dissolution structures including internal sediments (speleothems). In addition to these structural and textural distinctive features, there is also development of new non-sulfide mineral assemblages mainly of Pb-carbonates and Pb-sulfates deriving from in situ oxidation and replacement of the primary Pb-Zn-bearing sulfide mineralization.

The most impressive meteoric dissolution structures are mapped in the Touissit-Bou Beker district where part of the ore is clay hosted. This ore style is referred to by the mine geologists as "zone de collapse". The most productive orebodies are exemplified by Hassi Msidira and North Beddiane mines where the clay-hosted mineralization is strongly laminated, with contorted bands of reddish-brown to black illite and kaolinite accompanied by quartz and dolomite. Production data show that over 900,000 tons of lead-bearing ore grading ~20% Pb and 600 g/t Ag were extracted from these areas. Collapse zones are therefore interpreted as internal sediments formed by meteoric dissolution of the host dolostones, as evidenced by C-O data (Bouabdellah et al. 2012).

The resulting Pb, Zn, and Cu commodities are expressed mineralogically as carbonates (cerussite, smithsonite, malachite, azurite, hydrozincite), sulfates (anglesite) and silicates (willemite) intimately intergrown with iron oxide-hydroxides (goethite and hematite). In addition to these predominant phases, unusual mineral species, qualified as mineralogical curiosities, such as wulfenite and vanadinite, manganese oxides (coronadite, cryptomelane and hollandite; Jahn 2003), mottramite ($PbCu(VO_4)(OH)$), paralaurionite ($PbCl(OH)$) and phosgenite ($Pb_2CO_3Cl_2$) have been documented (Praszkier 2013). Gangue minerals consist predominantly of dolomite, calcite, aragonite, quartz and gypsum.

4.4.3. *Non-sulfide mineralization associated with MVT deposits of the calamine-bearing province of Central and Eastern High Atlas*

With a total production exceeding 1 million tons of calcined calamine with an average content of 35% ZnO, the Central and Eastern High Atlas is host to the largest calamine province, still active, in the Mediterranean basin. Ongoing artisanal mining is supervised by the Purchasing and Development Cooperative for the Tafilalat and Figuig Mining Region (CADETAF), a state institution in charge of the management of artisan miners, the exploitation of calcined calamine and the commercialization of lead, zinc and baryte ores. The CADETAF exploration license covers an area of over 60,000 km^2. Given the non-performance of the CADETAF's balance sheets, a new law (74-15) was enacted in order to encourage private mining companies to further contribute to the development of the CADETAF area. To this end, public–private partnerships have been set up to shed light and encourage investors on the attractiveness of the area in terms of mining potential including calamine.

Following this amendment, ManagemGroup, the largest private mining company in North Africa, acquired in 2005 a dozen of exploration and exploitation licenses. Between 2005 and 2019, ManagemGroup produced approximately 600,000 tons of calamine with an average grade of 16% ZnO. In this respect, it should be stressed that part of this production derived from the recovery of ancient ore dumps. The overall calamine resources are estimated at ~2 Mt ore grading 18% zinc (see: www.managemgroup.com).

Unlike supergene ore in Touissit-Bou Beker and Upper Moulouya districts, the supergene deposits and prospects of the Central and Eastern High Atlas are characterized by the wide development of calamine-bearing crusts and blankets overlying the primary Pb-Zn ± Cu sulfide ore. The resulting non-sulfide orebodies occur either as stratabound tabular blankets and pods of variable thickness, width and strike extent, oxide box works after sphalerite, and/or botryoidal concretions to columnar aggregate infills of interconnected karst cavities, vugs, veins and fractures. The non-sulfide paragenesis consists predominantly of smithsonite and cerussite with subordinate hydrozincite, anglesite, willemite, hemimorphite and Zn-enriched clays and iron oxide-hydroxides (goethite and hematite). Sauconite, greenockite, malachite, azurite and covellite complete the list of the mineral assemblages (Bouabdellah et al. 2012, 2021; Choulet et al. 2014, 2016, 2017). Gangue minerals are largely dominated by various proportions of calcite, aragonite, gypsum, quartz and baryte. The spatial distribution of these mineral assemblages delineates typical vertical zonation exemplified by the predominance of sulfides ore at the base and Zn-Pb ± Cu carbonates, sulfates and silicates in addition to iron oxide-hydroxide at the top of the alteration profile.

Based on the color of the exploited calamine, two types referred to as red and white calamines with contrasting mineralogical and geochemical attributes are recognized (Choulet et al. 2014; Bouabdellah et al. 2021). As inferred from their nomenclature, red calamine exhibits an ochre-red color due to the predominance of iron oxide-hydroxides, among which goethite and hematite are by far the most prominent phases. Conversely, white or gray calamine owes its name to its chalky-white color, talc-like aspect, and more importantly its mineralogical composition dominated by the occurrence of smithsonite and hydozincite combined with the overall absence of iron oxide-hydroxides due to the fixation of the latter at the level of the upper horizons of the exploited mines.

4.4.4. *Supergene mineralization associated with MVT deposits of the Tunisian "Diapir Zone"*

Similar to Moroccan MVTs, stratabound sulfide-bearing orebodies of the "Diapir Zone" in Tunisia (Fedj-El-Adoum, Bou Grine, Bou Jaber and Slata) (Figure 4.1) underwent the effects of variably advanced supergene oxidation giving rise to non-sulfide mineralization overlying the primary Pb-Zn ore (Garnit et al. 2018). Unlike the Moroccan counterpart, supergene mineralization in Tunisia is of lesser economic importance as the resulting non-sulfide ores were not exploited industrially so far. Where documented, supergene mineralization occurs as stalactitic incrustations lining wall cavities or vein and karst infills including internal sediments (speleothems) (Bruyère et al. 2010). Supergene mineral assemblages resulting from weathering of primary sulfides (sphalerite, galena, pyrite and chalcopyrite) consist of variable proportions of smithsonite, cerussite, hemimorphite and hydrozincite commonly associated with iron oxides and hydroxides (goethite, hematite) and to a lesser extent descloizite, woodruffite, brochantite and malachite, enclosed in a matrix of calcite, dolomite, quartz, kaolinite, fluorite, baryte and gypsum (Garnit et al. 2018).

4.4.5. *Discussion*

The carbon and oxygen isotope data (Choulet et al. 2017; Garnit et al. 2018; Bouabdellah et al. 2021) support a meteoric origin for the fluids responsible for supergene Pb-Zn mineralization that overlies the hypogene sulfide ores of the North African Atlas system. The oxidation of primary sulfides is assumed to have occurred at ambient temperatures decreasing with time from 30°C to 18°C (Bouabdellah et al. 2021). More importantly, the predominance of stromatolitic textures similar to those currently forming off the coast of Bahamas (Dill et al. 1986) coupled with the low $\delta^{13}C$ and $\delta^{34}S$ ratios provide strong support for the role of microbial activity in the supergene oxidation of sulfides and their subsequent replacement by low temperature assemblages in a sulfate-enriched medium (Bouabdellah et al. 2021). By contrast, Choulet et al. (2017) reported the occurrence of a supergene paragenesis with willemite variably enriched in Ge (up to 1,000 ppm) that is assumed to have formed at temperatures of about 130°C from a mixture of meteoric and hydrothermal fluids. It should be kept in mind that supergene oxidation is an ongoing process.

Moreover, the timing of supergene mineralization was constrained by combining indirect dating methods such as field mapping along with textural relationships, geochemical constraints with regionally well-correlated and radiochronologically dated deposits of the same type (Velasco et al. 2013). These approaches reveal that non-sulfide ores have formed through sulfide weathering during the Upper Miocene–Holocene interval, and even through the Quaternary. From a geodynamic and paleoclimatological perspectives, these time intervals coincide with the main tectonic uplift phases of the Atlas orogeny that presumably exhumed the sulfide-bearing mineralization in a regional setting of convergence between African and European plates. In term of exploration, the non-sulfide mineralization that overlies the North African Pb-Zn ± Cu MVTs is a promising target for the search of alternative sources to the conventional ore deposits of critical/strategic metals and related high-grade base metals. Future investigations combining climatic, neotectonic and hydrological approaches in addition to conventional geological studies would be welcomed for a better efficiency of the forthcoming mineral exploration campaigns.

4.5. Acknowledgments

The data presented in this study are undoubtedly the per value of a fruitful scientific collaboration that involves many worldwide MVT experts. In this respect, I express my gratitude to all those who warmly welcomed me into their host laboratories and generously provided the analytical facilities that allowed me to carry out my research. In this regard, I would like to thank the deceased D.F.

Sangster of the Geological Survey of Canada who was the first to introduce me to the MVT world, A.C. Brown from École polytechnique de Montréal, D.L. Leach of the US Geological Survey in Denver (Colorado), L. Fontboté from the University of Geneva (Switzerland), Y. Héroux from INRS-Georesources of Québec and E. Cardellach from the Universitat Autonoma de Barcelona Barcelona (Spain) for sharing their knowledge and expertise in the field of metallogeny. There were certainly divergent opinions concerning certain genetic interpretations and these discordant points of view did not impact our excellent academic relationships but instead was the stimulus for improving the MVT concepts. I cannot forget to address a special thank you to W. Boukirou and D. Margoum for the excellent illustrations they provided.

4.6. References

Aboutahir, N. (1999). Pétrographie et géochimie des dolomites du district de Touissit-Bou Beker : hôte de la minéralisation de type de la Vallée du Mississippi (Maroc Nord-Oriental). PhD Thesis, École Polytechnique, Montreal.

Adil, S., Bouabdellah, M., Grandia, F., Cardellach, E., Canals, A. (2004). Caractérisation géochimique des fluides associés aux minéralisations Pb-Zn de Bou Dahar (Maroc). *Comptes rendus geoscience*, 336, 1265–1272.

Affouri, H., Montacer, M., Disnar, J.R. (2012). Organic geochemistry of the Cenomanian–Turonian Bahloul formation petroleum source rock, central and northern Tunisia. *Resource Geol.*, 63(3), 262–287.

Agard, J. and Du, D.R. (1965). Les gîtes plombo-zincifères du Jbel Bou Dahar près de BeniTajjite (Haut Atlas oriental). *Notes Mém. Serv. Géol. Maroc.*, 181, 135–166.

Amouri, M. (1989). Les minéralisations Pb-Zn-Ba-F liées aux faciès carbonatés aptiens dans l'Atlas tunisien central. *Géologie méditerranéenne*, 16(2–3), 185–199.

Amrani, A., Said-Ahamed, W., Lewan, M.D., Aizenshtat, Z. (2006). Experiments of $\delta^{34}S$ mixing between organic and inorganic sulfur species during thermal maturation. *Geochim. Cosmochim. Acta*, 70, 5146–5161.

Anderson, G.M. (1991). Organic maturation and ore precipitation in Southeast Missouri. *Econ. Geol.*, 86, 909–926.

Anderson, G.M. (2015). Kerogen as a source of sulfur in MVT deposits. *Econ. Geol.*, 110, 837–840.

Annich, M. and Rahhali, M. (2002). Gisement de plomb de Zeida. In *Méthodes et techniques d'exploration minière et principaux gisements au Maroc*, Barodi, E.B. and Watanabe, Y. (eds). Projet JICA/BRPM, Bureau Recherche Participations Minières – BRPM, Rabat.

Appold, M.S. and Garven, G. (1999). The hydrology of ore formation in the southeast Missouri district: Numerical models of topography-driven fluid flow during the Ouachita orogeny. *Econ. Geol.*, 94, 913–935.

Appold, M.S. and Nunn, J.A. (2005). Hydrology of the western Arkoma Basin and Ozark Platform during the Ouachita Orogeny: Implications for Mississippi Valley-type ore formation in the Tri-State Zn-Pb district. *Geofluids*, 5, 308–325.

Auajjar, J. and Boulègue, J. (1999). Les minéralisations Pb-Zn-(Cu)-Ba du socle paléozoïque et de la plate-forme liasique du district du Tazekka (Taza, Maroc oriental) : une synthèse. *Chronique de la recherche minière*, 21–135, 536–537.

Auajjar, J. and Boulègue, J. (2002). Dolomites in the Tazekka Pb-Zn district, eastern Morocco: Poly-phase origin from hydrothermal fluids. *Terra Nova.*, 14, 175–182.

Bejaoui, J., Bouhlel, S., Sellami, A., Braham, A. (2014). Geology, mineralogy and fluid inclusion study of Oued Jebs Pb–Zn–Sr deposit; comparison with the Bou Grine deposit (diapirs zone, Tunisian atlas). *Arabian Journal of Geosciences*, 7, 2483–2497.

Belayouni, H. (1992). Oil seeps and associated phenomena in northern Tunisia. In *Journées de géologie Tunisienne appliquées à la recherche des hydrocarbures Field Trip*. ETAP, Tunis.

Benaissa, H. (1996). Les minéralisations stratiformes à Zn-Pb-F de la dorsale calcaire (Rif, Maroc) : étude gitologue et conditions de genèse des gisements d'Izilane, d'Adeldal et de Cadnar. PhD Thesis, Université Mohammed V, Rabat.

Bethke, C.M. and Marshak, S. (1990). Brine migrations across North America; the plate tectonics of groundwater. *Annu. Rev. Earth. Planet. Sci.*, 18, 287–315.

Bishop, W.F. (1988). Petroleum geology of the East-Central Tunisia. *A.A.P.G. Bull.*, 72, 1033–1058.

Boni, M. and Large, D.E. (2003). Nonsulfide zinc mineralization in Europe: An overview. *Econ. Geol.*, 98, 715–729.

Bouabdellah, M. and Sangster, D.F. (2016). Geology, geochemistry, and current genetic models for major Mississippi Valley-type Pb-Zn deposits of Morocco. In *Mineral Deposits of North Africa*, Bouabdellah, M. and Slack, J.F. (eds). Springer International Publishing, Berlin.

Bouabdellah, M., Héroux, Y., Brown, A.C. (1996). Pétrographie et altération de la matière organique du gisement de plomb-zinc-cuivre de Beddiane, district de Touissit-Bou Beker, Maroc. *Canad. Jour. Earth. Sci.*, 33, 1363–1374.

Bouabdellah, M., Héroux, Y., Chagnon, A. (2001). Zonation of organic matter reflectances and clay mineral assemblages around MVT deposits of Touissit-Bou Beker district, northeastern Morroco. In *Mineral Deposits at the Beginning of the 21st Century*, Piestrzynski, A., Pasava, J., Gize, A., Sass-Gustkiewicz, M., Leach, D., Muchez, P., Oszczepalski, S., Brown, A.C., Blundell, D.J., Kotlinski, R. et al. (eds). CRC Press, Boca Raton.

Bouabdellah, M., Sangster, D.F., Leach, D.L., Brown, A.C., Johnson, C.A., Emsbo, P. (2012). Genesis of the Touissit-Bou Beker Mississippi Valley-type district (Morocco–Algeria) and its relationships to the Africa–Europe collision. *Econ. Geol.*, 107, 117–146.

Bouabdellah, M., Castorina, F., Bodnar, R.J., Banks, D., Jébrak, M., Prochaska, W., Lowry, D., Klügel, H.K. (2013). Petroleum migration, fluid mixing, and halokinesis as the main ore-forming processes at the peridiapiric Jbel Tirremi fluorite-barite hydrothermal deposit, Northeastern Morocco. *Econ. Geol.*, 108, 1223–1256.

Bouabdellah, M., Niedermann, S., Velasco, F. (2015). The Touissit-Bou Beker Mississippi Valley-type district of northeastern Morocco: Relationships to the Messinian salinity crisis, Late Neogene-Quaternary alkaline magmatism, and buoyancy-driven fluid convection. *Econ. Geol.*, 110, 1455–1484.

Bouaziz, S., Barrier, E., Soussi, M., Turki, M.M., Zouari, H. (2002). Tectonic evolution of the northern African margin in Tunisia from paleostress data and sedimentary record. *Tectonophysics*, 357, 227–253.

Bouhlel, S. (1987). Mineralized carbonates in fluorite and Pb-Zn ore deposits: Jurassic and Upper Cretaceous of northern Tunisia. In *Excursion Guidebook I.A.S. 8th Regional Meeting of Sedimentology*. International Association of Sedimentology, Habib Belayouni, Tunis.

Bouhlel, S. (1993). Gîtologie, minéralogie et essai de modélisation des minéralisations de Pb–Zn–Sr–Ba–F–(S°) associées aux carbonates jurassiques et crétacés et aux diapirs triasiques (Tunisie septentrionale). PhD Thesis, Université Tunis II.

Bouhlel, S., Johnson, C.A., Leach, D.L. (2007). The peridiapiric-type Pb-Zn deposit at Fedj el Adoum, Tunisia: Geology, petrography, and stable isotopes. In *Digging Deeper*, Andrew, C., Annesley, I., Archibald, S., Beaudoin, G., Bierlein F.P., Borg G., Boyce A., Broughton, D., Canby, V.M., Cleverley, J. (eds). Ninth Biennial SGA Meeting, Dublin.

Bouhlel, S., Leach, D.L., Johnson, C.A., Marsh, E., Salmi-Laouar, S., Banks, D.A. (2016). A salt diapir-related Mississippi Valley-type deposit: The Bou Jaber Pb-Zn-Ba-F deposit, Tunisia: Fluid inclusion and isotope study. *Mineralium Deposita*, 51, 749–780.

Bouladon, J. (1949). Les gisements de plomb et de zinc en Afrique du Nord. Service Géologique du Maroc. *Notes Mém. Serv. Géol. Maroc.*, 1, 123–133.

Boutaleb, A. (2001). Les minéralisations Pb-Zn du domaine Sétifien-Hodna : gîtologie, pétrographie des dolomies, microthermométrie et implications métallogénique. PhD Thesis, Université Houari Boumedienne, Alger.

Boutaleb, A., Afalfiz, A., Aissa, D.E., Kolli, O., Marignac, C., Touahri, B. (2000). Métallogénie et évolution géodynamique de la chaine tellienne en Algérie. *Bulletin du service géologique d'Algérie*, 11(1), 3–27.

Bouzenoune, A., Rouvier, H., Thibieroz, J. (1995). Trias de l'Ouenza : contexte diapirique, zonation mineralogique et conséquences métallogéniques. *Bulletin du service géologique d'Algérie*, 6(1), 3–24.

Bouzenoune, A., Rouvier, H., Thibieroz, J. (2006). Chronologie relative et conditions de mise en place des minéralisations du massif de l'Ouenza (Algérie NE). *Bulletin du service géologique d'Algérie*, 17(1), 3–27.

Braham, A., Hatira, N., Hammami, M., Kadri, A., Arfaoui, M., Chihi, L., Mansouri, A., Souid, R. (2002). Le diapir de Bou Jaber (Tunisie Centre Ouest) : précisions biostratigraphiques et conséquences structurales. *Serv. Géol. Tunisie*, 69, 93–103.

Bruyère, D., De Putter, T., Decrée, S., Dupuis, C., Fuchs, Y., Jamoussi, F., Perruchot, A., Arbey, F. (2010). Miocene karsts and associated Fe–Zn-rich minerals in Aïn Khamouda (Central Tunisia). *J. Afr. Earth Sci.*, 57, 70–78.

Burollet, P.F. (1956). Contribution à l'étude stratigraphique de la Tunisie centrale. *Annales des mines et de la géologie (Tunis)*, 18, 350.

Burollet, P.F. and Stainfeld, P. (1956). Notice explicative de la carte géologique de la Tunisie au 50.000ème, feuille 44, El Kef, avec une étude sommaire des microfaunes de la région d'El Kef. Report, Serv. Minist. Indust. Energ., Tunis.

Butler, I.B., Boettcher, M.E., Rickard, D., Oldroyd, A. (2004). Sulfur isotope partitioning during experimental formation of pyrite via the polysulfide and hydrogen sulfide pathways; implications for the interpretation of sedimentary and hydrothermalpyrite isotope records. *Earth Planet. Sci. Letters*, 228, 495–509.

Campbell, A.E. and Stafleu, J. (1992). Seismic modeling of an Early Jurassic, drowned carbonate platform: Djebel Bou Dahar, High Atlas, Morocco. *A.A.P.G. Bull.*, 76, 1760–1777.

Charef, A. and Sheppard, S.M.F. (1987). Pb-Zn mineralization associated with diapirism: Fluid inclusion and stable isotope (H, C, O) evidence for the origin and evolution of the fluids at Fedj-El-Adoum, Tunisia. *Chem. Geol.*, 61, 113–134.

Chazot, G., Abbassene, F., Maury, R.C., Déverèche, F., Bellon, H., Ouabadi, A., Bosch, D. (2017). An overview on the origin of post-collisional Miocene magmatism in the Kabylies (northern Algeria): Evidence for crustal stacking, delamination and slab detachment. *J. Afri. Earth Sci.*, 125, 27–41.

Chihaoui, A., Jaillard, E., Latil, J.L., Zghal, I., Susperregui, A.S., Touir, J., Ouali, J. (2010). Stratigraphy of the Hameima and lower Fahdene Formations in the Tadjerouine area (northern Tunisia). *J. Afri. Earth Sci.*, 58, 387–399.

Choulet, F., Charles, N., Barbanson, L., Branquet, Y., Sizaret, S., Ennaciri, A., Badra, L., Chen, Y. (2014). Non-sulfide zinc deposits of the Moroccan High Atlas: Multi-scale characterization and origin. *Ore Geol. Rev.*, 56, 115–140.

Choulet, F., Buatier, M., Barbanson, L., Guégan, R., Ennaciri, A. (2016). Zinc-rich clays in supergene non-sulfide zinc deposits. *Mineralium Deposita*, 51, 467–490.

Choulet, F., Barbanson, L., Martine, B., James, R., Torsten, V., Ennaciri, A., Zouhair, M. (2017). Characterization and origin of low-T willemite (Zn_2SiO_4) mineralization: The case of the Bou Arhous deposit (High Atlas, Morocco). *Mineralium Deposita*, 52, 1085–1102.

Claveau, J., Paulhac, J., Pellerin, J. (1952). The lead and zinc deposits of the Bou-Beker-Touissit area, Eastern French Morocco. *Econ. Geol.*, 47, 481–493.

Claypool, G.E., Holser, W.T., Kaplan, T.R., Sakai, H., Zak, I. (1980). The age curves of sulfur and oxygen isotopes in marine sulfate and their mutual interpretations. *Chem. Geol.*, 28, 199–260.

Coulon, C., Megartsi, M., Fourcade, S., Maury, R.C., Bellon, H., Louni-Hacini, A., Cotten, J., Coutelle, A., Hermitte, D. (2002). Post-collisional transition from calc-alkaline to alkaline volcanism during the Neogene in Oranie (Algeria): Magmatic expression of a slab breakoff. *Lithos.*, 62, 87–110.

Crevello, P.D. (1990). Stratigraphic evolution of Lower Jurassic carbonate platforms: Record of rift tectonics and eustatsy, central and eastern High Atlas, Morocco. PhD Thesis, Colorado School of Mines, Golden.

Cumming, G.L. and Richards, J.R. (1975). Ore lead isotope ratios in a continuously changing Earth. *Earth Planet. Sci. Letters*, 28, 155–171.

Dagallier, G. and Macaudiere, J.M. (1987). Contrôles tectoniques des concentrations Pb-Ba en milieu carbonaté de Mibladen (Maroc). *Bulletin de la société géologique de France*, 8, 387–394.

Davis, D.W., Lowenstein, T.K., Spencer, R.J. (1990). Melting behavior of fluid inclusions in laboratory grown halite crystals in the systems NaCl-H_2O, NaCl-KCl-H_2O, NaCl-$MgCl_2$-H_2O and NaCl-$CaCl_2$-H_2O. *Geochim. Cosmochim. Acta*, 54, 591–601.

Decrée, S., Marignac, C., Liégeois, J.P., Yans, J., Ben Abdallah, R., Demaiffe, D. (2014). Miocene magmatic evolution in the Nefza district (Northern Tunisia) and its relationship with the genesis of polymetallic mineralizations. *Lithos*, 92–195, 240–258.

Dill, R.F., Shinn, E.A., Jones, A.T., Kelly, K., Steinen, R.P. (1986). Giant subtidal stromatolites forming in normal salinity water. *Nature*, 324, 55–58.

Dubois, M. and Marignac, C. (1997). The H_2O-NaCl-$MgCl_2$ ternary phase diagram with special application to fluid inclusion studies. *Econ. Geol.*, 92, 114–119.

Dubourdieu, G. (1959). Exquise géologique du Dj. Mesloula. *Pub. Serv. Carte Géol. Algér. Bull.*, 21, 162.

Duggen, S., Hoernle, K., van den Bogaard, P., Garbe-Schönberg, D. (2005). Post collisional transition from subduction- to intraplate-type magmatism in the westernmost Mediterranean: Evidence from continental-edge delamination of subcontinental lithosphere. *Jour. Petrology*, 46, 1115–1201.

Duggen, S., Hoernle, K., Hauff, F., Klügel, A., Bouabdellah, M., Thirlwall, M.F. (2009). Flow of Canary mantle plume material through a subcontinental lithospheric corridor beneath Africa to the Mediterranean. *Geology*, 37, 283–286.

Dupuy, J.J. and Touray, J.C. (1986). Multi-stage ore deposition at the Oued Mekta strata-bound lead deposit, Touissit-Bou Beker district, eastern Morocco. *Econ. Geol.*, 181, 1558–1561.

Eisenlohr, B.N., Tompkins, L.A., Cathles, L.M., Barley, M.E., Groves, D.I. (1994). Mississippi Valley-type deposits: Products of brine expulsion by eustatically induced hydrocarbon generation? An example from northwestern Australia. *Geology*, 22, 315–318.

Elmi, S. (1973). Décrochements et mouvements atlasiques dans la région frontalière algéro-marocaine (Monts Rhar Roubane). *Comptes rendus de l'académie des sciences*, 276, 1521–1524.

Emberger, A. (1965). Caractères polygénétiques des minéralisations plombifères de la Haute Moulouya (gisements d'Aouli, Mibladen et Zeïda, Maroc). *Comptes rendus de l'académie des sciences*, 260(9), 3433–3436.

Espitalié, J., Laporte, J.L., Madec, M., Marquis, F., Leplat, P., Paulet, J., Boutefeu, A. (1977). Méthode rapide de caractérisation des roches mètres, de leur potentiel pétrolier et de leur degré d'évolution. *Revue de l'institut français du petrole*, 32, 23–42.

Espitalié, J., Deroo, G., Marquis, F. (1985). La pyrolyse Rock-Eval et ses applications. *Revue de l'institut français du petrole*, 40(6), 755–784.

Faccenna, C. and Becker, T.W. (2020). Topographic expressions of mantle dynamics in the Mediterranean. *Earth. Sci. Rev.*, 209, 103327.

Fedan, B., Laville, E., Mezgueldi, A.E. (1989). Le bassin jurassique du Moyen Atlas (Maroc) : exemple de bassin sur relais de décrochements. *Bulletin de la société géologique de France*, 5(6), 1123–1136.

Fiechtner, L., Friedrichsen, H., Hammerschmid, K. (1992). Geochemistry and geochronology of early Mesozoic tholeiites from central Morocco. *Geol. Rundschau.*, 81, 45–62.

Filali, F., Guiraud, M., Burg, J.P. (1999). Nouvelles données petro-structurales sur la boutonnière d'Aouli (Haute Moulouya) : leurs conséquences sur la géodynamique hercynienne du Maroc. *Bulletin de la société géologique de France*, 4, 435–450.

François, T. (1970). Les mines et la region d'Annaba. *Géocarrefour*, 45(1), 31–59.

Frizon de Lamotte, D., Saint-Bezar, B., Bracene, R., Mercier, E. (2000). The two main steps of the Atlas building and geodynamics of the western Mediterranean. *Tectonics*, 19, 740–761.

Frizon de Lamotte, D., Michard, A., Saddiqi, O. (2006). Quelques développements récents sur la géodynamique du Maghreb. *Comptes rendus geosciences*, 338, 1–10.

Frizon de Lamotte, D., Leturmy, P., Missenard, Y., Khomsi, S., Ruiz, G., Saddiqi, O., Guillocheau, F., Michard, A. (2009). Mesozoic and Cenozoic vertical movements in the Atlas system (Algeria, Morocco, Tunisia): An overview. *Tectonophysics*, 475, 9–28.

Garnit, H., Boni, M., Buongiovanni, G., Arfè, G., Mondillo, N., Joachimski, M., Bouhlel, S., Balasson, G. (2018). C–O stable isotopes geochemistry of Tunisian nonsulfide zinc deposits: A first look. *Minerals*, 8, 13.

Garven, G., Ge, S., Person, M.A., Sverjensky, D.A. (1993). Genesis of stratabound ore deposits in the Midcontinent basins of North America. The role of regional groundwater flow. *Am. J. Sci.*, 293, 497–568.

Ge, S. and Garven, G. (1992). Hydromechanical modeling of tectonically driven groundwater flow with application to the Arkoma foreland basin. *J. Geophys. Res.*, 97(6), 9119–9144.

Gharbi, R.A., Chihi, L., Hammami, M., Soumaya, A., Kadri, A. (2005). Manifestations tectono diapiriques synsédimentaires et polyphasées d'âge Crétacé supérieur – Quaternaire dans la région de Zag Et Tir (Tunisie centre-nord). *Comptes rendus geoscience*, 337, 1293–1300.

Gomez-Rivas, E., Corbella, M., Martín-Martín, J.D., Stafford, S.L., Teixell, A., Bons, P.D., Griera, A., Cardellach, E. (2014). Reactivity of dolomitizing fluids and Mg source evaluation of fault-controlled dolomitization at the Benicàssim outcrop analogue (Maestrat Basin, E Spain). *Mar. Petrol. Geol.*, 55, 26–42.

Haddouche, O., Boutaleb, A., Hébert, R., Picard, D., Sami, L. (2004). Les minéralisations à Pb-Zn, Fe, Ba (Sr) d'El Ouasta (Algérie Nord Oriental). Typologie et apport des études d'inclusions fluides. *Bulletin du service géologique d'Algérie*, 15(2), 87–105.

Haddouche, O., Boutaleb, A., Benhamoud, I. (2014). Contexte structural des minéralisations liées à la bordure nord des Monts des Aures (NE de l'Algérie) et des régions voisines : exemple des gisements à Ba-Pb (Zn-Cu) d'Ichmoul et d'Ain Mimoun. *Bulletin du service géologique d'Algérie*, 25(1), 3–19.

Haddouche, O., Boutaleb, A., Chamam, M., Ysbaa, S., Hammouche, H., Boubaya, D. (2016). Pb-Zn (Ba) deposits of the oriental Saharan Atlas (north-eastof Algeria): Distribution, control and implications for mining exploration. *Arab. Jour. Geosci.*, 9(42). doi: 10.1007/s12517-016-2406-x.

Hanor, J.S. (1980). Dissolved methane in sedimentary brines: Potential effect on PVT properties of fluid inclusion. *Econ. Geol.*, 75, 603–609.

Hanor, J.S. (1994). Origin of saline fluids in sedimentary basins. *Geol. Soc. Spec. Pub.*, 78, 151–174.

Harmand, C. and Cantagrel, J.M. (1984). Le volcanisme alcalin Tertiaire et Quaternaire du moyen Atlas (Maroc) : chronologie K/Ar et cadre géodynamique. *J. Afr. Earth Sci.*, 2, 51–55.

Heijlen, W., Muchez, P., Banks, D.A. (2001). Origin and evolution of high-salinity Zn-Pb mineralizing fluids in the Variscides of Belgium. *Miner. Depos.*, 36, 165–176.

Hsü, K.J., Cita, M.B., Ryan, W.B.F. (1973). The origin of the Mediterranean evaporites. Report, Deep Sea Drilling Project.

Ibouh, H. and Mouguina, E.M. (2012). Aguerd n'Tazoult, a Zn-Pb Mine in the Central High Atlas. In *Les principales mines du Maroc*, Mouttaqi, A., Rjimati, E.C., Maacha, L., Michard, A., Soulaimani, A., Ibouh, H. (eds). Éditions du service géologique du Maroc, Rabat.

Jackson, S.A. and Beales, F.W. (1967). An aspect of sedimentary basin evolution: The concentration of Mississippi Valley-type ores during the late stages of diagenesis. *Bull. Canad. Petrol. Geol.*, 15, 393–433.

Jaillard, E., Dumont, T., Ouali, J., Bouillin, J.P., Chihaoui, A., Latil, J.L., Arnaud, H., Arnaud-Vanneau, A., Zghal, I. (2013). The Albian tectonic "crisis" in central Tunisia: Nature and chronology of the deformations. *J. Afri. Earth Sci.*, 85, 75–86.

Jébrak, M., Marcoux, E., Nasloubi, M., Zahraoui, M. (1998). From sandstone- to carbonate-hosted stratabound deposits: An isotope study of galena in the Upper-Molulouya district (Morocco). *Miner. Depos.*, 33, 406–415.

Jemmali, N., Souissi, F., Carranza, E.J.M., Bouabdellah, M. (2013). Lead and sulfur isotope constraints on the genesis of the polymetallic mineralization at Oued Maden, Jebel Hallouf and FedjHassene carbonate-hosted Pb–Zn (As–Cu–Hg–Sb) deposits, northern Tunisia. *Jour. Geochem. Explor.*, 132, 6–14.

Jenkyns, H.C., Jones, C.E., Grocke, D.R., Hesselbo, S.P., Parkinson, D.N. (2002). Chemostratigraphy of the Jurassic system: Applications, limitations and implications for palaeooceanography. *Jour. Geol. Soc.*, 159, 351–378.

Jones, G.D., Whitaker, F.F., Smart, P.L., Sanford, W.E. (2004). Numerical analysis of seawater circulation in carbonate platforms: II. The dynamic interaction between geothermal and brine reflux circulation. *Amer. Jour. Sci.*, 304, 250–284.

Juteau, M., Michard, A., Albarede, F. (1986). The Pb–Sr–Nd isotope geochemistry of some recent circum-Mediterranean granites. *Contrib. Miner. Petrol.*, 92, 331–340.

Kampschulte, A. and Strauss, H. (2004). The sulfur isotope evolution of Phanerozoic seawater based on the analysis of structurally substituted sulfate in carbonates. *Chem. Geol.*, 204, 255–286.

Keller, J. and Hoefs, J. (1995). Stable isotope characteristics of recent natrocarbonatite from Oldoinyo Lengai. In *Carbonatite Volcanism: Oldoinyo Lengai and the Petrogenesis of Natrocarbonatites*, Bell, K. and Keller, J. (eds). IAVCE I, Proceedings in Volcanology, Springer, Berlin, Heidelberg.

Kesler, S.E., Martini, A.M., Appold, M.S., Walter, L.M., Huston, T.J., Furman, F.C. (1996). Na-Cl-Br systematics of fluid inclusions from Mississippi Valley-type deposits, Appalachian basin: Constraints on solute origin and migration paths. *Geochimica et cosmochimica acta*, 60, 225–233.

Kowalski, W.M. and Hamimed, M. (2000). Diapirisme polyphasé ou glacier de sel Albien ? Dilemme du matériel triasique des confins algéro-tunisiens. *Bulletin du service géologique d'Algérie*, 11(1), 29–60.

Krijgsman, W., Langereis, C.G., Zachariasse, W.J., Boccaletti, M., Moratti, G., Gelati, R., Iaccarino, S., Papani, G., Villa, G. (1999). Late Neogene evolution of the Taza-Guercif basin (Rifian corridor, Morocco) and implications for the Messinian salinity crisis. *Mar. Geol.*, 153, 147–160.

Laatar, E. (1980). Gisements de plomb-zinc et diapirisme du Trias salifère en Tunisie septentrionale : les concentrations péridiapiriques du district minier de Nefate-Fedj el Adoum. PhD Thesis, Université Pierre et Marie Curie, Paris.

Laffitte, P., Permingeat, F., Routhier, P. (1965). Cartographie métallogénique, métallotecte et géochimie régionale. *Bulletin de minéralogie*, 88(1), 3–6.

Laouar, R., Salmi-Laouar, S., Sami, L., Boyce, A.J., Kolli, O., Boutaleb, A., Fallick, A.E. (2016). Fluid inclusion and stable isotope studies of the Mesloula Pb-Zn-Ba ore deposit, NE Algeria: Characteristics and origin of the mineralizing fluids. *J. Afri. Earth Sci.*, 121, 119–135.

Layeb, M. and Belayouni, H. (2000). Paléogéographie de la Formation Bahloul (passage Cénomanien–Turonien). *Ann. Min. Géol. Tunisie.*, 40, 21–44.

Leach, D.L. and Rowan, E.L. (1986). Genetic link between Ouachita foldbelt tectonism and the Mississippi Valley-type lead-zinc deposits of the Ozarks. *Geology*, 14, 931–935.

Leach, D.L. and Sangster, D.F. (1993). Mississippi Valley type lead-zinc deposits. *Geological Association of Canada*, 40, 289–314.

Leach, D.L., Bradley, D.C., Huston, D., Pisarevsky, S.A., Taylor, R.D., Gardoll, S.J. (2010a). Sediment-hosted lead-zinc deposits in Earth history. *Econ. Geol.*, 10., 593–625.

Leach, D.L., Taylor, R.D., Fey, D.L., Diehl, S.F., Saltus, R.W. (2010b). A deposit model for Mississippi Valley-type lead-zinc ores. In *Mineral Deposit Models for Resource Assessment. Critical Reviews of Oxidative Stress and Aging*, Cutler, R.G. and Boucher, D. (eds). World Scientific Publishing, Singapore [Online]. Available at: http://www.usgs.gov/pubprod.

Lüning, S., Kolonic, S., Belhadj, E.M., Belhadj, Z., Cota, L., Baric, G., Wagner, T. (2004). Integrated depositional model for the Cenomanian–Turonian organic-rich strata in North Africa. *Earth Sci. Rev.*, 64, 51–117.

Lustrino, M. and Wilson, M. (2007). The circum-Mediterranean anorogenic Cenozoic igneous province. *Earth Sci. Rev.*, 81, 1–65.

Machel, H.G., Krouse, H.R., Sassen, R. (1995). Products and distinguishing criteria of bacterial and thermochemical sulfate reduction. *App. Geochem.*, 10, 373–389.

Makhoukhi, S., Marignac, C., Pironon, J., Schmitt, J.M., Marrakchi, C., Bouabdelli, M., Bastoul, A. (2003). Aqueous and hydrocarbon inclusions in dolomite from Touissit-Bou Beker district, eastern Morocco: A Jurassic carbonate hosted Pb-Zn (Cu) deposit. *Jour. Geochem. Explor.*, 78–79, 545–551.

Monciardini, C. and Orgeval, J.J. (1987). Apport de la biostratigraphie à la connaissance du gisement (Zn-Pb) de Bou Grine (Tunisie). Report, BRGM, Principal Sci. Tech. Results.

Montacer, M. (1989). Étude de la matière organique de la formation Bahloul (Cénomano-Turonien) dans l'environnement sédimentaire du gisement Zn-Pb de Bou Grine (Nord-Ouest de la Tunisie). PhD Thesis, Université d'Orléans.

Montacer, M., Disnar, J.R., Orgeval, J.J. (1986). Étude préliminaire de la matière organique de la formation Bahloul dans l'environnement du gisement Zn-Pb de Bou Grine (Tunisie). Report, BRGM, Principal Sci. Tech. Results.

Montacer, M., Disnar, J.R., Orgeval, J.J., Trichet, J. (1988). Relationship between Zn-Pb ore and oil accumulation processes: Example of the Bougrine deposit (Tunisia). *Org. Geochem.*, 13, 423–431.

Monthel, J., Lagny, P., Orgeval, J.J. (1986). Inventaire des gisements et indices de la zone des Dômes de l'Atlas Tunisien et de sa bordure septentrionale. Report, BRGM 86 TUN 1127 GMX, 37.

Mouguina, E.M., Ibouh, H., Alansari, A. (2012). The Ali-ou-Daoud zinc mine (Imilchil, Central High Atlas). In *Les principales mines du Maroc*, Mouttaqi, A., Rjimati, E.C., Maacha, L., Michard, A., Soulaimani, A., Ibouh, H. (eds). Éditions du service géologique du Maroc, Rabat.

Noronha, F., Auajjar, J., Guedes, A., Banks, D. (2007). Pb-Zn-Ba mineralization from the Tazekka Pb-Zn district, eastern Morocco. In *Digging Deeper Proceedings of the Ninth Biennial*, Andrew, C., Annesley, I., Archibald, S., Beaudoin, G., Bierlein F.P., Borg G., Boyce A., Broughton, D., Canby, V.M., Cleverley, J. (eds). SGA Meeting, Dublin.

Ohmoto, H. (1972). Systematics of sulfur and carbon isotopes in hydrothermal ore deposits. *Econ. Geol.*, 67, 551–578.

Ohmoto, H. and Rye, R.O. (1979). Isotopes of sulfur and carbon. In *Geochemistry of Hydrothermal Ore Deposits*, Barnes, H.L. (ed.). Wiley, New York.

Oliver, J. (1986). Fluids expelled tectonically from orogenic belts: Their role in hydrocarbon migration and other geologic phenomena. *Geology*, 14, 99–102.

Orgeval, J.J. (1994). Peridiapiric metal concentration: Example of the Bougrine deposit (Tunisian Atlas). In *Sediment-Hosted Zn–Pb Ores*, Fontboté, L. and Boni, M. (eds). Springer, Berlin.

Orgeval, J.J. (1995). Peridiapiric metal concentration at Bou Grine (Tunisian Atlas): Some geochemical characteristics. In *Mineral Deposits: From Their Origin to Their Environmental Impacts*, Pasava, J., Kribek, B., Zak, K. (eds). Balkema, Rotterdam.

Orgeval, J.J., Giot, D., Karoui, J., Monthel, J., Sahli, R. (1986). Le gisement de Zn–Pb de Bou Grine (Atlas tunisien). Description et historique de la découverte. *Chron. Rech. Min.*, 485, 5–32.

Orr, W.L. (1974). Changes in sulfur content and isotopic ratios of sulfur during petroleum maturation; a study of big horn basin Paleozoic oils. *A.A.P.G. Bull.*, 58(11), 2295–2318.

Perthuisot, V. (1978). Dynamique et pétrogenèse des extrusions triasiques en Tunisie septentrionale. PhD Thesis, École Normale Supérieure, Paris.

Perthuisot, V. (1981). Diapirism in northern Tunisia. *J. Struct. Geol.*, 3, 231–235.

Perthuisot, V. (1992). Les diapirs du Maghreb central et oriental : des diapirs variés, résultats d'une évolution structurale et pétrogénétique complexe. *Bulletin de la société géologique de France*, 163(6), 751–760.

Perthuisot, V. (1994). Structures et géométrie des diapirs maghrébins. Essai de synthèse. *Bulletin du service géologique d'Algérie*, 6, 153–159.

Perthuisot, V., Rouvier, H., Smati, A. (1998). Style et importance des déformations anté-vraconiennes dans le Maghreb oriental : exemple du diapir du Jebel Slata. *Bulletin de la société géologique de France*, 8, 389–398.

Potra, A., Garmon, W.T., Samuelsen, J.R., Wulff, A., Pollock, E.D. (2018). Lead isotope trends and metal sources in the Mississippi Valley-type districts from the mid-continent United States. *Jour. Geochem. Explor.*, 192, 174–186.

Prochaska, W., Kolli, O., Boutaleb, A., Sami, L., Haddouche, O., Laouar, R. (2007). Investigations on the chemistry of mineralizing fluids of vein-type mineralizations related to the Diapiric zone in Northen Algeria. In *Digging Deeper Proceedings of the Ninth Biennial*, Andrew, C., Annesley, I., Archibald, S., Beaudoin, G., Bierlein F.P., Borg G., Boyce A., Broughton, D., Canby, V.M., Cleverley, J. (eds). SGA Meeting, Dublin.

Rahhali, M. (2002). Gisements stratiformes de Mibladen. In *Méthodes et techniques d'exploration minière et principaux gisements au Maroc*, Barodi, E.B. and Watanabe, Y. (eds). Projet JICA/BRPM, Bureau Recherche Participations Minières – BRPM, Rabat.

Rajlich, P. (1983). Geology of Oued Mekta, a Mississippi Valley-type deposit, Touissit-Bou Beker region, eastern Morocco. *Econ. Geol.*, 78, 1239–1254.

Rddad, L. and Bouhlel, S. (2016). The Bou Dahar Jurassic carbonate-hosted Pb–Zn–Ba deposits (Oriental High Atlas, Morocco): Fluid inclusion and C–O–S–Pb isotope studies. *Ore Geol. Rev.*, 72, 1072–1087.

Rddad, L., Mouguina, E.M., Muchez, P., Darling, R.S. (2018). The genesis of the Ali Ou Daoud Jurassic carbonate Zn-Pb Mississippi Valley type deposit, Moroccan central High Atlas: Constraints from bulk stable C-OS, in situ radiogenic Pb isotopes, and fluid inclusion studies. *Ore Geol. Rev.*, 99, 365–379.

Rddad, L., Jemmali, N., Sośnicka, M., Cousens, B. (2019). The genesis of the salt diapir-related Mississippi Valley-type Ba-Pb-(± Zn) ore of the Slata District, Tunisia: The role of halo-kinesis, hydrocarbon migration, and alpine orogenesis. *Econ. Geol.*, 114(8), 1599–1620.

Roedder, E. (1984). Fluid inclusions. *Rev. Miner. Geochem.*, 12, 644.

Rouvier, H., Perthuisot, V., Mansouri, A. (1985). Deposits and salt bearing diapirs in southern Europe and North Africa. *Econ. Geol.*, 80, 666–687.

Sainfeld, P. (1952). Les gîtes plombo-zincifères de Tunisie. *Annales des mines et géologie de Tunis*, 9, 285.

Saintilan, N.J., Spangenberg, J.E., Chiaradia, M., Chelle-Michou, C., Stephens, M.B., Fontboté, L. (2019). Petroleum as source and carrier of metals in epigenetic sediment-hosted mineralization. *Scientific Reports*, 9, 8283. doi: 10.1038/s41598-019-44770-7.

Salmi-Laouar, S. (2004). Contribution à l'étude géologique et géochimie des isotopes stables (S, O, C) des minéralisations polymétalliques (Zn-Pb-F-Ba-Fe-Hg) de la zone des diapirs du nord de Tebessa (NE Algérien). PhD Thesis, Université de Annaba.

Salmi-Laouar, S., Laouar, R., Boyce, A.J., Zerdazi, A., Arrouche, Y. (2004). Rapports isotopiques du soufre, de l'oxygène et du carbone dans le massif de Boujaber, NE algérien : origine des minéralisations à Pb-Zn-Ba et source des fluides. *Bulletin de service géologique de l'Algérie*, 15(1), 3–25.

Salmi-Laouar, S., Laouar, R., Boyce, A.J., Boutaleb, A., Lamouroux, C. (2007). Premières données isotopiques sur la mer triasique dans l'Atlas Saharien Oriental (Algérie). *Bulletin du service géologique de l'Algérie*, 18(3), 315–323.

Sami, L. (2011). Caractérisation géochimique des minéralisations à Pb-Zn, F, Ba, Cu, Fe et Hg des confins Algéro-tunisiens. PhD Thesis, Université d'Alger, Algiers.

Samson, P. (1973). *Un gisement plombo-zincifère en milieu récifal : Touissit (Maroc oriental)*. Éditions du service géologique du Maroc, Rabat.

Schmidt, S.C. (1999). Re-activation of the Bougrine Mine, Tunisia. In *Caglary 99–101st Ann. General Meeting Can. Inst. Min. Metal Petrol, Techn. Progr. CD-ROM*. Major Canadian Overseas Projects.

Sheppard, S.M.F. (1986). Characterization and isotopic variations in natural waters. In *Stable Isotopes in High Temperature Geological Processes*, Valley, J.W., Taylor, H.P., O'Neil, J.R. (eds). De Gruyter, Berlin.

Sheppard, S.M.F., Charef, K., Bouhlel, S. (1996). Diapirs and Pb–Zn mineralization: A general model based on Tunisian (N. Africa) and Gulf Coast (U.S.A.) deposits. In *Carbonate-hosted Lead Zinc Deposits*, Sangster, D.F. (ed.). Society of Economic Geologists, Littleton.

Smati, A. (1986). Les gisements de Pb-Ba et de Fe du Jebel Slata (Tunisie du Centre-Nord) : minéralisations épigénétiques dans le Crétacé néritique de la bordure d'un diapir de Trias. Gisements de Sidi Amor Ben Salem et de Slata-Fer. PhD Thesis, Université Pierre et Marie Curie, Paris.

Spirakis, C.S. and Heyl, A.V. (1996). Fracturing radioactive basement rocks as the energy source to drive mineralization in the Upper Mississippi Valley zinc-lead district. In *Carbonate-hosted Lead Zinc Deposits*, Sangster, D.F. (ed.). Society of Economic Geologists, Littleton.

Stacey, J.S. and Kramers, J.C. (1975). Approximation of terrestrial lead isotope evolution by a two-stage model. *Earth Planet. Sci. Letters*, 26, 207–221.

Taylor, H.P., Frechen, J., Degens, E.T. (1967). Oxygen and carbon isotope studies of carbonatites from the Laacher See district, West Germany and the Alnö district, Sweden. *Geochimica et cosmochimica acta*, 31, 407–430.

Teixell, A., Arboleya, M.L., Charroud, M., Julivert, M. (2003). Tectonic shortening and topography in the central High Atlas (Morocco). *Tectonics*, 22(5), 1051.

Teixell, A., Ayarza, P., Zeyen, H., Fernandez, M., Arboleya, M.L. (2005). Effects of mantle upwelling in a compressional setting: The Atlas Mountains of Morocco. *Terra Nova.*, 17, 456–461.

Tompkins, L.A., Rayner, M.J., Groves, D.I., Roche, M.T. (1994). Evaporites: In-situ sulfur source for rhythmically banded ore in the Cadjebut MVT Zn-Pb deposit, Western Australia. *Econ. Geol.*, 89, 467–492.

Torbi, A. and Gélard, J.P. (2000). Paléocontraintes identifiées dans la couverture mésocénozoïque du Maroc nord oriental : relations avec l'ouverture de l'Atlantique et rapprochement Afrique-Europe. *Comptes rendus de l'académie des sciences*, 330, 853–858.

Tornos, F., Velasco, F., Menor-Salván, C., Delgado, A., Slack, F.F., Escobar, J.M. (2013). Formation of recent Pb-Ag-Au mineralization by potential sub-surface microbial activity. *Nature Communications*, 5, 46.

Tosdal, R.M., Wooden, J.L., Bouse, R.M. (1999). Pb isotopes, ore deposits, and metallogenic terranes. *Rev. Econ. Geol.*, 12, 1–28.

Touahri, B. (1991). *Géochimie et métallogénie des minéralisations à Pb et Zn du nord de l'Algérie*. Office national de géologie, Algiers.

Ulrich, M.R. and Bodnar, R.J. (1988). Systematics of stretching of fluid inclusions II: Barite at 1 atm confining pressure. *Econ. Geol.*, 83, 1037–1046.

Vauchez, A. (1976). Les déformations anté-triasiques dans la boutonnière d'Aouli-Mibladen. *Comptes rendus de l'académie des sciences*, 282, 425–428.

Veizer, J., Ala, D., Azmy, K., Bruskschen, P., Buhl, D., Bruhn, F., Carden, G.A.F., Diener, A., Ebneth, S., Godderis, Y. et al. (1999). $^{87}Sr/^{86}Sr$, $\delta^{13}C$, and $\delta^{18}O$ evolution of Phanerozoic seawater. *Chem. Geol.*, 161, 59–88.

Viets, J.G., Hofstra, A.H., Emsbo, P. (1996). Solute composition of fluid inclusions in sphalerite from North American and European Mississippi Valley-type ore deposits: Ore fluids derived from evaporated seawater. *Geol. Soc. Spec. Pub.*, 4, 465–482.

Vila, J.M. (1980). La chaîne alpine d'Algérie nord-orientale et des confins algéro-tunisiens. PhD Thesis, Université Pierre et Marie Curie, Paris.

Vila, J.M., Youssef, M., Charrière, A., Chikhaoui, M., Ghanmi, M., Kamoun, F., Peybernès, B., Saadi, J., Souquet, P., Zarbout, M. (1994). Découverte en Tunisie au SW du Kef de matériel triasique interstratifié dans l'Albien : extension du domaine à "glaciers de sel" sous-marin des confins algéro-tunisiens. *Comptes rendus de l'académie des sciences*, 318, 109–116.

Vila, J.M., Ben Youssef, M., Chikhaoui, M., Ghanmi, M. (1996). Deuxième étude de surface d'un "glacier de sel" sous-marin albien (250 km^2?) ; les masses triasiques du "diapir" de Ben Gasseur et de l'anticlinal du Kef (Nord-Ouest tunisien). *Bulletin de la société géologique de France*, 167, 235–246.

Voirin, J. (1965). Géologie du gisement plombo-zincifère de Bou Beker. In *Colloque sur les gîtes stratiformes de plomb, zinc et manganèse du Maroc 1962*. Éditions du service géologique du Maroc, Rabat.

Vydrine, A. (1972). Géologie et lois de répartition des gisements de métaux non ferreux de l'Algérie du Nord. Report, Entreprise Nationale de Recherche Minière, Boumerdès.

Wittig, N., Pearson, D.G., Baker, J., Duggen, S., Hoernle, K. (2010). A major element, PGE and Re-Os isotope study of Middle Atlas (Morocco) peridotite xenoliths: Evidence for coupled introduction of metasomatic sulphides and clinopyroxene. *Lithos*, 115, 15–26.

Wood, S.A., Crerar, D.A., Borcsik, M.P. (1987). Solubility of the assemblage pyrite-pyrrhotite-magnetite-sphalerite-galena-gold-stibnite-bismuthinite-argentite-molybdenite in H (sub 2) O-NaCl-CO (sub 2) solutions from 200° to 350°C. *Econ. Geol.*, 82, 1864–1887.

Ysbaa, S., Haddouche, O., Boutaleb, A., Chemam, A., Sadaoui, M. (2019). Mineral deposits of northeastern Algeria (southern Medjerda mounts and diapiric zone): Regional-scale structural controls, spatial distribution, and importance of geophysical lineaments. *Arab Jour. Geosci.*, 12, 482. doi: 10.1007/s12517-019-4611-x.

Yukler, M.A., Moumen, A., Daadouch, I., Bouhlel, H., Meskini, A., Saidi, M., Jarraya, H. (1994). Quantitative evaluation of the geologic and hydrocarbon potential of the Gulf of Gabès. In *Proceedings of the 4th Tunisian Petroleum Exploration Conference*, Tunis Memoir, Tunis.

Zartman, R.E and Doe, B.R. (1981). Plumbotectonics – The model. *Tectonophysics*, 75, 35–162.

Zheng, Y.F. (1999). Oxygen isotope fractionation in carbonate and sulfate minerals. *Geochem. Journal*, 33(2), 109–126.

5

West African Leo-Man Shield Metallogenic Province

Aurélien EGLINGER[1], Anne-Sylvie ANDRÉ-MAYER[1], Nicolas THÉBAUD[2] and Quentin MASUREL[2]

[1] *GeoRessources, University of Lorraine, Vandœuvre-lès-Nancy, France*

[2] *University of Western Australia, Crawley, Australia*

5.1. Introduction

5.1.1. *Definition of a mineral system*

An ore deposit is the local expression of a series of geological processes occurring at various time and space scales, allowing the concentration of elements of economic interest in a volume compatible with their exploitation (Meyer 1988; Barley and Groves 1992; Groves et al. 2005; Kerrich et al. 2005; Groves and Bierlein 2007; Bierlein et al. 2009; Cawood and Hawkesworth 2015; Jenkin et al. 2015). By definition, ore deposits involve small volumes of rock compared to the surface and thickness of the continental crust, even if they result from geological and geodynamic processes that occur and can be described on a crustal or lithospheric scale (McCuaig and Hronsky 2014).

Mobilization, transfer and deposition of metallic elements in the lithosphere can be integrated in the concept of "mineral systems" including "all the geological factors that control the generation and preservation of mineral deposits" (McCuaig and Hronsky 2014) (Figure 5.1). Despite the variety of interpretations of what a mineral

---

For a color version of all figures in this chapter, see www.iste.co.uk/decree/metallic2.zip.

Metallic Resources 2,
coordinated by Sophie DECRÉE.

system is (Wyborn et al. 1994; Barnicoat 2007; McCuaig et al. 2010; Huston et al. 2012), most of them conclude that some factors are essential, quite critical, that must combine in nested scales in space and time for the formation of giant ore deposit. These four critical elements include whole lithosphere architecture, transient favorable geodynamics, fertility and preservation of the primary depositional zone. The mineral systems concept is related to the "source-transport-deposit" triad used in the oil industry for the exploration of hydrocarbon resources (Lord et al. 2001) and is now considered as a valuable approach in mineral exploration compared to approaches that use deposit models (McCuaig and Hronsky 2014).

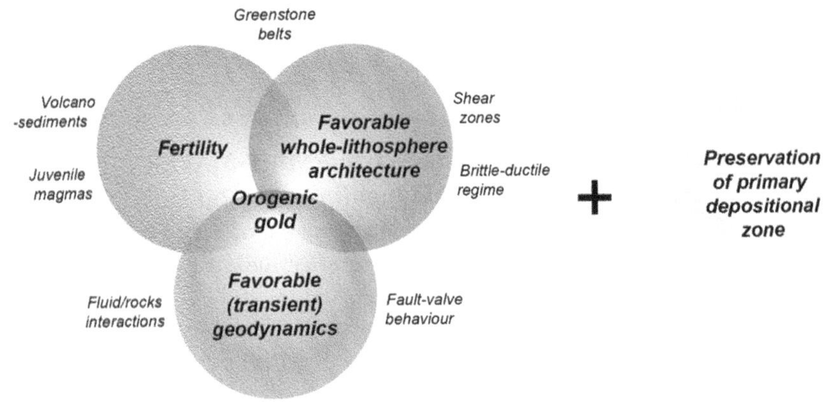

Figure 5.1. *Key elements of a mineral system: example of Precambrian orogenic gold systems. Modified from McCuaig and Hronsky (2014)*

5.1.2. *Metallogenic period and province*

The concept of metallogenic period and province was introduced by Louis de Launay (1913). It covers the idea of duration and synchronism in the formation of mineralizations. The ore deposits are indeed distributed heterogeneously in space and time (Lindgren 1909; Barley and Groves 1993; Titley 1993; Groves et al. 2005; Kerrich et al. 2005; Groves and Bierlein 2007; Bierlein et al. 2009; Goldfarb et al. 2010; Maier and Groves 2011; Cawood and Hawkesworth 2015; Jenkin et al. 2015). This heterogeneous distribution is related to the following three main factors: (a) the evolution of the hydrosphere-atmosphere; (b) the secular changes in the Earth's heat flow; (c) the evolution of the tectonic processes and orogenic cycles that marked the geological history of the Earth. Ore deposits can also be considered as tracers, records of the evolution of the Earth system during specific geological periods. Indeed, their temporal distribution can be used to decipher the secular trends of the Wilson cycles, defined by the fragmentation and periodical amalgamation of mega- and supercontinents (Kerrich et al. 2005; Goldfarb et al. 2010), but also to reflect the succession of various geodynamic contexts within the same orogenic cycle (Eglinger 2013; Eglinger et al. 2013; Turlin 2017; Perret 2021). Cawood and Hawkesworth (2015) argue for a potential gap in these secular records linked to the surface effects related to the resulting material erosion and orogenic reworking. They also highlight the existence of a systematic bias in the preservation of segments of the continental crust potentially mineralized during various Wilson cycles.

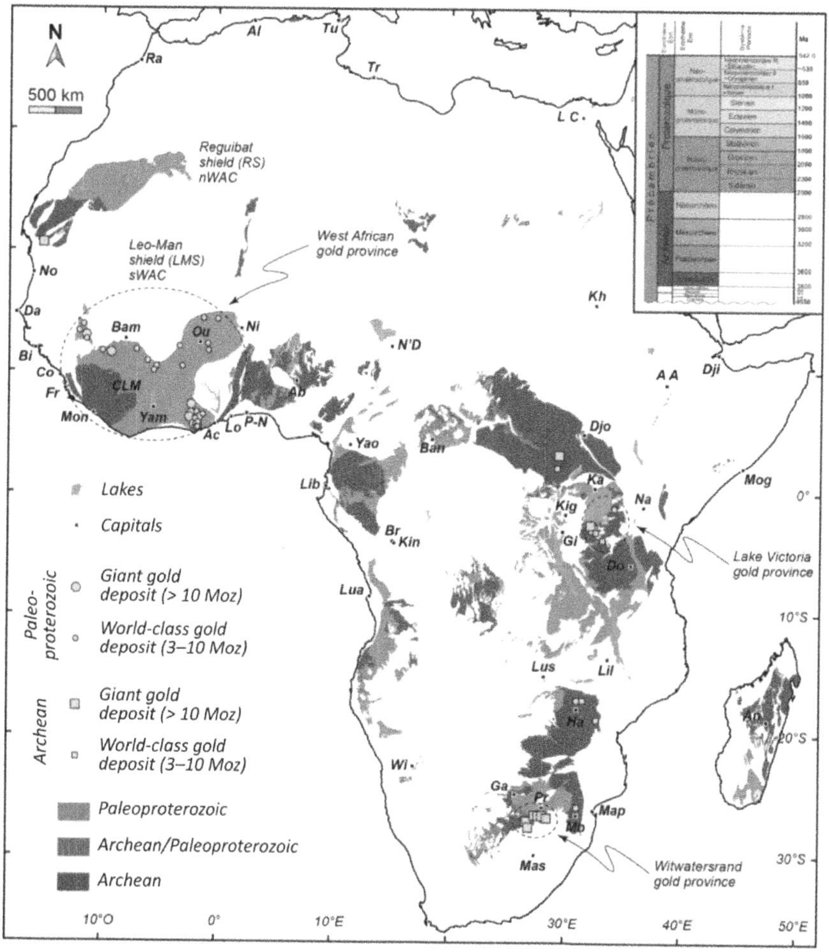

Figure 5.2. *Simplified geological map at 1:10 million scale of Archean and Paleoproterozoic gold provinces in Africa. Modified from Thiéblemont et al. (2016)*

A great example of period and associated metallogenic province is the African continent and its gold mineral systems formed in the Archean, between ca. 3.8 and 2.5 Ga (orogenic gold-type deposits in Tanzania, Victoria Lake gold province and paleoplacer gold-type deposits in South Africa, Witwatersrand gold province) and in Paleoproterozoic, between ca. 2.5 and 1.6 Ga (gold deposits of the Leo-Man Shield (LMS) in southern West Africa) (Figure 5.2). The main objective of this chapter is to illustrate the Precambrian geodynamic evolution of what is known as the West African Craton (WAC) or more precisely, the LMS, through the distribution of ore deposits, with a specific interest in gold mineralization, as tracers of geological and geodynamic processes. On the other hand, the purpose of this chapter is to show how the geodynamic evolution of West Africa was conducive to the formation and preservation of ore deposits and showings of various mineral resources. The African continent features large areas of Precambrian crust (<0.5 Ga), notably of Archean and Paleoproterozoic ages (Figure 5.2) that indicate a complex geodynamic evolution. This history is expressed by the presence of Archean *nuclei* stabilized around 2.5 Ga and surrounded by Proterozoic mobile orogenic belts emplaced around 2.0 Ga. These orogenic systems are known at worldwide scale and reflect the formation of the Columbia supercontinent between ca. 2.1 and 1.8 Ga. The terms Eburnean orogeny (2.0–1.8 Ga), Kibaran orogeny (1.4–1.0 Ga) and Pan-African orogeny (0.7–0.5 Ga) are used specifically to describe these orogenic systems on the African continent. These Precambrian orogenies are spatially and temporally associated with many metallogenic, especially auriferous, provinces (Figure 5.2).

5.2. Geology of LMS

The WAC of Precambrian age, as defined by Rocci (1965), crops out in three windows of unequal surface area, as they are covered by a large sedimentary cover of Neoproterozoic to Paleozoic age, forming the Taoudéni basin. The two main outcropping zones are, respectively, located to the north, in the Reguibat rise (Morocco, Algeria and Mauritania), and to the south, in the Leo or Man Rise (Sierra Leone, Guinea, Liberia, Ivory Coast, Mali, Burkina Faso and Ghana) (Figure 5.2). Between these two main zones, the Kédougou-Kéniéba inlier outcrops (Mali and Senegal). The following terms will be used in this chapter: Reguibat Shield (RS) for the northern part of the West African Craton (nWAC) and Leo-Man Shield (LMS) for its southern part (sWAC), rather than the obsolete term "rise", which may be confusing. These two continental shields are both composed of an Archean *nucleus* in their W-SW, on which Paleoproterozoic orogenic belts were accreted. This chapter focuses on the geodynamic and metallogenic evolution of LMS bounded by the Panafrican orogenic belts of Dahomeyide, to the east, and of Rockelide and Bassaride to the west (Villeneuve 2008).

The Archean part mainly outcrops to the W-SW of the LMS and covers an area of over 300,000 km^2 (Rollinson 2016). Historically, this domain had several names, such as "Sierra Leone Ivory Coast *nucleus*" (Clifford 1970); "the Liberian province" (Hurley et al. 1971); "the Kénéma-Man domain" (Thiéblemont et al. 2001); "the Man shield" (Barth et al. 2002); "the Man craton" (Skinner et al. 2004) or "the Leo-Man Shield" (De Waele et al. 2015). The name that will be used in this chapter is Archean Kénéma-Man domain. Accreted on this Archean *nucleus* and forming the eastern part of the craton, Paleoproterozoic greenstone belts and granitoids are organized along large structural lineaments and define the Baoulé-Mossi domain. These two domains, the Archean Kénéma-Man domain and the Paleoproterozoic Baoulé-Mossi domain, are separated by the Sassandra shear zone (Feybesse and Milesi 1994).

5.2.1. *The Archean Kénéma-Man domain: genesis of a continental crust*

The Archean rocks mainly outcrop in Liberia and Sierra Leone, but cover also the south-eastern part of Guinea and the western part of the Ivory Coast in the southern part of the WAC thus defining the Kénéma-Man domain (Figure 5.2). Thanks to the 1:200,000 surveys conducted by the French geological survey (Bureau de Recherches Géologiques et Minières (BRGM)) in Guinea between 1998 and 2003, orthogneiss and metagabbros have been discovered and dated at ca. 3.5 Ga, Paleoarchean age, using zircon U-Pb dating (Thiéblemont et al. 2001). The plutonic protoliths of these metamorphic rocks are therefore considered as evidence of the first phase of magmatic accretion and of the genesis of the West African Precambrian continental crust (Thiéblemont et al. 2001). Identical ages of ca. 3.5–3.4 Ga have also been obtained by zircon U-Pb dating and recorded by sodic orthogneiss, defined as forming the tonalite-trondhjemite-granodiorite (TTG) series in the RS in Mauritania (Potrel et al. 1996). The formation and evolution of this Archean crustal segment is characterized by several tectono-magmatic and metamorphic periods defining three cycles: Paleoarchean (ca. 3.5–3.4 Ga), Leonian (ca. 3.2–3.0 Ga) and Liberian (ca. 2.9–2.8 Ga) (Thiéblemont et al. 2004). However, in a more recent synthesis, Rollinson (2016) suggests that this division must be refined by considering the new geochronological records, undoubtedly involving poorly constrained sub-cycles. He insists on the poorly documented and underexplored character of this Archean domain with a strong metallogenic potential, especially for gold, iron, diamond, chromium and critical metals, such as niobium and tantalum.

The first Paleoarchean cycle is defined by the emplacement of magmatic TTG-type series (presently represented by more or less migmatitic orthogneiss). These TTGs are silica-rich with high Na_2O contents and exhibit characteristic trace element signatures with high contents of light rare earth element (LREE) and depleted contents of heavy rare earth element (HREE). Some mafic complexes (presently represented by granulitic metagrabbros) whose crystallization ages vary between ca. 3.6 and 3.5 Ga have also been emplaced during this cycle. The presence of a continental protocrust formed in this period is also confirmed by the presence of zircon xenocrysts dated between ca. 3.6 and 3.4 Ga in Neoarchean intrusions (Barth et al. 2002; Thiéblemont et al. 2004) and also Paleoproterozoic intrusions (Eglinger et al. 2017). The model ages calculated by various isotopic systems in South-Eastern Guinea (zircon Lu-Hf isotope analyses) (Eglinger et al. 2017) and the Western Ivory Coast (Sm-Nd garnet-whole rock geochronology) (Kouamelan et al. 1997) are in agreement with a Paleoarchean mantle extraction between ca. 3.7 and 3.3 Ga. At the same time, eclogite xenoliths sampled in kimberlites near the Sierra Leone-Guinea border were dated at ca. 3.4 Ga by whole rock Re-Os isotope analyses (Barth et al. 2002). These rocks, among which protolith was interpreted based on their subchondritic chemical and isotopic composition as basalt, are the oldest witnesses of a juvenile magmatism in Western Africa (Thiéblemont 2005). These basalts, evidence of an oceanic crust, had presumably been buried and metamorphosed at eclogite facies conditions in the Paleoarchean and then recycled in the subcontinental lithospheric mantle (Rollinson 2010). Various authors proposed that during subduction a part of these basalts melted and generated the magmatic TTG series dated at ca. 3.2–3.0 Ga (Leonian cycle), leaving an eclogitic residue in the mantle (Thiéblemont 2005; Rollinson 2016). While TTG formation is still under heated scientific debate (Bédard 2006; Moyen and Martin 2012), it has been proposed that these West African TTGs of Leonian age are derived from the partial melting of mafic rocks, probably a hydrated metabasalt, in the garnet stability field (1.2–2.5 GPa). For some authors, the emplacement of TTG series represents the early stages of plate tectonics with the emplacement of proto-subduction zones since ca. 3.2–3.0 Ga (Thiéblemont 2005; Shirey and Richardson 2011).

The second Leonian cycle (ca. 3.2–3.0 Ga) is characterized by the formation of greenstone belts (komatiites and tholeiitic basalts), the emplacement of magmatic rocks with intermediate to acid compositions and of volcano-sedimentary sequences (banded iron formation, graywacke, pelite and conglomerate). A subduction environment, as discussed above, is usually proposed to explain this period during which the metabasalts outcropping at the core of greenstone belts had presumably formed in extensive context in back-arc basins before deformation and metamorphism (Thiéblemont 2005). However, the temporal and geochemical relations that may exist between the eclogite xenoliths (melting residue) in the sub-continental lithospheric mantle and the emplacement of Leonian TTG series need to be consolidated through other geochemical and isotopic studies in order to test the alternative Archean petrogenetic and geodynamic models (Rollinson 2016). Among these alternative models, the mantle activity associated with upwelling mantle plumes was proposed to explain the formation of ultramafic to mafic lavas (komatiites and tholeiitic basalts) (Arndt et al. 2009).

The third and last period that characterizes the crustal evolution of the Archean Kénéma-Man domain is known as the Liberian cycle (ca. 2.9–2.8 Ga). This episode is marked by a high-temperature metamorphism and the emplacement of large batholits composed of potassic felsic granitoids (granodiorite, granite) and enriched in incompatible elements, containing some microgranular, gabbroic and mafic dioritic enclaves (Thiéblemont 2005). The existence of a high-grade metamorphic event in the granulite facies is evidenced by the presence of charnockites (granite with orthopyroxene), dated at ca. 2.8 Ga by Pb-Pb and U-Pb dating on monazite and zircon, and the formation of associated migmatites. The geodynamic environment of this high temperature event at crustal scale is still poorly understood (Rollinson 2016). It is important to note that these Liberian plutons often contain Paleoarchean TTG gneiss enclaves and/or Paleo- to Mesoarchean zircon xenocrysts. This Liberian episode is therefore considered as a phase of crustal reworking and differentiation (Kouamelan et al. 1997) rather than a phase of magmatic accretion, crustal growth and recycling, as defined for the first two cycles, Paleoarchean and Leonian (Thiéblemont 2005).

5.2.2. *Columbia supercontinent: stabilization of LMS*

Understanding a mineral system, such as the one associated with gold mineralization, requires a multiscale approach, from the lithospheric scale to the mineral scale (McCuaig and Hronsky 2014), but also to constrain the temporal relationship between geodynamic processes and mineralizations. It is therefore essential to understand the tectono-metamorphic evolution of the metallogenic LMS province during the Paleoproterozoic, following the formation of the Archean *nucleus* detailed in the previous section. Although there are ambiguities about the Precambrian Earth's tectonic mode, there is broad consensus that the early Paleoproterozoic era marks the period in which plate tectonics became the dominant tectonic regime on Earth (see Brown et al. (2020) and the references therein). This is reflected by the supercontinent cycle from the breakup of a few supercratons to the collision of these continental segments/terranes leading to formation of the Columbia supercontinent, including Nuna (Rogers and Santosh 2002). Indeed, the paleomagnetic data obtained on various cratons show large horizontal displacements since ca. 2.2 Ga, which is considered as the beginning of the assembly of Columbia (see Meert and Santosh (2017) and the references therein). Moreover, these horizontal displacements are also supported by the abundance of passive margin at ca. 2.2–1.8 Ga (Brown et al. 2020). Finally, subduction-related magmatism and bimodal metamorphism recording prove the development of orogenic activity at the craton margin throughout the globe, between ca. 2.1 and 1.8 Ga (Zhao et al. 2002). These orogenic systems represent markers of the geological evolution of cratons that will be presented in detail in the following sections for the LMS case. From a geodynamic point of view, it has been admitted that the main driving force of plate tectonics is the traction of the slab in the subduction zones, known as the slab-pull force (Forsyth and Uyeda 1975). From a geochemical point of view, subduction zones represent privileged sites for the generation of continental crust (Davidson and Arculus 2006). Paradoxically, these sites where the oceanic lithosphere (the subducting plate) sinks back into the mantle under an overriding continental or oceanic plate are also places of crustal recycling (associated sediments and volatile elements) in the mantle (Rüpke et al. 2004). From a metallogenic point of view, this geodynamic context represents specific environments (e.g. back-arc basins) favorable to the fertilization (e.g. enrichment in large-ion lithophile elements (LILE)) of the lithosphere with various metals (Bierlein et al. 2009). Thus, convergent boundaries represent a setting of tectonic boundaries where continental crust is formed and oceanic lithosphere is recycled yielding a wide range of geological processes such as deformation, metamorphism, magmatism and mineralization. LMS is rightly part of this "dance of the continents" through the Eburnean orogenic cycle, involving the amalgamation of Archean *nuclei*, such as the Kénéma-Man domain, and of orogenic mobile belts of Paleoproterozoic age defining the Baoulé-Mossi domain during the Eburnean orogenic cycle (Ledru et al. 1994) (Figure 5.3). The South-American equivalent involves the Archean Imataca and Amapa domains and the Paleoproterozoic Maroni-Itacaiunas orogenic belts formed during the Transamazonian orogenic cycle. Both domains partly compose the Amazonian craton (see Grenholm (2019) and the references therein) (Figure 5.3).

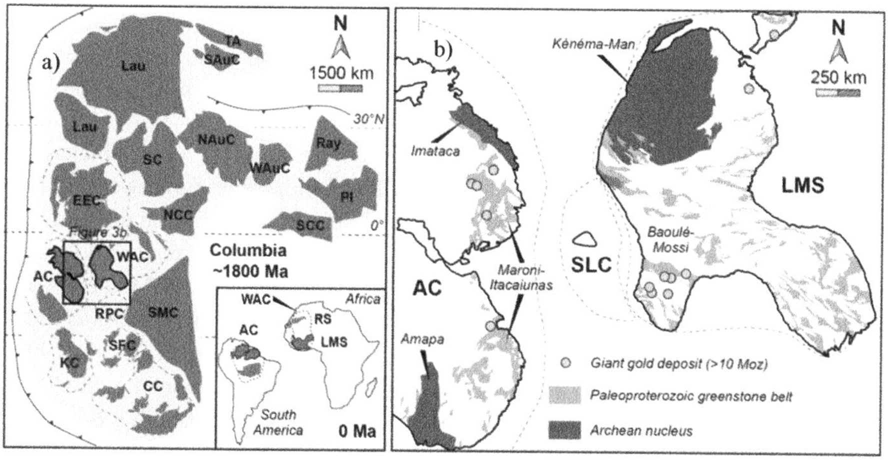

Figure 5.3. *Reconstruction of the Columbia supercontinent at ca. 1.8 Ga. Modified from Grenholm (2019)*

COMMENT ON FIGURE 5.3.– *Abbreviations: PI – Peninsular India; AC – Amazonian Craton; WAC – West African Craton; CC – Congo Craton; EEC – East European Craton; LMS – Leo-Man Shield; NAuC – North Australian Craton; NCC – North China Craton; KC – Kalahari Craton; RS – Reguibat Shield; RPC – Rio de la Plata Craton; SC – Siberian Craton; SAuC – South Australian Craton; SCC – South China Craton; SFC – Sao Francisco Craton; SLC – Sao Luis Craton; Lau – Laurentia; SMC – Saharan Metacraton; Ray – Rayner; TE –Terre Adélie.*

5.2.3. *The Paleoproterozoic Baoulé-Mossi domain*

5.2.3.1. *Lithologies and geodynamic environments*

Various geodynamic models have been published in the literature to explain the formation, evolution and deformation of the greenstone belts and the associated plutonic suites and also sedimentary sequences, all of which compose the Baoulé-Mossi domain (Abouchami et al. 1990; Feybesse et al. 2006; Lompo 2009; Vidal et al. 2009; Baratoux et al. 2011; Ganne et al. 2012; de Kock et al. 2012; Cooper and Miller 2014; Ganne et al. 2014; Eglinger et al. 2017; Gapais 2018; Wane et al. 2018; McFarlane et al. 2019a; Masurel et al. 2021). Supracrustal rocks that mainly compose the greenstone belts are predominantly composed of mafic volcanic rocks and volcaniclastic (pyroclastic flows and epiclastic deposits) and detritic sediments, all these lithologies having been deformed and metamorphosed. These metavolcanic and metasedimentary rocks were distinguished into two lithostratigraphic units with metavolcano-sediments forming the Birimian Supergroup (from the Birim river in Ghana) at the base and clastic metasediments belonging to the Tarkwaian Group (from the name of a city in Ghana) at the top.

The base of the Birimian stratigraphic unit ("Lower Birimian") is characterized by the presence of rare ultramafic (UM) rocks: lherzolite, harzburgite, wehrlite and dunite (Attoh et al. 2006; Dampare et al. 2020; Labou et al. 2020), and mainly of mafic (M) rocks: dominant basalts with several gabbros (Baratoux et al. 2011). The geodynamic environments associated with the emplacement of UM-M rocks with predominantly tholeiitic signature, more or less enriched in Th, are either a context of plume-related oceanic plateau (Abouchami et al. 1990; Boher et al. 1992; Augustin and Gaboury 2017), or a context of mid-ocean ridge (Pouclet et al. 2006; Labou et al. 2020). The age of this juvenile magmatism (material with a depleted-mantle-like isotopic signature) remains poorly constrained because few studies were conducted on these UM-M complexes. Several Sm-Nd isochron ages (Abouchami et al. 1990) or recent Nd model ages (Dampare et al. 2020; Sakyi et al. 2020) indicate an extraction between ca. 2.5 and 2.2 Ga for these juvenile magmas (characterized by positive εNd, therefore with supra-chondritic signatures). The transition from mafic to intermediate magmatism is dated at ca. 2.19–2.16 Ga. Bimodal volcanic series of tholeiitic to calc-alkaline composition, with an interlayer of volcano-derived sediments (mainly pyroclastites), succeed these UM-M complexes. Their emplacement is expressed by a basaltic to andesitic volcanism, more locally rhyolitic, in volcanic arc systems (Baratoux et al. 2011) and/or in back-arc basins (Vidal and Alric 1994; Grenholm et al. 2019), witnessing subduction settings (Eglinger et al. 2017). For example, based on geochemical (major and trace elements) and isotopic data (Sr and Nd), Labou et al. (2020) identify three distinct magmatic series for the formation of these UM-M complexes sampled along the Mako belt (Senegal): tholeiitic series, called type 1 ($^{87}Sr/^{86}Sr_{2,1Ga}$ = 0.701882 ± 10 and $\varepsilon Nd_{2,1Ga}$ = + 3.6 ± 0.01) result from a more juvenile source, while tholeiitic series, type 2 ($^{87}Sr/^{86}Sr_{2,1Ga}$ = 0.703214 ± 9 and $\varepsilon Nd_{2,1Ga}$ = + 3.4 ± 0.01) and calc-alkaline series ($^{87}Sr/^{86}Sr_{2,1Ga}$ = 0.702039 ± 9 and $\varepsilon Nd_{2,1Ga}$ = + 2.8 ± 0.02) derive from sources respectively more enriched and whose the positive $\varepsilon Nd_{2,1Ga}$ values underline the absence of contamination of the mantle source by a crustal component. The authors then propose a geodynamic evolution since the opening of an oceanic domain (tholeitiic series, type 1 whose geochemical composition is close to that of Mid Ocean Ridge Basalt (MORB)); intra-oceanic subduction involving the formation of an immature island arc (tholeiitic series, type 2) and maturation of the arc (calc-alkaline series related to the sinking of the subducting plate and to the thickening of the crustal part of the arc). This type of arc magmatism in the context of intra-oceanic subduction was also described in other greenstone belts through the WAC, such as the Ashanti belt (Dampare et al. 2020) and the Lawra belt (Sakyi et al. 2020) in Ghana. The top of the stratigraphic unit of the Birimian (Upper Birimian) is characterized by flysch-type sedimentary sequences that include sandstones, siltstones and argillites with several carbonates and minor volcaniclastic layers, chemical sedimentary and volcanic rocks. These thick deep-water rock sequences were deposited between ca. 2.14 and 2.10 Ga (e.g. the Kumasi basin in Ghana). About the depositional setting, some authors favor syn-orogenic foreland basin sedimentation, whereas others favor back-arc and fore-arc basin sedimentation controlled by extensional detachments or transtensional deformation during lateral escape of the crust against Archean indenters (see Masurel et al. (2021) and the references therein).

Closely associated with these greenstone belts, TTG plutonic suites occurred as early as ca. 2.25 Ga, but became widespread across the southern part of the WAC (sWAC) after ca. 2.20 Ga, with peak in magma flux between ca. 2.16 and 2.15 Ga. While there is a consensus on the source of these TTG magmas commonly interpreted as derived from partial melting and/or fractional crystallization of hydrous basalt, their geodynamic setting is still extensively debated (Laurent et al. 2020). Indeed, various hypotheses were advanced to explain the formation of these sodic plutonic rocks, characterized by low K_2O/Na_2O (<0.5), particularly due to their geochemical diversity in trace elements. Some TTG are depleted in heavy rare earth elements (HREE), involving the fractionation of residual garnet in the petrogenesis of these magmas (minerals having a first-order pressure-sensitive behavior in mafic system), and therefore their genesis in the garnet stability field, at high pressure (greater than ca. 1.2 GPa, or greater than ca. 40 km in depth). However, according to recent studies, the same geochemical signatures may have been generated by fluid-fluxed crustal melting at relatively shallow depths (25–35 km) (Pourteau et al. 2020) or through fractionation of melts derived from metasomatically enriched lithospheric mantle (Smithies et al. 2019). Finally, other authors proposed that the geochemical diversity of TTG does not necessarily result from processes related to the genesis of deep magmas, but rather from their crystallization dynamics in the mid-/upper crust. Hence, the high Sr/Y and Na/K ratios of TTG may equally be explained through amphibole-dominated fractionation (Liou and Guo 2019) or by the differentiation of a single, tonalitic parent liquid formed in the lower crust by basalt melting and/or crystallization at <40 km depth (Laurent et al. 2020). These various interpretations of the geochemical signature of TTG lead to equally varied geodynamic interpretations, with actualistic hypotheses proposing that these magmas could have formed at convergent plate margins by melting of over-thickened arc roots, subducted oceanic crust or fractionation of hydrous arc basalts (see Laurent et al. (2020) and the references therein). By contrast, non-uniformitarian scenarios proposed that TTG suites were emplaced during the protracted differentiation of thick basaltic plateaus above long-lived mantle upwellings (Bédard 2018).

This magmatic event is followed by a period of crustal deformation marked by the arc(s)-continent collision between these Paleoproterozoic magmatic arcs, involving the closing and inversion of associated sedimentary basins, and the Archean Kénéma-Man *nuclei* (Baratoux et al. 2011; Eglinger et al. 2017). These tectonic events are dated between ca. 2.11 and 2.05 Ga and are characterized by the formation of intermountain and flexural basins whose filling is dominated by molasse-type synorogenic sedimentary series mainly dated between ca. 2.11 and 2.09 Ga and known as stratigraphic Tarkwa units. The base of these units is underlined by polygenic conglomerates, composed of quartz pebbles and clasts of Birimian metabasite alternating with sandstone. The top of this unit is rather composed of mudstone and pelite. All these sedimentary units were deposited and deformed during incremental NW-SE-directed bulk crustal shortening towards the culmination of the Eburnean orogeny (see Masurel et al. (2021) and the references therein). This period is followed by crustal reworking of the thickened Birimian crust coeval with transcurrent tectonics (e.g. the shear zone of Greenville-Ferkessédougou in Ivory Coast) (Chardon et al. 2020; Masurel et al. 2021). The transitions between syn-, late- to post-collisional stages are still vague and poorly determined in terms of absolute chronology due to a diachronism between east and west at the WAC scale. The best record of the transition from the collisional stage to late- to post-collisional stages remain of magmatic order, marked by the emplacement of granodiorite, monzogranite and granite intrusions whose geochemical signatures can evolve from potassic calc-alkaline to shoshonitic (Eglinger et al. 2017). The presence of peraluminous magmatism is also well documented and represented by leucogranite, mainly by muscovite-bearing granite (Parra-Avila et al. 2017). This magmatic activity is dated between ca. 2.09 and 2.05 Ga and mainly expressed to the west of the WAC, along the margin of the Archean Kénéma-Man domain (Mériaud et al. 2020). Several geodynamic drivers have been suggested to explain this elevated geothermal gradient associated with the production of these magmas and recorded by granulite facies conditions (Pitra et al. 2010), such as: syn-collisional lateral extrusion during transtensional deformation, thus explaining the mafic volcanism with local injection of juvenile magmas (Grenholm et al. 2019); syn- to late-collisional crustal thickening and then differential exhumation of the partially molten lower crust accommodated by low-angle detachments (McFarlane et al. 2019a); syn-collisional lateral flow is also suggested to have compensated part of the crustal thickening of the soft Birimian crust (Chardon et al. 2020); late- to post-collisional lithospheric delamination involving the partial melting of both the crust and enriched lithospheric mantle (Eglinger et al. 2017). The various magmatic and sedimentary lithologies presented in this section were formed, deformed and metamorphosed during the Eburnean orogenic cycle, as defined by Bonhomme (1962). This chapter does not employ the term Eoeburnean cycle, as it does not correspond to a supercontinent cycle resulting in periodic aggregation of most of the extant continental lithosphere into a single entity as defined by Wilson (1966). This word was rather used to describe a major and specific period of magmatism dated between ca. 2.19 and 2.14 Ga (Perrouty et al. 2012) or to characterize a local deformation phase (de Kock et al. 2011). Moreover, the terminology used in the literature to define different deformation events having affected the Birimian and Tarkwaian lithologies is somewhat complex and even confusing. Indeed, the deformation phase approach used by many geologists could be misleading and lead to several deformation phases (Chardon et al. 2022). Most of the structural interpretations at the sWAC-scale have been published involving polyphase deformation, for which the term "deformation phase" is used to characterize the discrete and well-defined period during which the orientation and relative magnitudes of principal stresses and instantaneous stretching axes should be constant (Ledru et al. 1991; Feybesse et al. 2006; de Kock et al. 2012; Perrouty et al. 2012; Masurel et al. 2021). However, this approach is certainly debatable, as the structures identified, most often at the deposit-scale, correspond to a local deformation regime extrapolated at the scale of the metallogenic province. It is worth considering that, for the same deformation event known as progressive, the spatial kinematic diversity of localities and the composite structural patterns and overprinting relation can form by progressive single-phase deformation, coherent at the regional-scale (Fossen et al. 2019). Moreover, ductile versus brittle deformation is clearly scale-sensitive and the coexistence of both structures may be possible during progressive deformation and tend to be ignored in the structural phase approach (Chardon et al. 2020; Perret et al. 2020). In the absence of sufficiently precise geochronological constraints, it is difficult to attest or not the diachronism between the composite structural patterns described in the field. Finally, there is diversity in the nomenclature of deformation phases described across the sWAC. For example, folded and metamorphosed early Birimian volcano-sediments are overlain with marked unconformity by late Tarkwaian sediments. This first crustal shortening event affecting the Birimian units was differently defined depending on the locality where it has been observed: "Burkinian cycle" in Ivory Coast, "phase D1" (Milési et al. 1991) or "Eoeburnean tectonic cycle" in Ghana (de Kock et al. 2011, 2012) or "Tangean event" in Burkina Faso (Tshibubudze et al. 2009). To explain this heterogeneity in terms of deformation phases and their names, some authors published syntheses and comparisons at the craton-scale (Perrouty et al. 2012; Masurel et al. 2021).

5.2.3.2. *Tectonic models*

The Archean-Paleoproterozoic boundary is recognized as being a period of major geodynamic, geothermal, petrological and atmospheric transition marking the transition from an early tectonic mode (stagnant or sluggish lid) to plate tectonics (Gapais et al. 2009). Indeed, plate tectonics can be demonstrated on Earth since the early Paleoproterozoic (since ca. 2.2 Ga), but before the Proterozoic Earth's tectonic style remains contentious (see Brown et al. (2020) and the references therein). These two tectonic modes are opposed especially because the deformation mode of continental lithospheres is closely related to the Moho temperature and the associated rheological behavior of the lithospheric mantle (Chardon et al. 2009; Gapais et al. 2009). Modern plate tectonics involve stiff lithospheres and the crustal thickening is localized at plate boundaries following classical thrusting models. Archaic tectonics is possible due to the hotter and weaker rheological state of Archean lithospheres. In this case, a weak lithospheric mantle will favor dominant distributed shortening and vertical motions.

Concerning the LMS in Western Africa, the models supporting a distributed crustal thickening rely mainly on a litho-structural approach (Vidal et al. 2009; Lompo 2010; Chardon et al. 2020), while those that support a localized crustal thickening advance tectono-metamorphic and geochemical arguments (Feybesse et al. 2006; Baratoux et al. 2011; Ganne et al. 2012; de Kock et al. 2012; Eglinger et al. 2017; McFarlane et al. 2019a). The question then is: are the Paleoproterozoic rocks from the LMS the youngest witnesses of archaic tectonics or the oldest witnesses of modern tectonics, or are they marking the transition between these two extreme tectonic modes? In the following paragraphs, the different tectonic (structural patterns), metamorphic (geothermobarometric gradients) and magmatic (geochemical signatures) markers will be briefly described.

In the case of the Eburnean orogenic system in West Africa, various tectonic models have been published. While the presence of an anastomosed shear zone network related to late strike-slip deformation is no longer discussed, the same cannot be said for the previous tectonic features (see Chardon et al. (2020) and the references therein). Indeed, the mainly steep fabrics in greenstone belts (foliation/lineation and axial plane/hinge line) remain ambiguous and are interpreted in one of the following ways:

– As originally shallow fold and thrust fabrics that were steepened and/or transposed by late strike-slip deformation (Feybesse and Milesi 1994).

– As "dome and basin" structures in which the granitoids emplaced by diapirism and the supracrustal rocks were buried by sagduction (Vidal et al. 2009). The juxtaposition of domes and basins is accommodated by the existence vertical shear zones (Lompo 2010).

– As transpressional anastomosed shear zones with burial of supracrustal rocks within soft crust along vertical deformation zones (pop-down tectonics) (Chardon et al. 2020).

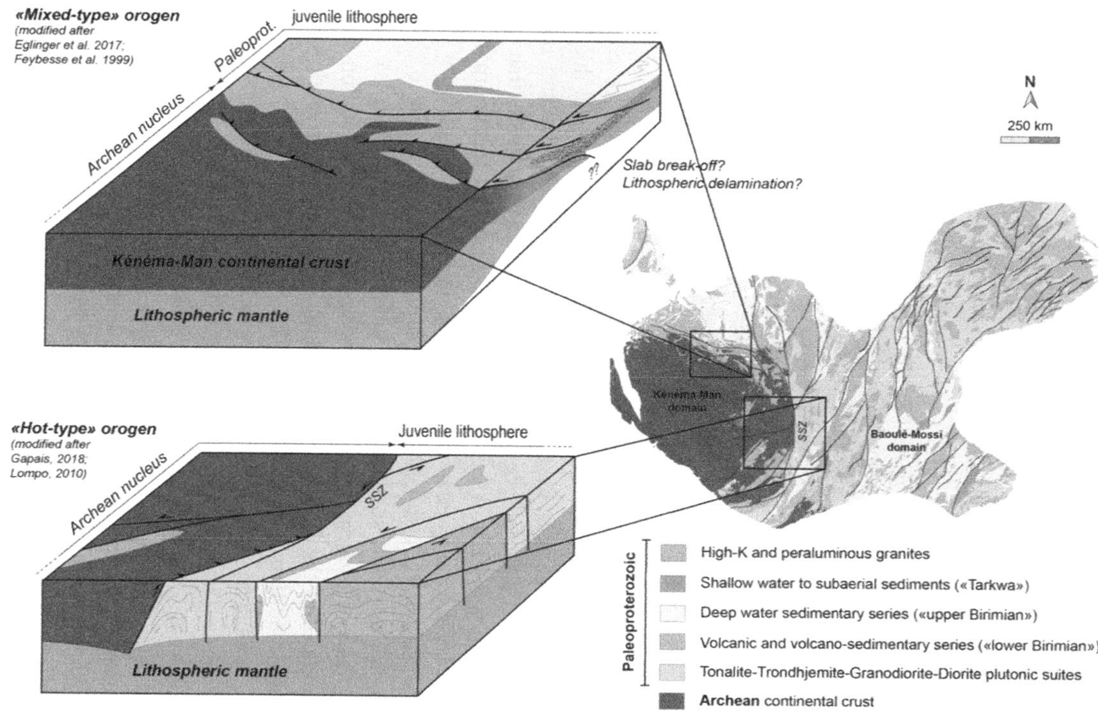

Figure 5.4. *Lihospheric-scale cross-sections along the northern and eastern margin of the Kénéma-Man nucleus illustraiting two modes of collision tectonics between this Archean domain and the Baoulé-Mossi juvenile lithosphere: mixed-type orogen for the N-S cross-section versus hot-type orogen for the E-W cross-section. Abbreviation: SSZ – Sassandra shear zone*

All of these interpretations suggest various geodynamic models to explain the crustal formation and evolution of the Paleoproterozoic greenstone belts and related granitoids involving either a "cold-type" orogen (Ganne et al. 2012), "hot-type" orogen (Lompo 2010; Gapais 2018; Chardon et al. 2020) or "mixed-type" orogen (Feybesse et al. 1999; Eglinger et al. 2017). In the case of a hot Eburnean orogen, the transpressional anastomosed network accommodated long-lasting distributed shortening and lateral flow of the weak Birimian lithosphere in a compressive setting between the Paleoproterozoic Baoulé-Mossi domain and the Archean Kénéma-Man and Congo *nuclei* (Chardon et al. 2020). In this tectonic mode, crustal thickening is moderate, distributed and accommodated along mainly transpressive subvertical shear zones (Cagnard et al. 2006). Lateral flow is interpreted to have compensated part of the crustal thickening (Chardon et al. 2020). Moreover, the predominance of regional HT-LP metamorphism associated with partial melting and juvenile magmatism are parameters in favor of hot orogens (Chardon et al. 2009). All these characteristics seem to indicate that the Birimian lithosphere was hotter, more ductile, and therefore certainly weaker than the current stable lithospheres (Rey et al. 2003).

These characteristics contrast with those involving collision between cold and stiff lithospheres where compression induces localized strains and crustal thickening is accommodated along large-scale thrusts (Gapais et al. 2009). Modern orogens may be composed of belts with contrasting types of metamorphism that record apparent thermal gradients (Brown et al. 2020). Classic paired metamorphic belts are composed of HT-LP belt juxtaposed along tectonic structure against a HP-LT belt (Brown et al. 2020). Relics of HP-LT metamorphism, recording low to very low thermal gradients due to subduction at convergent plate margins during pre-collisional phase of orogenic cycle, has been locally described within specific greenstone lithologies of the Baoulé-Mossi domain. For example, in Burkina-Faso, blueschist-facies conditions were recorded ($P > 1$ GPa, $T < 450°C$) by metasediments, associated with the formation of Columbia by subduction-driven collisional orogenesis (Ganne et al. 2012). In modern-type tectonics, exhumation of lower crust generally involves crustal-scale normal shearing, leading to the juxtaposition of high temperature rocks against low-grade supracrustal units. Some evidence in Ghana of post-thickening extensional detachments have been proposed to explain the exhumation of the Birimian partially molten lower crust (Block et al. 2015, 2016; McFarlane et al. 2019a).

A progressive transition in space and time from a distributed deformation, as expressed in hot-type orogens, to a more localized deformation, as observed for cold-type orogens, with a variety of possible orogenic styles between these two extremes has been suggested by Chardon et al. (2009). For example, an intermediate orogen building mode, called mixed-type orogen, has been proposed for Proterozoic collisional belts involving thrusting of a hot and juvenile crust over adjacent continental stiff lithosphere. Thrust tectonics and high-grade metamorphic conditions (up to 6 GPa and 800°C) have been recorded by metasediments along the southern margin of the Siguiri basin in Guinea (Feybesse et al. 1999). These metasediments have been interpreted as a part of an intensively deformed granulite-facies nappe that was thrust onto the Archean Kénéma-Man craton during the Eburnean orogeny (Eglinger et al. 2017; Figure 5.4). A possibility is that this subduction zone at the northern margin of the Kénéma-Man domain has absorbed the long-lived convergence between the different Paleoproterozoic domains and Archean *nuclei*. Thus, the Eburnean orogen shares deformation modes intermediate between hot weak orogens and cold stiff ones.

5.3. Spatiotemporal distribution of gold in LMS

5.3.1. *Introduction*

Until the 1990s, Western Africa has been underexplored, geological studies being very limited. Indeed, with the exception of several world-class gold deposits, essentially located in Ghana (for example, the Obuasi deposit that has been quasi-continuously exploited since 1897), the numerous gold anomalies presented no economic interest until the increase in gold price starting in 2005 (Goldfarb et al. 2017).

WAC is part of the regions in which the mining activity was revived in the past two decades. The individual study of several deposits located in various West African countries gave rise to many scientific publications dealing with gold metallogeny, from deposit-scale studies to district-scale syntheses, and even at the metallogenic province scale (Markwitz et al. 2016; Goldfarb et al. 2017; Thébaud et al. 2020; Masurel et al. 2021).

The sWAC hosts the world's premier Paleoproterozoic gold province with an overall endowment >10,000 metric tons, estimated in 2017 (Goldfarb et al. 2017).

Gold resources are mainly concentrated in the Paleoproterozoic greenstone belts of the Baoulé-Mossi domain.

Most gold deposits are classified as orogenic type, though there are also other types of deposits, such as paleoplacer, porphyry, skarn or intrusion-related (Markwitz et al. 2016; Goldfarb et al. 2017; Thébaud et al. 2020; Masurel et al. 2021).

The Paleoproterozoic greenstone belts containing the most significant gold deposits are mainly located in South-Western Ghana, North-Eastern and Western Burkina Faso, Southern Mali and North-Eastern Guinea, and also along the Senegal-Mali border (Figure 5.5).

5.3.2. *Archean gold*

The Archean terrains of the LMS and RS received little attention in terms of exploration during the past decades, therefore they are a relatively unknown part of the WAC with an underestimated metallogenic potential.

The metamorphic grade described by rare studies (Heron et al. 2015; Rollinson 2016) corresponds to amphibolite facies, which may first be considered as too high PT conditions to be conducive to the presence of orogenic gold (Foster and Piper 1993), given the greenschist metamorphic facies considered as optimal for the formation of this type of gold deposit (Groves 1993; Kerrich et al. 2000; Goldfarb et al. 2005; Goldfarb and Groves 2015).

However, the discovery and exploitation of the giant gold deposit of Tasiast in the RS, indirectly dated at 2,840 ± 36 Ma (U-Pb dating on hydrothermal zircon) (Heron et al. 2015), indicate that the Archean terrains of the WAC must also be considered for the exploration of economically important occurrences (Goldfarb et al. 2017). Indeed, new world-class Archean deposits with amphibolite facies gold mineralization have been discovered in the past decade, as the Éléonore deposit, in James Bay, Superior Province, Canada (Fontaine et al. 2018).

Gunn et al. (2018) have also described the gold potential of Archean terrains in Liberia where lode-gold mineralizations have been described in association with a wide variety of greenstone belt lithologies with metamorphic grade ranging from lower greenschist to upper amphibolite facies. The best known and economically most important is the New Liberty deposit, which is hosted by Archean mafic and ultramafic rocks, representing a relict greenstone belt. This deposit is structurally controlled by a first-order steeply dipping silicified shear zone, parallel to the orientation of greenstone belts. Many other major gold occurrences occur along the same structural corridor extending north-east from New Liberty. These Liberian gold mineralizations are similar to those described by Barrie and Touret (1999) in Sierra Leone, particularly at the Yirisen deposit in the Sula-Kangari greenstone belt (Figure 5.5).

5.3.3. *Paleoproterozoic gold*

The current production of Paleoproterozoic gold of WAC (~350 t in 2019) is comparable to that of other world-class gold provinces such as Western Australia, Canada, Russia and South Africa. Ghana remains the most productive region (~130 t in 2019), but Mali, Burkina Faso and, to a lesser extent, the Ivory Coast and Guinea have also become important producers (Goldfarb et al. 2017).

Historically, there are two types of gold mineralization in Western Africa: Birimian orogenic gold deposits, such as Obuasi, and the paleoplacer deposits expressed as gold-rich deformed conglomerates (Junner and Wild 1935), such as Tarkwa or Damang.

However, more in-depth research, particularly conducted by the BRGM teams in the 1980–1990 period, highlighted a wider diversity in the Paleoproterozoic gold metallogeny of the sWAC. Indeed, based on a phase structural approach, they identified a three-phase evolution to decipher the gold mineralization history across the sWAC: (1) pre-orogenic (pre-D_1) deposits related to early extension zones and represented by stratiform Au-bearing tourmalinite of Loulo in Mali; (2) syn-orogenic (post-D_1 to syn-D_2/D_3) deposits with the formation of disseminated gold deposits such as Yaouré in the Ivory Coast or the paleoplacer type with the gold-bearing conglomerates of Tarkwa; and finally (3) late-orogenic (post-D_2/D_3) deposits, which is the most economically interesting period with the emplacement of quartz-(carbonate) veins with free gold, representing the most high-grade ore shoots, as a classical example, Obuasi in Ghana (Milési et al. 1992).

COMMENT ON FIGURE 5.5.– *This map presents the typology of deposits, their tonnage (data published by mining companies) and their age, if known (either directly: absolute dating of the mineralization, or indirectly: relative chronology). Abbreviations: BEN – Bénin; BUR – Burkina Faso; CIV – Ivory Coast; GHA – Ghana; GUI – Guinea; LBR – Liberia; MLI – Mali; MTN – Mauritania; NIG – Niger; SEN – Senegal; SLE – Sierra Leone; TGO – Togo.*

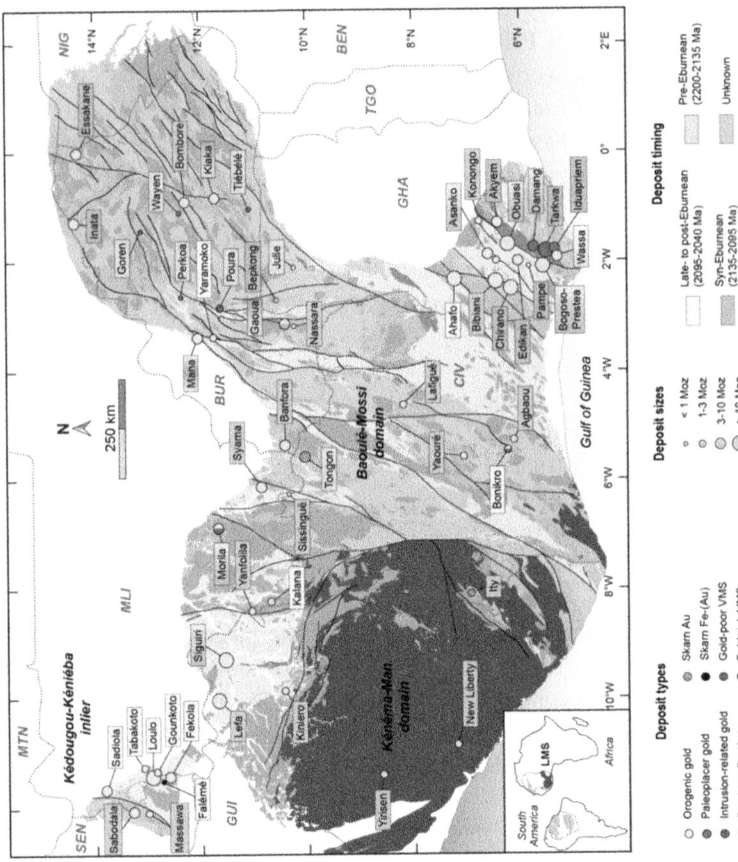

Figure 5.5. *Gold metallogenic map of the Paleoproterozoic Baoulé-Mossi domain. Modified from Markwitz et al. (2016) and Gunn et al. (2018)*

More recently, from 2006, in the framework of the West African Exploration Initiative (WAXI), the gold mineral system has been reevaluated in order to enhance the exploration potential of the WAC. The WAXI is a unique international and collaborative research and training program, in close relation with mining companies and local geological surveys, focusing on the link between tectonics and ore minerals. This multidisciplinary approach gathering geophysical, structural, metamorphic, geochemical and isotopic data made it possible to describe the metallogenic and geodynamic evolution in three stages during the Eburnean orogenic cycle (Masurel et al. 2021). These three stages of gold endowment are thus defined, by considering the syn-Eburnean period as the paroxysm of this cycle, as the main period of continental crust thickening.

1) The pre-Eburnean (also known as Eoeburnean) gold deposits are formed between ca. 2.20 and 2.13 Ga, during a period of convergence and magmatic accretion between the juvenile Baoulé-Mossi domain and an Archean *nucleus* no longer outcropping, which was certainly located at the current position of Togo-Bénin (Petersson et al. 2016). The few deposits formed during this period are mostly emplaced in magmatic arc settings and within the related fore-, intra- and back-arc basins. They are preserved only in the eastern part of the Baoulé-Mossi domain, including Mana (I) and Kiaka (I) in Burkina Faso and Julie and Wassa (I) in Ghana (Figure 5.6(a)). This period represents approximately 10% of the gold endowment of the sWAC (Thébaud et al. 2020).

2) The early syn-Eburnean deposits were formed between ca. 2.13 and 2.10–2.09 Ga, during a period of oblique shortening, with inversion of the newly formed crustal terranes, tectonic accretion and arc(s)-continent collision (McFarlane et al. 2019a, 2019b), including, in particular the collision with the Archean Kénéma-Man domain (Eglinger et al. 2017). The vast majority of these gold deposits are orogenic types, with free gold in quartz veins and refractory gold in arsenopyrite and/or pyrite and alteration halos of carbonate, sericite, chlorite and albite. These deposits are represented by Obuasi (I), Chirano and Edikan in Ghana, and Essakane, Inata, Mana (II) and Kiaka (II) in Burkina Faso. The other predominance in terms of gold mineralization types that characterize this period of crustal thickening is represented by paleoplacers well expressed in Ghana with Tarkwa, Damang and Iduapriem deposits. These deposits are associated with gold-bearing conglomerates deposited in intra-orogenic sedimentary basins (Figure 5.6(b)). The formation of these conglomerate-hosted gold deposits then involves the remobilization of an early gold source (the pre-Eburnean and the early syn-Eburnean mineralizations), with or without a new syn-Eburnean gold input. This period represents approximately 50–70% of the gold budget of the sWAC (Thébaud et al. 2020).

3) The late- to post-Eburnean gold deposits were formed between ca. 2.10–2.09 and 2.04 Ga, after the complete collision between the Paleoproterozoic and Archean domains. This period was marked by transcurrent tectonics and crustal reworking as

attested by the emplacement of peraluminous and K-rich granitoids formed through melting of older crustal lithologies (Eglinger et al. 2017; Masurel et al. 2021). The gold deposits formed during this period have a wider variety of styles (gold associated with intrusions, orogenic type and/or skarn type), geological settings and metal assemblages. They are described nearly all over across the sWAC, even though they are predominantly expressed in the western part with the deposits of Loulo, Fekola and Sadiola in Mali, Siguiri in Guinea, while they are less abundant or less well preserved in the south-west of Ghana, for example, the Ahafo, Obuasi (II) and Damang (II) deposits (Figure 5.6(c)). This gold period represents about 20–40% of the gold budget (Thébaud et al. 2020).

This synthesis highlights the entire spatiotemporal complexity of the gold system in Western Africa. The world-class deposit of Obuasi is considered as the most important in the gold province of West Africa with resources estimated at 130.74 Mt at 7.39 g/t for 31.04 Moz of gold for a past production of approximately 32 Moz (AngloGold Ashanti 2019). This deposit is a good example of the complexity of mineralization in terms of (re)mobilization-(re)precipitation of gold during deformation (Fougerouse et al. 2016a). It is hosted in a flysch-type metasedimentary series, known as the Kumasi group, representing the top of the Birimian stratigraphic unit ("Upper Birimian"). This series lies unconformably on the Sefwi group ("Lower Birimian"), represented by metamorphosed volcano-sedimentary series. The metasediments of the Kumasi group were the witnesses of a back-arc basin formed between two magmatic arc systems, possibly insular, then deformed by the collision of these arcs, nowadays represented by the Ashanti and Sefwi greenstone belts, respectively, from east to west (McFarlane et al. 2019b).

Two styles of economic mineralization are clearly described at Obuasi: a mineralization characterized by the presence of arsenopyrite grains disseminated in metasediments and a mineralization carried by native gold quartz veins (Oberthur 1994; Oliver et al. 2020). The first style of mineralization has the form of auriferous arsenopyrite grains that may contain up to 2,000 ppm of so-called invisible or refractory gold – trapped in the crystal structure – that crystallize in parallel to a penetrative fabric. This fabric forms during a coaxial deformation, developing an axial plane schistosity under *P-T* conditions of upper greenschist facies. Arsenopyrite is described as proximal alteration sulfide of boudinaged quartz veins in shear zones (Figure 5.6(b)). The second style of mineralization is expressed by visible gold, also known as free gold, present in microfractures in the same system of boudinaged quartz veins that are folded during a subsequent deformation phase. A fine petro-structural analysis coupled with high spatial resolution chemical mapping tools, reaching atomic scale, allowed the role of deformation in gold remobilization by dissolution-recrystallization processes and pseudomorphism phenomena affecting early auriferous sulfides to be highlighted (Fougerouse et al. 2017). Fougerouse et al. (2016b) prove that arsenopyrite grains are affected by plastic deformation and recrystallize locally, leading to the formation of subgrains. Gold is then released from the sulfide-bearing matrix and mobilized over distances from several micrometers to several dozen meters and reprecipitate in microfractures mainly in quartz veins.

These two mineralization events were interpreted as formed either by a single-phase progressive deformation (Blenkinsop et al. 1994), or by a polyphased deformation model with two separated, at least, deformation phases (Allibone et al. 2002). Unfortunately, geochronology does not allow us to decide with certainty between these two deformation models, as a single age could be obtained by Re-Os dating on arsenopyrite, at 2,045 ± 40 Ma, interpreted as the recrystallization age of arsenopyrite grains (Le Mignot et al. 2017a).

The Kiaka deposit in Burkina Faso and the Wassa deposit in Ghana express the overlapping of two gold events witnessing a polyphase of two pre-Eburnean systems and one syn-Eburnean hydrothermal system (Figure 5.6(a)). The Kiaka deposit is hosted in a volcano-sedimentary sequence, metamorphosed in the amphibolite greenschist facies, belonging to the Manga-Fada N'Gourma belt and located at the intersection of the Markoye shear zone, interpreted as a first-order structure, and second-order shear zones (Fontaine et al. 2017). Resources are estimated at 257.5 Mt at 0.7–0.8 g/t for 7 Moz of gold (B2Gold 2021). The host rocks are essentially conglomerates, metagraywackes and metapelites interbedded with garnet-orthopyroxene schists and mafic metavolcanites. Several magmatic rocks are described in this sector, an amphibole-biotite diorite dated at 2,140 ± 7 Ma using zircon U-Pb dating, as well as an undeformed biotite granite dated between ca. 2.12 and 2.10 Ga (Fontaine et al. 2017 and cited references). The auriferous mineralization of Kiaka is expressed in two forms: a disseminated ante-Eburnean mineralization (Figure 5.6(a)) and a syn-Eburnean vein-type of mineralization (Fontaine et al. 2017). The disseminated mineralization, rich sulfides with an average grade of 1–3 g/t, was dated at 2,157 ± 24 Ma by Re-Os dating on pyrrhotite (Le Mignot et al. 2017a) and thus interpreted as more or less coeval with the emplacement of amphibole-biotite diorite. This mineralization consists of an assemblage of pyrrhotite (60–80%), pyrite (10–20%), chalcopyrite (5–10%), arsenopyrite (5–10%) and sparse löllingite mainly disseminated in a sericite-bearing metagraywacke and less frequently in a mafic volcanic structure with a hydrothermal biotite-titanite-clinozoisite assemblage. Gold is expressed as nano-inclusions in pyrite, arsenopyrite and löllingite. The second type of high-grade (50–60 g/t) gold mineralization is expressed as free gold in quartz veins or breccias, markers of a deformation phase considered as syn-Eburnean. Mineralization consists of pyrrhotite, chalcopyrite, arsenopyrite, electrum and native gold (2–20 µm), and to a lesser extent, grains of tellurobismuthinite (Gauriau et al. 2020).

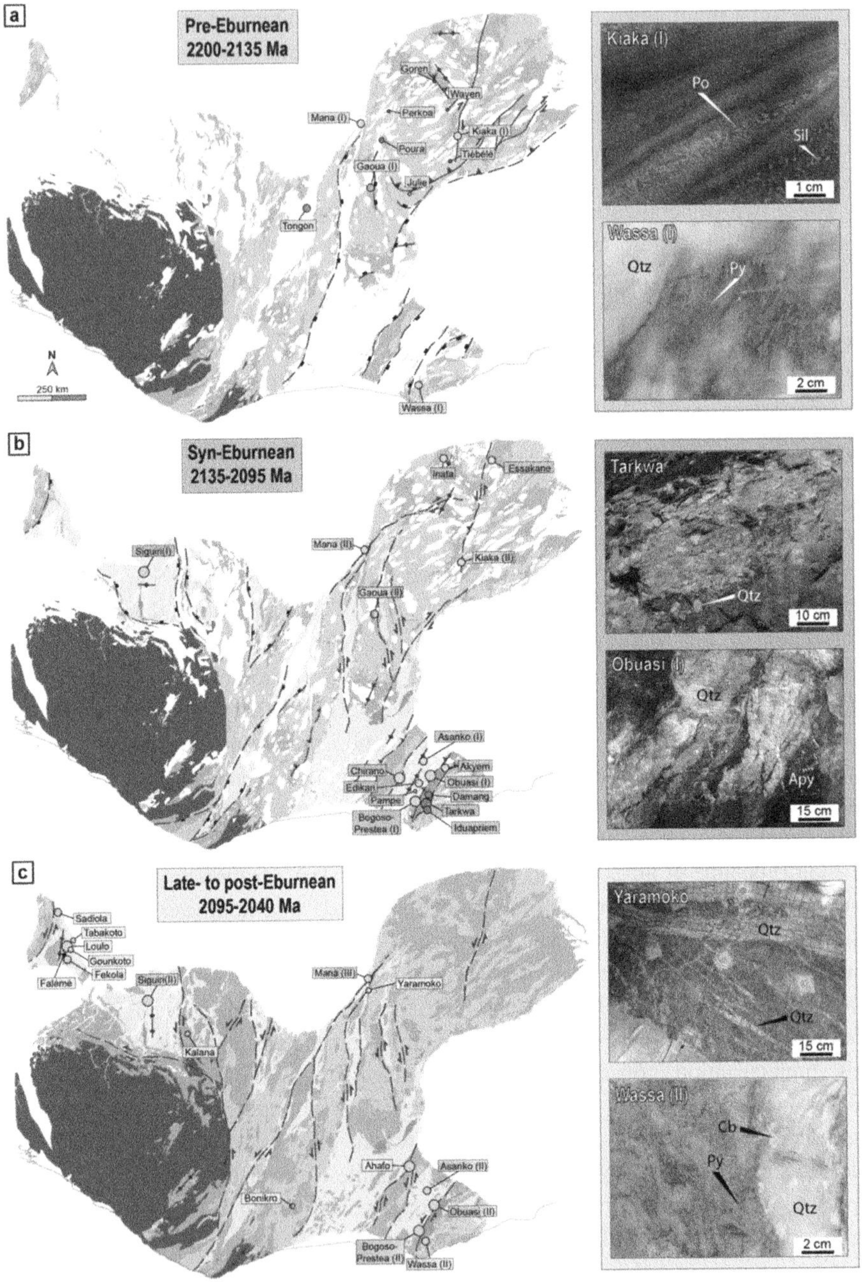

Figure 5.6. *Time and space evolving litho-structural maps highlighting the polyphase gold system of the LMS. Modified from Masurel et al. (2021)*

COMMENT ON FIGURE 5.6.– *These maps present the location of the main gold deposits known as (a) pre-Eburnean: for example, Kiaka (I): disseminated pyrrhotite ± gold mineralization in metasediments (Fontaine* et al. *2017); Wassa (I): folded quartz-pyrite ± gold veins (Perrouty et al. 2015); (b) syn-Eburnean: for example, Tarkwa: gold-bearing conglomerate pebbles (Masurel et al. 2021); Obuasi (I) sheared quartz veins developing a proximal arsenopyrite ± gold alteration with (Fougerouse et al. 2017) and (c) late- to post-Eburnean: for example, Yaramoko: quartz ± gold fault-infill veins and associated extension veins (Hein 2016); Wassa (II): quartz-carbonate ± gold veins developing a proximal pyrite alteration (Perrouty et al. 2015).*

The Wassa deposit is located in the Ashanti greenstone belt in Ghana. The resources are estimated at 29.3 Mt at 3.8 g/t for 3.5 Moz of gold (Golden Star Ltd 2020). The Wassa deposit is hosted in a metamorphosed volcano-sedimentary sequence of Sefwi group, intruded by mafic volcanites and felsic intrusions (Perrouty et al. 2015). The Wassa deposit is affected by several deformation phases from pre-Eburnean to late-Eburnean stages and thus represents a good example of a polyphased mineral system (Perrouty et al. 2015). Out of the six generations of veins described at Wassa by these authors, two are gold-bearing: the first generation of auriferous veins is predominant at the deposit scale, the transposition of these veins within S_1 schistosity indicates an pre-Eburnean emplacement (Figure 5.6(a)), with associated pyrite grains dated at 2,164 ± 14 Ma (Le Mignot et al. 2017a); the veins associated with the second gold event crosscutting the S_1 (Figure 5.6(b)), in a brittle deformation regime, were dated at 2,055 ± 14 Ma by Re-Os dating on pyrite (Le Mignot et al. 2017a).

5.3.4. *Alluvial and eluvial deposits*

Deep and intense alteration from early Miocene (23 Ma), as well as persistent erosion from Neogene in moderate to low relief zones led to a regolithic cover in the form of lateritic paleosurfaces and favored the development of eluvial and alluvial gold deposits throughout the WAC, which is reflected by intense activity of artisanal exploitation of alluvial and eluvial gold in the craton (Chardon et al. 2018; Chevrillon-Guibert et al. 2019; Sawadogo et al. 2020).

The density of eluvial and alluvial gold deposits in the WAC indicates a significant potential of auriferous mineralization of the bedrock. This potential was transformed into many exploitations in the Paleoproterozoic terrains and can offer useful indications for exploration targeting in these Archean and Panafrican terrains, marked for centuries by intense artisanal mining (see Barrie and Touret (1999) and Gunn et al. (2018) for the Archean part of Liberia; Adjo et al. (2019) for the Panafrican belt of Togo-Bénin).

5.4. Spatiotemporal distribution of other LMS metallogenic systems

In addition to gold mining, in the past 20 years the WAC has become a target for the exploration and exploitation of other commodities: iron, diamond, manganese, base metal and nickel, among others (Markwitz et al. 2016) (Figure 5.7).

5.4.1. *Iron*

Significant iron deposits and anomalies are present in the Archean to Panafrican WAC system. The most significant iron deposits are *Banded Iron Formations* located in the Archean part of the craton. The world-class deposit of Simandou in South-East Guinea extends over 7.5×0.5 km^2 and includes three banded iron layers, two of which reach contents of 65 wt% over 250 m in thickness (Cope et al. 2008).

Ferriferous skarn deposits are also described in the Falémé district, where the magnetite ± apatite mineralization is hosted in the Falémé batholite, a potassium calc-alkaline magmatic complex, intrusive in the Kofi metasedimentary series (including marble levels), and whose emplacement ages (determined by zircon U-Pb dating) (Allibone et al. 2020) range between ca. 2.09 and 2.07 Ga.

Other iron ore deposits related to magmatic intrusions are described in the Oursi region, north-east of Burkina Faso, where the dykes carrying Fe-V-Ti cut the Proterozoic mafic and ultramafic intrusions (Neybergh et al. 1980). Iron-Oxide-Gold-Copper (IOGC) deposits were described in Mauritania, for example, the Guelb Moghrein deposit dated at 2,492 ± 9 Ma using monazite and xenotime U-Pb dating (Kolb et al. 2006; Meyer et al. 2006).

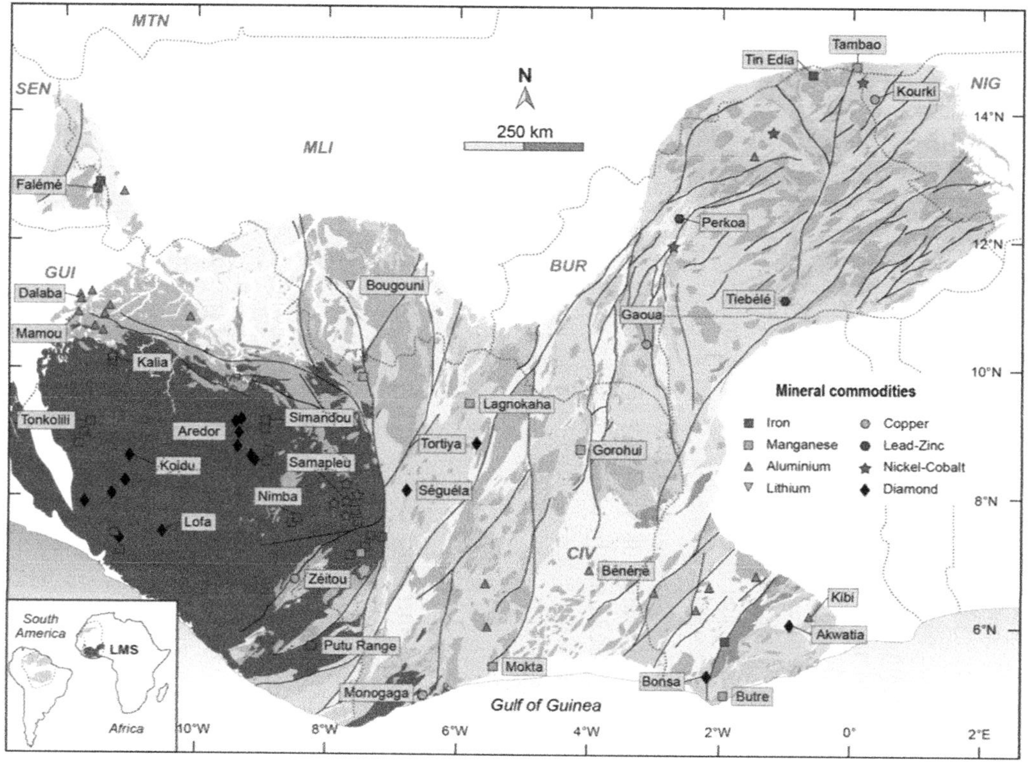

Figure 5.7. *Metallogenic map that highlights the distribution of the main mineral commodities other than gold of the sWAC. Modified from Markwitz et al. (2016). Abbreviations: See Figure 5.5*

5.4.2. *Nickel and chrome*

Nickel mineralization is described both during Eburnean and Panafrican orogenic cycles. In the Ivory Coast, Ni-Cu-PGE (Platinum Group Elements) mineralization is closely associated with Eburnean mafic to ultramafic magmatism described within the granulitic gneisses along the Archean Kénéma-Man margin (Gouedji et al. 2014). This type of mineralization is illustrated by the Samapleu ore deposit in association with mafic to ultramafic intrusions with layers of websterites, peridotites and gabbronorites mineralized in Ni-bearing and copper-bearing sulfides and in platinum group minerals (PGM). The contact with the gneissic host rocks is marked by a hybrid zone a few meters thick, composed of plagioclase-orthopyroxene metabasites having locally saphirine-cordierite-sillimanite-spinelle ± rutile metamorphic paragenesis dated at ca. 2.09 Ga by U-Pb dating on rutile. This age is interpreted as the age of contact metamorphism related to the emplacement of the ultramafic-mafic complex (Gouedji et al. 2014). The emplacement of such dykes can be closely linked to the beginning of the late-Eburnean geodynamic history. In the Panafrican, the anomalies of Ni and Cr concentrations, described in South-East Ghana and in Togo, are hosted by ultramafic and mafic lithologies interpreted as the exhumed root of an island arc or litho-tectonic units of an oceanic lithosphere subducted and exhumed during the formation of the Dahomeyide orogen (Guillot et al. 2019).

5.4.3. *Copper, zinc and lead*

There are few base metal deposits known in West Africa, and their economic importance is relatively minor. They were essentially formed in relation with the early magmatic accretion of the Eburnean orogenic cycle.

Pb-Zn volcanogenic massive sulfides (VMS) are associated with the emplacement of a mafic volcanic series, with tholeiitic signature forming the base of the Birimian lithostratigraphic unit. The main currently exploited VMS are those of Perkoa, located in the Paleoproterozoic Boromo-Goren belt in Burkina Faso. Production started in 2013, with reserves estimated at 5.7 Mt at 18.2 wt% Zn, 10 wt% $BaSO_4$ and 26 g/t Ag (Schwartz and Melcher 2003). VMS are also known in the Panafrican segment of Mali, where the Tessalit deposit produces zinc, lead and copper (Leblanc and Sauvage 1986).

Small copper-bearing porphyries are also described in the Boromo-Goren belt, particularly in the Gaoua and Goren districts (Le Mignot et al. 2017b). These porphyries associated with the pre-Eburnean period (ca. 2.20–2.13 Ga) were significantly reworked in the later stages of the Eburnean orogenic cycle and thus express a polyphased history of deformation and mineralizing events. For example, porphyry copper mineralization of the Gaoua district is contemporary with the emplacement of its dioritic host rock, dated at 2,165 ± 24 Ma by U-Pb dating on zircon (Le Mignot et al. 2017b). This Cu-(Au) mineralization has been overprinted by an orogenic gold event at ca. 2.12 Ga (Le Mignot et al. 2017b). This Cu-(Au) porphyry system remains quite rare in the Paleoproterozoic terrains, as the exposed terrains have probably not been preserved during their exhumation and associated erosion.

5.4.4. *Diamond*

Western Africa diamonds have been exploited from (i) primary kimberlitepipes, mainly of Mesozoic age in relation with the Pangea breakup between the Jurassic and the Cretaceous, and (ii) secondary (alluvial) deposits in Sierra Leone, Ghana, Guinea and Liberia. The structure represents approximately 1% of the world production. The diamond production is essentially linked to small-size artisanal operations on paleoplacers or alluvial placers (Dampare et al. 2005; Markwitz et al. 2016).

5.4.5. *Other mineral substances*

A wide range of other metals and industrial minerals are found in the WAC (see Milesi et al. (1992), Markwitz et al. (2016), Gunn et al. (2018) and the references therein), including:

– world-class rutile resources in the alluvial-fluvial deposits in the Rockelide belt of South-West Sierra Leone that have been exploited since the 1960s;

– manganese deposits located mainly in the Paleoproterozic Baoulé-Mossi domain, in the Dahomeyide and Rockelide, as well as on the selvages of the Volta Basin – these are in situ supergene enrichment deposits of iron and manganese-rich Cretaceous sedimentary rocks, with 17–52% MnO_2 contents (Markwitz et al. 2016);

– sedimentary phosphate deposits in coastal and intracratonic basins in Togo, Senegal, Guinea-Bissau, Burkina Faso, Niger and Mauritania;

– alluvial deposits of columbite-tantalite and cassiterite, associated with the erosion of granitoids, worked by artisanal miners in Sierra Leone, at the border between Senegal and Mali and in Burkina Faso. Nigerian anorogenic magmatic complexes are also known for their tin mineralization feeding alluvial deposits.

5.5. Conclusion

This synopsis on the evolution of WAC mineral systems highlights the importance of mineralization as a marker of the tectonic, magmatic and underlying metamorphic and hydrothermal processes. The scientific approach to describe and decipher complex mineral systems involves various observations in time and space and at different scales: from a geological point of view that involves: lithospheric (up to several hundred km); crustal (from several dozen km to km); regional (km to m); macroscopic (dm to cm); microscopic (mm to μm) and crystallographic (<hundreds of μm) scales, or from a metallogenic point of view, the following scale-related terms are used: the metallogenic province (up to several hundred km), the mining district (100–10 km) and camp (10–1 km), the deposit (km to m), the ore shoot (dozen dm to several mm) and the ore mineral (<1 mm in general).

The VMS and porphyry deposits are characteristic for the period of magmatic accretion in convergent context and the formation and maturation of magmatic arc(s) and related basins, for example. The orogenic and paleoplacer deposits indicate the thickening of this segment of continental crust involving the collision of arc(s) and continental *nuclei* during the Eburnean orogeny.

Finally, other deposits described as more hybrid, that can be of orogenic gold type and/or associated with intrusions, may highlight crustal reworking during transcurrent tectonics. Lithospheric mantle delamination at the craton margin could be the trigger of the large-scale tectono-magmatic activity generating a protracted period of hydrothermal fluid flux. In conclusion, the metallogenic features of the WAC, expressed by the diversity of mineralization styles and their nonrandom distribution in time and space, offer an opportunity to observe and better understand the formation of the Columbia supercontinent during the Paleoproterozoic era and to have interesting discussions on its tectono-magmatic evolution.

5.6. References

Abouchami, W., Boher, M., Michard, A., Albarede, F. (1990). A major 2.1 Ga event of mafic magmatism in west Africa: An early stage of crustal accretion. *Journal of Geophysical Research*, 95(11), 17605.

Adjo, F.B., Bolarinwa, A.T., Adissin Glodji, L., Dongmo, F.W.N., Olajide-Kayode, J.O. (2019). A review of the current state of knowledge on gold mineralisation in Benin Republic, West Africa. *Applied Earth Science*, 128(1), 2–14.

Allibone, A.H., McCuaig, T.C., Harris, D., Etheridge, M., Munroe, S., Byrne, D., Amanor, J., Gyapong, W. (2002). Structural controls on gold mineralization at the Ashanti deposit, Obuasi, Ghana. In *Integrated Methods for Discovery: Global Exploration in the Twenty-First Century*, Goldfarb, R.J. and Nielsen, R.L. (eds). Society of Economic Geologists, Littleton.

Allibone, A.H., Lawrence, D., Scott, J., Fanning, M., Lambert-Smith, J., Stenhouse, P., Harbidge, R., Vargas, C., Turnbull, R., Holliday, J. (2020). Paper 7: Paleoproterozoic gold deposits of the Loulo District, Western Mali. *Society of Economic Geologists*, Special Publication.

Arndt, N.T., Coltice, N., Helmstaedt, H., Gregoire, M. (2009). Origin of Archean subcontinental lithospheric mantle: Some petrological constraints. *Lithos*, 109(1–2), 61–71.

Attoh, K., Evans, M.J., Bickford, M.E. (2006). Geochemistry of an ultramafic-rodingite rock association in the Paleoproterozoic Dixcove greenstone belt, southwestern Ghana. *Journal of African Earth Sciences*, 45(3), 333–346.

Augustin, J. and Gaboury, D. (2017). Paleoproterozoic plume-related basaltic rocks in the Mana gold district in western Burkina Faso, West Africa: Implications for exploration and the source of gold in orogenic deposits. *Journal of African Earth Sciences*, 129, 17–30.

Baratoux, L., Metelka, V., Naba, S., Jessel, M.W., Grégoire, M., Ganne, J. (2011). Juvenile Paleoproterozoic crust evolution during the Eburnean orogeny (~2.2-2.0 Ga), western Burkina Faso. *Precambrian Research*, 191, 18–45.

Barley, M.E. and Groves, D.I. (1992). Supercontinent cycles and the distribution of metal deposits through time. *Geology*, 20(4), 291–294.

Barnicoat, A.C. (2007). Mineral systems and exploration science: Linking fundamental controls on ore deposition with the exploration process. In *Digging Deeper. Proceedings of the Ninth Biennial SGA Meeting*, Andrews C.J. et al. (eds). SGA Meeting, Dublin.

Barrie, I.J. and Touret, J.L.R. (1999). Fluid inclusion studies of gold-bearing quartz veins from the Yirisen deposit, Sula Mountains greenstone belt, Masumbiri, Sierra Leone. *Ore Geology Reviews*, 14(3), 203–225.

Barth, M.G., Rudnick, R.L., Carlson, R.W., Horn, I., McDonough, W.F. (2002). ReÃ/Os and UÃ/Pb geochronological constraints on the eclogite–tonalite connection in the Archean Man Shield, West Africa. *Precambrian Research*, 118(3–4), 267–283.

Bédard, J.H. (2006). A catalytic delamination-driven model for coupled genesis of Archaean crust and subcontinental lithospheric mantle. *Geochimica et Cosmochimica Acta*, 70(5), 1188–1214.

Bédard, J.H. (2018). Stagnant lids and mantle overturns: Implications for Archaean tectonics, magmagenesis, crustal growth, mantle evolution, and the start of plate tectonics. *Geoscience Frontiers*, 9(1), 19–49.

Bierlein, F.P., Groves, D.I., Cawood, P.A. (2009). Metallogeny of accretionary orogens – The connection between lithospheric processes and metal endowment. *Ore Geology Reviews*, 36(4), 282–292.

Blenkinsop, T., Schmidt Mumm, A., Kumi, R., Sangmor, S. (1994). Structural geology of the Ashanti gold mine. *Geologisches Jahrbuch D*, 100, 131–153.

Block, S., Ganne, J., Baratoux, L., Zeh, A., Parra-Avila, L.A., Jessell, M., Ailleres, L., Siebenaller, L. (2015). Petrological and geochronological constraints on lower crust exhumation during Paleoproterozoic (Eburnean) orogeny, NW Ghana, West African craton. *Journal of Metamorphic Geology*, 33(5), 463–494.

Block, S., Jessell, M., Aillères, L., Baratoux, L., Bruguier, O., Zeh, A., Bosch, D., Caby, R., Mensah, E. (2016). Lower crust exhumation during Paleoproterozoic (Eburnean) orogeny, NW Ghana, West African Craton: Interplay of coeval contractional deformation and extensional gravitational collapse. *Precambrian Research*, 274, 82–109.

Boher, M., Abouchami, W., Michard, A., Albarede, F., Arndt, N.T. (1992). Crustal growth in West Africa at 2.1 Ga. *Journal of Geophysical Research*, 97(1), 345.

Bonhomme, M. (1962). Contribution à l'étude géochronologique de la plateforme de l'ouest africain. Thesis, Université de Clermont-Ferrand.

Brown, M., Johnson, T., Gardiner, N.J. (2020). Plate tectonics and the Archean Earth. *Annual Review of Earth and Planetary Sciences*, 48(1), 291–320.

Cagnard, F., Durrieu, N., Gapais, D., Brun, J.P., Ehlers, C. (2006). Crustal thickening and lateral flow during compression of hot lithospheres, with particular reference to Precambrian times. *Terra Nova*, 18(1), 72–78.

Cawood, P.A. and Hawkesworth, C.J. (2015). Temporal relations between mineral deposits and global tectonic cycles. *Geological Society, London, Special Publications*, 393(1), 9–21.

Chardon, D., Gapais, D., Cagnard, F. (2009). Flow of ultra-hot orogens: A view from the Precambrian, clues for the Phanerozoic. *Tectonophysics*, 477(3–4), 105–118.

Chardon, D., Grimaud, J.L., Beauvais, A., Bamba, O. (2018). West African lateritic pediments: Landform-regolith evolution processes and mineral exploration pitfalls. *Earth-Science Reviews*, 179, 124–146.

Chardon, D., Bamba, O., Traoré, K. (2020). Eburnean deformation pattern of Burkina Faso and the tectonic significance of shear zones in the West African craton. *Bulletin de la Société Géologique de France*, 191(2), 1–18.

Chevrillon-Guibert, R., Gagnol, L., Magrin, G. (2019). Les ruées vers l'or au Sahara et au nord du Sahel. Ferment de crise ou stabilisateur ? *Hérodote*, 172(1), 193.

Clifford, T.N. (1970). The structural framework of Africa. In *African Magmatism and Tectonics*, Clifford, T.N. and Gass, I. (eds). Oliver and Boyd, Edinburgh.

Cooper, C.M. and Miller, M.S. (2014). Craton formation: Internal structure inherited from closing of the early oceans. *Lithosphere*, 6(1), 35–42.

Cope, I.L., Wilkinson, J.J., Boyce, A.J., Chapman, J.B., Herrington, R.J., Harris, C.J. (2008). Genesis of the pic de fon iron oxide deposit, Simandou Range, Republic of Guinea, West Africa. In *Banded Iron Formation-Related High-Grade Iron Ore*, Hagemann, S., Rosière, A.C., Gutzmer, J., Beuker, J.N. (eds). Society of Economic Geologists, Littleton.

Dampare, S., Asiedu, D., Banoeng-Yakubo, B. (2005). Heavy mineral analysis of alluvial aedinents from the Akwatia area of the Birim diamonndierous field, Ghana. Report, Okayama University Earth Science.

Dampare, S., Shibata, T., Asiedu, D.K., Okano, O., Osae, S.K.D., Atta-Peters, D., Sakyi, P.A. (2020). Ultramafic–mafic and granitoids supra-subduction magmatism in the southern Ashanti volcanic belt, Ghana: Evidence from geochemistry and Nd isotopes. *Geological Journal*, 55(4), 2495–2531.

Davidson, J.P. and Arculus, R. (2006). The significance of Phanerozoic arc magmatism in generating continental crust. In *Evolution and Differentiation of the Continental Crust*, Brown, M. and Rushmer, T. (eds). Cambridge University Press.

De Waele, B., Lacorde, M., Vergara, F., Chan, G. (2015). New insights on proterozoic tectonics and sedimentation along the peri-Gondwanan West African margin based on zircon U–Pb SHRIMP geochronology. *Precambrian Research*, 259, 156–175.

Eglinger, A. (2013). Cycle de l'uranium et évolution tectono-métamorphique de la ceinture orogénique Pan-Africaine du Lufilien (Zambie). PhD Thesis, Université de Lorraine, Nancy.

Eglinger, A., André-Mayer, A.S., Vanderhaeghe, O., Mercadier, J., Cuney, M., Decrée, S., Feybesse, J.L., Milesi, J.P. (2013). Geochemical signatures of uranium oxides in the Lufilian belt: From unconformity-related to syn-metamorphic uranium deposits during the Pan-African orogenic cycle. *Ore Geology Reviews*, 54, 197–213.

Eglinger, A., Thébaud, N., Zeh, A., Davis, J., Miller, J., Parra-Avila, L.A., Loucks, R., McCuaig, C., Belousova, E. (2017). New insights into the crustal growth of the Paleoproterozoic margin of the Archean Kéména-Man domain, West African craton (Guinea): Implications for gold mineral system. *Precambrian Research*, 292, 258–289.

Feybesse, J.L. and Milesi, J. (1994). The Archaean/Proterozoic contact zone in West Africa: A mountain belt of décollement thrusting and folding on a continental margin related to 2.1 Ga convergence of Archaean cratons? *Precambrian Research*, 69, 199–227.

Feybesse, J.L., Bangoura, A., Billa, M., Costea, A., Diabaté, B., Diaby, S., Diallo, A.B., Egal, E., Freyssinet, P., Gaye, F. et al. (1999). Notice explicative de la carte géologique de la Guinée à 1/200 000. Report, Ministère des Mines de la Géologie et de l'Environnement.

Feybesse, J.L., Billa, M., Guerrot, C., Duguey, E., Lescuyer, J.L., Milesi, J.P., Bouchot, V. (2006). The paleoproterozoic Ghanaian province: Geodynamic model and ore controls, including regional stress modeling. *Precambrian Research*, 149(3–4), 149–196.

Fontaine, A., Eglinger, A., Ada, K., André-Mayer, A.S., Reisberg, L., Siebenaller, L., Le Mignot, E., Ganne, J., Poujol, M. (2017). Geology of the world-class Kiaka polyphase gold deposit, West African Craton, Burkina Faso. *Journal of African Earth Sciences*, 126, 96–122.

Fontaine, A., Dubé, B., Malo, M., Turcotte, J., Doucet, D. (2018). Geology of the Cheechoo gold property, Eeyou Istchee Baie-James, Superior Province, northern Quebec. Report, Open File 8403, Natural Resources Canada.

Forsyth, D. and Uyeda, S. (1975). On the relative importance of the driving forces of plate motion. *Geophysical Journal International*, 43(1), 163–200.

Fossen, H., Cavalcante, G.C.G., Pinheiro, R.V.L., Archanjo, C.J. (2019). Deformation – Progressive or multiphase? *Journal of Structural Geology*, 125, 82–99.

Foster, R.P. and Piper, D.P. (1993). Archaean lode gold deposits in Africa: Crustal setting, metallogenesis and cratonization. *Ore Geology Reviews*, 8(3), 303–347.

Fougerouse, D., Micklethwaite, S., Tomkins, A.G., Mei, Y., Kilburn, M., Guagliardo, P., Fisher, L.A., Halfpenny, A., Gee, M., Paterson, D. et al. (2016a). Gold remobilisation and formation of high grade ore shoots driven by dissolution-reprecipitation replacement and Ni substitution into auriferous arsenopyrite. *Geochimica et Cosmochimica Acta*, 178, 143–159.

Fougerouse, D., Micklethwaite, S., Halfpenny, A., Reddy, S.M., Cliff, J.B., Martin, L.A.J., Kilburn, M., Guagliardo, P., Ulrich, S. (2016b). The golden ark: Arsenopyrite crystal plasticity and the retention of gold through high strain and metamorphism. *Terra Nova*, 28(3), 181–187.

Fougerouse, D., Micklethwaite, S., Ulrich, S., Miller, J., Godel, B., Adams, D.T., McCuaig, T.C. (2017). Evidence for two stages of mineralization in West Africa's largest gold deposit: Obuasi, Ghana. *Economic Geology*, 112(1), 3–22.

Ganne, J., De Andrade, V., Weinberg, R.F., Vidal, O., Dubacq, B., Kagambega, N., Naba, S., Baratoux, L., Jessell, M., Allibon, J. (2012). Modern-style plate subduction preserved in the Palaeoproterozoic West African craton. *Nature Geoscience*, 5(1), 60–65.

Ganne, J., Gerbault, M., Block, S. (2014). Thermo-mechanical modeling of lower crust exhumation – Constraints from the metamorphic record of the Palaeoproterozoic Eburnean orogeny, West African Craton. *Precambrian Research*, 243, 88–109.

Gapais, D. (2018). Tectonics-mineralisation relationships within weak continental lithospheres: A new structural framework for Precambrian cratons. *BSGF – Earth Sciences Bulletin*, 189(3), 14.

Gapais, D., Cagnard, F., Gueydan, F., Barbey, P., Ballèvre, M. (2009). Mountain building and exhumation processes through time: Inferences from nature and models. *Terra Nova*, 21(3), 188–194.

Gauriau, J., Harlaux, M., André-Mayer, A.S., Eglinger, A., Richard, A., Fontaine, A., Lefebvre, M.G., Béziat, D., Villeneuve, J., Lemarchand, D. (2020). Chemical and boron isotope composition of tourmaline from the Kiaka orogenic gold deposit (Burkina Faso, West African Craton) as a proxy for ore-forming processes. *Mineralium Deposita*, 57, 581–600.

Goldfarb, R.J. and Groves, D.I. (2015). Orogenic gold: Common or evolving fluid and metal sources through time. *Lithos*, 233, 2–26.

Goldfarb, R.J., Baker, T., Dubé, B., Groves, D.I., Hart, C.J.R., Gosselin, P. (2005). Distribution, character and genesis of gold deposits in metamorphic terranes. *Economic Geology*, 100, 407–450.

Goldfarb, R.J., Bradley, D., Leach, D.L. (2010). Secular variation in economic geology. *Economic Geology*, 105(3), 459–465.

Goldfarb, R.J., André-Mayer, A.S., Jowitt, S.M., Mudd, G.M. (2017). West Africa: The world's premier Paleoproterozoic gold province. *Economic Geology*, 112(1), 123–143.

Gouedji, F., Picard, C., Coulibaly, Y., Audet, M.A., Auge, T., Goncalves, P., Paquette, J.L., Ouattara, N. (2014). The Samapleu mafic-ultramafic intrusion and its Ni-Cu-PGE mineralization: An Eburnean (2.09 Ga) feeder dyke to the Yacouba layered complex (Man Archean craton, western Ivory Coast). *Bulletin de la Société Géologique de France*, 185(6), 393–411.

Grenholm, M. (2019). The global tectonic context of the ca. 2.27-1.96 Ga Birimian Orogen – Insights from comparative studies, with implications for supercontinent cycles. *Earth-Science Reviews*, 193, 260–298.

Grenholm, M., Jessell, M., Thébaud, N. (2019). A geodynamic model for the Paleoproterozoic (ca. 2.27–1.96 Ga) Birimian Orogen of the southern West African Craton – Insights into an evolving accretionary-collisional orogenic system. *Earth-Science Reviews*, 192, 138–193.

Groves, D.I. (1993). The crustal continuum model for late-Archaean lode-gold deposits of the Yilgarn Block, Western Australia. *Mineralium Deposita*, 28(6), 366–374.

Groves, D.I. and Bierlein, F.P. (2007). Geodynamic settings of mineral deposit systems. *Journal of the Geological Society*, 164(1), 19–30.

Groves, D.I., Condie, K.C., Goldfarb, R.J., Hronsky, J.M.A., Vielreicher, R.M. (2005). Secular changes in global tectonic processes and their influence on the temporal distribution of gold-bearing mineral deposits. *Economic Geology*, 100, 203–224.

Guillot, S., Agbossoumondé, Y., Bascou, J., Berger, J., Duclaux, G., Hilairet, N., Ménot, R.P., Schwartz, S. (2019). Transition from subduction to collision recorded in the Pan-African arc complexes (Mali to Ghana). *Precambrian Research*, 320, 261–280.

Gunn, A.G., Dorbor, J.K., Mankelow, J.M., Lusty, P.A.J., Deady, E.A., Shaw, R.A., Goodenough, K.M. (2018). A review of the mineral potential of Liberia. *Ore Geology Reviews*, 101, 413–431.

Hein, K.A.A. (2016). The Bagassi gold deposits on the eastern margin of the Houndé greenstone belt, Burkina Faso. *Ore Geology Reviews*, 78, 660–666.

Heron, K., Jessell, M., Benn, K., Harris, E., Crowley, Q.G. (2015). The tasiast deposit, Mauritania. *Ore Geology Reviews*, 78, 564–572.

Hurley, P.W., Leo, G.W., White, R.W., Fairbairn, H.W. (1971). Liberian age province (about 2,700 m.y.) and adjacent provinces in Liberia and Sierra Leone. *Geological Society of America Bulletin*, 82(12), 3483–3490.

Huston, D.L., Blewett, R.S., Champion, D.C. (2012). Australia through time: A summary of its tectonic and metallogenic evolution. *Episodes*, 35(1), 23–43.

Jenkin, G.R., Lusty, P.A., McDonald, I., Smith, M.P., Boyce, A.J., Wilkinson, J.J. (2015). Ore deposits in an evolving Earth: An introduction. *Geological Society*, 393(1), 1–8.

Junner, N.R. and Wild, R.P. (1935). *Gold in the Gold Coast.* Government Printer, Accra.

Kerrich, R., Goldfarb, R.J., Groves, D., Garwin, S., Jia, Y. (2000). The characteristics, origins, and geodynamic settings of supergiant gold metallogenic provinces. *Science in China Series D: Earth Sciences*, 43(S1), 1–68.

Kerrich, R., Goldfarb, R.J., Richards, J.P. (2005). Metallogenic provinces in an evolving geodynamic framework. *Economic Geology*, 13, 1097–1136.

de Kock, G.S., Armstrong, R.A., Siegfried, H.P., Thomas, E. (2011). Geochronology of the Birim supergroup of the West African craton in the Wa-Bolé region of west-central Ghana: Implications for the stratigraphic framework. *Journal of African Earth Sciences*, 59, 1–40.

de Kock, G.S., Théveniaut, H., Botha, P.M.W., Gyapong, W. (2012). Timing the structural events in the Palaeoproterozoic Bolé–Nangodi belt terrane and adjacent Maluwe basin, West African craton, in central-west Ghana. *Journal of African Earth Sciences*, 65, 1–24.

Kolb, J., Sakellaris, G.A., Meyer, F.M. (2006). Controls on hydrothermal Fe oxide–Cu–Au–Co mineralization at the Guelb Moghrein deposit, Akjoujt area, Mauritania. *Mineralium Deposita*, 41(1), 68–81.

Kouamelan, A.N., Delor, C., Peucat, J.J. (1997). Geochronological evidence for reworking of Archean terrains during the Early Proterozoic (2.1 Ga) in the western Côte d'Ivoire (Man Rise-West African Craton). *Precambrian Research*, 86(3), 177–199.

Labou, I., Benoit, M., Baratoux, L., Grégoire, M., Ndiaye, P.M., Thebaud, N., Béziat, D., Debat, P. (2020). Petrological and geochemical study of Birimian ultramafic rocks within the West African Craton: Insights from Mako (Senegal) and Loraboué (Burkina Faso) lherzolite/harzburgite/wehrlite associations. *Journal of African Earth Sciences*, 162, 103677.

Launay, L. (1913). *Traité de métallogénie : gîtes minéraux et métallifères.* Charles Béranger, Brussels.

Laurent, O., Björnsen, J., Wotzlaw, J.F., Bretscher, S., Silva, M.P., Moyen, J.F., Ulmer, P., Bachmann, O. (2020). Earth's earliest granitoids are crystal-rich magma reservoirs tapped by silicic eruptions. *Nature Geoscience*, 13(2), 163–169.

Le Mignot, E., Reisberg, L., André-Mayer, A.S., Bourassa, Y., Fontaine, A., Miller, J. (2017a). Re-Os geochronological evidence for multiple paleoproterozoic gold events at the scale of the West African Craton. *Economic Geology*, 112(1), 145–168.

Le Mignot, E., Siebenaller, L., Béziat, D., André-Mayer, A.S., Reisberg, L., Salvi, S., Velasquez, G., Zimmermann, C., Naré, A., Franceschi, G. (2017b). The paleoproterozoic copper-gold deposits of the Gaoua District, Burkina Faso: Superposition of orogenic gold on a porphyry copper occurrence? *Economic Geology*, 112(1), 99–122.

Leblanc, M. and Sauvage, J.F. (1986). Un gisement polymétallique en milieu volcanosédimentaire du Protérozoïque supérieur : Tessalit (Adrar des Iforas, Mali). *Journal of African Earth Sciences*, 5(3), 201–207.

Ledru, P., Pons, J., Milesi, J.P., Feybesse, J.L., Johan, V. (1991). Transcurrent tectonics and polycyclic evolution in the lower proterozoic of Senegal-Mali. *Precambrian Research*, 50(3), 337–354.

Ledru, P., Johan, V., Milési, J.P., Tegyey, M. (1994). Markers of the last stages of the Palaeoproterozoic collision: Evidence for a 2 Ga continent involving circum-South Atlantic provinces. *Precambrian Research*, 69(1), 169–191.

Lindgren, W. (1909). Metallogenetic epochs. *Economic Geology*, 4(5), 409–420.

Liou, P. and Guo, J. (2019). Generation of Archaean TTG gneisses through amphibole-dominated fractionation. *Journal of Geophysical Research: Solid Earth*, 124(4), 3605–3619.

Lompo, M. (2009). Geodynamic evolution of the 2.25-2.0 Ga Palaeoproterozoic magmatic rocks in the Man-Leo Shield of the West African Craton. A model of subsidence of an oceanic plateau. *Geological Society*, 323(1), 231–254.

Lompo, M. (2010). Paleoproterozoic structural evolution of the Man-Leo Shield (West Africa). Key structures for vertical to transcurrent tectonics. *Journal of African Earth Sciences*, 58(1), 19–36.

Lord, D., Etheridge, M., Willson, M., Hall, G., Uttley, P.J. (2001). Measuring exploration success: An alternative to the discovery-cost-per-ounce method of quantifying exploration success. *Society of Economic Geologists Newsletter*, 45, 10–16.

Maier, W.D. and Groves, D.I. (2011). Temporal and spatial controls on the formation of magmatic PGE and Ni–Cu deposits. *Mineralium Deposita*, 46(8), 841–857.

Markwitz, V., Hein, K.A.A., Miller, J. (2016). Compilation of West African mineral deposits: Spatial distribution and mineral endowment. *Precambrian Research*, 274, 61–81.

Masurel, Q., Eglinger, A., Thébaud, N., Allibone, A., André-Mayer, A.S., McFarlane, H., Miller, J., Jessel, M., Aillères, L., Vanderhaeghe, O. et al. (2021). Paleoproterozoic gold events in the southern West African Craton: Review and synopsis. *Mineralium Deposita*, 57, 513–537.

McCuaig, T.C. and Hronsky, J.M. (2014). The mineral system concept: The key to exploration targeting. *Society of Economic Geologist*, 18, 153–175.

McCuaig, T.C., Beresford, S., Hronsky, J. (2010). Translating the mineral systems approach into an effective exploration targeting system. *Ore Geology Reviews*, 38(3), 128–138.

McFarlane, H.B., Ailleres, L., Betts, P., Ganne, J., Baratoux, L., Jessell, M.W., Block, S. (2019a). Episodic collisional orogenesis and lower crust exhumation during the Palaeoproterozoic Eburnean orogeny: Evidence from the Sefwi Greenstone Belt, West African Craton. *Precambrian Research*, 325, 88–110.

McFarlane, H.B., Thébaud, N., Parra-Avila, L.A., Armit, R., Spencer, C., Ganne, J., Aillères, L., Baratoux, L., Betts, P.G., Jessell, M.W. (2019b). Onset of the supercontinent cycle: Evidence for multiple oceanic arc accretion events in the Paleoproterozoic Sefwi Greenstone Belt of the West African Craton. *Precambrian Research*, 335, 105450.

Meert, J.G. and Santosh, M. (2017). The Columbia supercontinent revisited. *Gondwana Research Golden Jubilee Volume*, 50, 67–83.

Mériaud, N., Thébaud, N., Masurel, Q., Hayman, P., Jessell, M., Kemp, A., Evans, N.J., Fisher, C.M., Scott, P.M. (2020). Lithostratigraphic evolution of the Bandamian Volcanic Cycle in central Côte d'Ivoire: Insights into the late Eburnean magmatic resurgence and its geodynamic implications. *Precambrian Research*, 347, 105847.

Meyer, C. (1988). Ore deposits as guides to geologic history of the earth. *Annual Review of Earth and Planetary Sciences*, 16(1), 147–171.

Meyer, F.M., Kolb, J., Sakellaris, G.A., Gerdes, A. (2006). New ages from the Mauritanides Belt: Recognition of Archean IOCG mineralization at Guelb Moghrein, Mauritania. *Terra Nova*, 18(5), 345–352.

Milési, J.P., Ledru, P., Ankrah, P., Johan, V., Marcoux, E., Vinchon, C. (1991). The metallogenic relationship between Birimian and Tarkwaian gold deposits in Ghana. *Mineralium Deposita*, 26, 228–238.

Milési, J.P., Ledru, P., Feybesse, J.L., Dommanget, A., Marcoux, E. (1992). Early proterozoic ore deposits and tectonics of the Birimian orogenic belt, West Africa. *Precambrian Metallogeny Related to Plate Tectonics*, 58(1), 305–344.

Moyen, J.F. and Martin, H. (2012). Forty years of TTG research. *Lithos*, 148, 312–336.

Neybergh, H., Laduron, D., Martin, H., Verkaeren, J. (1980). The vanadiferous magnetite deposits of the Oursi region, Upper-Volta. *Economic Geology*, 75(7), 1042–1052.

Oberthür, T., Vetter, U., Schmidt Mumm, A., Weiser, T., Amanor, J.A., Gyapong, W.A., Kumi, R., Blenkinsop, T.G. (1994). The Ashanti gold mine at Obuasi, Ghana: Mineralogical, geochemical, stable isotope and fluid inclusion studies on the metallogenesis of the deposit. *Geologisches Jahrbuch*, 100, 31–129.

Oliver, N.H.S., Allibone, A., Nugus, M.J., Vargas, C., Jongens, R., Peattie, R., Chamberlain, V.A. (2020). Paper 6: The supergiant, high-grade, paleoproterozoic metasedimentary rock- and shear vein-hosted Obuasi (Ashanti) gold deposit, Ghana, West Africa. *Society of Economic Geologists*, 121–140.

Parra-Avila, L.A., Kemp, A.I.S., Fiorentini, M.L., Belousova, E., Baratoux, L., Block, S., Jessell, M., Bruguier, O., Begg, G.C., Miller, J. et al. (2017). The geochronological evolution of the Paleoproterozoic Baoulé-Mossi domain of the Southern West African Craton. *Precambrian Research*, 300, 1–27.

Perret, J. (2021). Répartition spatio-temporelle du système métallogénique de l'or panafricain au sein du bouclier arabo-nubien : étude multiscalaire le long de la suture de Keraf (Soudan). PhD Thesis, Université de Lorraine, Nancy.

Perret, J., Eglinger, A., André-Mayer, A.S., Aillères, L., Feneyrol, J., Hartshorne, C., Abanyin, E., Bosc, R. (2020). Subvertical, linear and progressive deformation related to gold mineralization at the Galat Sufar South deposit, Nubian Shield, NE Sudan. *Journal of Structural Geology*, 135, 104032.

Perrouty, S., Aillères, L., Jessell, M.W., Baratoux, L., Bourassa, Y., Crawford, B. (2012). Revised Eburnean geodynamic evolution of the gold-rich southern Ashanti Belt, Ghana, with new field and geophysical evidence of pre-Tarkwaian deformations. *Precambrian Research*, 204–205, 12–39.

Perrouty, S., Jessell, M.W., Bourassa, Y., Miller, J., Apau, D., Siebenaller, L., Velásquez, G., Baratoux, L., Aillères, L., Béziat, D., Salvi, S. (2015). The Wassa deposit: A poly-deformed orogenic gold system in southwest Ghana – Implications for regional exploration. *Journal of African Earth Sciences*, 112, 536–547.

Petersson, A., Scherstén, A., Kemp, A.I.S., Kristinsdóttir, B., Kalvig, P., Anum, S. (2016). Zircon U–Pb–Hf evidence for subduction related crustal growth and reworking of Archaean crust within the Palaeoproterozoic Birimian terrane, West African Craton, SE Ghana. *Precambrian Research*, 275, 286–309.

Pitra, P., Kouamelan, A.N., Ballèvre, M., Peucat, J.J. (2010). Palaeoproterozoic high-pressure granulite overprint of the Archean continental crust: Evidence for homogeneous crustal thickening (Man Rise, Ivory Coast). *Journal of Metamorphic Geology*, 28(1), 41–58.

Potrel, A., Peucat, J.J., Fanning, C.M., Auvray, B., Burg, J.P., Caruba, C. (1996). 3.5 Ga old terranes in the West African Craton, Mauritania. *Journal of the Geological Society*, 153(4), 507–510.

Pouclet, A., Doumbia, S., Vidal, M. (2006). Geodynamic setting of the Birimian volcanism in central Ivory Coast (western Africa) and its place in the Palaeoproterozoic evolution of the Man Shield. *Bulletin de la Société Géologique de France*, 177(2), 105–121.

Pourteau, A., Doucet, L.S., Blereau, E.R., Volante, S., Johnson, T.E., Collins, W.J., Li, Z.X., Champion, D.C. (2020). TTG generation by fluid-fluxed crustal melting: Direct evidence from the Proterozoic Georgetown Inlier, NE Australia. *Earth and Planetary Science Letters*, 550, 116548.

Rey, P.F., Philippot, P., Thébaud, N. (2003). Contribution of mantle plumes, crustal thickening and greenstone blanketing to the 2.75–2.65Ga global crisis. *Precambrian Research*, 127(1–3), 43–60.

Rocci, G. (1965). Essai d'interpretation de mesures geochronologiques. La structure de l'Ouest Africain. *Sciences Terre Nancy*, 10, 461–479.

Rogers, J.J.W. and Santosh, M. (2002). Configuration of Columbia, a mesoproterozoic supercontinent. *Gondwana Research*, 5(1), 5–22.

Rollinson, H. (2010). Coupled evolution of Archean continental crust and subcontinental lithospheric mantle. *Geology*, 38(12), 1083–1086.

Rollinson, H. (2016). Archaean crustal evolution in West Africa: A new synthesis of the Archaean geology in Sierra Leone, Liberia, Guinea and Ivory Coast. *Precambrian Research*, 281, 1–12.

Rüpke, L.H., Morgan, J.P., Hort, M., Connolly, J.A.D. (2004). Serpentine and the subduction zone water cycle. *Earth and Planetary Science Letters*, 223(1), 17–34.

Sakyi, P.A., Su, B.X., Manu, J., Kwayisi, D., Anani, C.Y., Alemayehu, M., Malaviarachchi, S.P.K., Nude, P.M., Su, B.C. (2020). Origin and tectonic significance of the metavolcanic rocks and mafic enclaves from the Palaeoproterozoic Birimian Terrane, SE West African Craton, Ghana. *Geological Magazine*, 157(8), 1349–1366.

Sawadogo, B., Bamba, O., Chardon, D. (2020). Landform-regolith mapping in the West African context. *Ore Geology Reviews*, 126, 103782.

Schwartz, M.O. and Melcher, F. (2003). The Perkoa zinc deposit, Burkina Faso. *Economic Geology*, 98(7), 1463–1485.

Shirey, S.B. and Richardson, S.H. (2011). Start of the Wilson cycle at 3 Ga shown by diamonds from subcontinental mantle. *Science*, 333(6041), 434–436.

Skinner, E.M.W., Apter, D.B., Morelli, C., Smithson, N.K. (2004). Kimberlites of the Man craton, West Africa. *Selected Papers from the Eighth International Kimberlite Conference. The C. Roger Clement Volume*, 76(1), 233–259.

Smithies, R.H., Lu, Y., Johnson, T.E., Kirkland, C.L., Cassidy, K.F., Champion, D.C., Mole, D.R., Zibra, I., Gessner, K., Sapkota, J. et al. (2019). No evidence for high-pressure melting of Earth's crust in the Archean. *Nature Communications*, 10(1), 5559.

Thébaud, N., Allibone, A., Masurel, Q., Eglinger, A., Davis, J., André-Mayer, A.S., Miller, J., Ouedrago, M.F., Jessell, M. (2020). Paper 34: The paleoproterozoic (Rhyacian) gold deposits of West Africa. *Society of Economic Geologist*, 735–752.

Thiéblemont, D. (2005). Geology and petrology of Archean rocks in Guinea: A regional contribution to the knowledge of continental crust formation. French Geological Survey, Université de Bretagne Occidentale, Brest.

Thiéblemont, D., Delor, C., Cocherie, A., Lafon, J.M., Goujou, J.C., Baldé, A., Bah, M., Sané, H., Mark Fanning, C. (2001). A 3.5 Ga granite–gneiss basement in Guinea: Further evidence for early archean accretion within the West African Craton. *Precambrian Research*, 108(3), 179–194.

Thiéblemont, D., Goujou, J.C., Egal, E., Cocherie, A., Delor, C., Lafon, J.M., Fanning, C.M. (2004). Archean evolution of the Leo Rise and its Eburnean reworking. *Journal of African Earth Sciences*, 39(3–5), 97–104.

Thiéblemont, D., Liégeois, J.P., Fernandez-Alonso, M., Ouabadi, A., Le Gall, B., Maury, R., Vidal, M., Ouattara, G., Tchameni, R., Michard, A. et al. (2016). Geological map of Africa at 1:10M scale, CGMW-BRGM 2016. *35th International Geological Congress, Cape Town, 29 August–2 September 2016*. Commission of the Geological Map of the World.

Titley, S.R. (1993). Characteristics of porphyry copper occurrence in the American Southwest. *Mineral Deposit Modelling: Geological Association of Canada*, Special Paper, 40, 433–464.

Tshibubudze, A., Hein, K.A.A., Marquis, P. (2009). The Markoye shear zone in NE Burkina Faso. *Journal of African Earth Sciences*, 55(5), 245–256.

Turlin, F. (2017). Les granites pegmatitiques enrichis en éléments de terres rares légères marqueurs des processus de croissance et de différenciation crustale : exemple de la Province Protérozoïque de Grenville, Québec. PhD Thesis, Université de Lorraine, Nancy.

Vidal, M. and Alric, G. (1994). The palaeoproterozoic (Birimian) of Haute-Comoé in the West African craton, Ivory Coast: A transtensional back-arc basin. *Precambrian Research*, 65(1–4), 207–229.

Vidal, M., Gumiaux, C., Cagnard, F., Pouclet, A., Ouattara, G., Pichon, M. (2009). Evolution of a Paleoproterozoic "weak type" orogeny in the West African Craton (Ivory Coast). *Tectonophysics*, 477(3–4), 145–159.

Villeneuve, M. (2008). Review of the orogenic belts on the western side of the West African craton: The Bassarides, Rokelides and Mauritanides. *Geological Society*, 297(1), 169–201.

Wane, O., Liégeois, J.P., Thébaud, N., Miller, J., Metelka, V., Jessell, M. (2018). The onset of the Eburnean collision with the Kenema-Man craton evidenced by plutonic and volcano sedimentary rock record of the Masssigui region, southern Mali. *Precambrian Research*, 305, 444–478.

Wilson, J.T. (1966). Did the Atlantic close and then re-open? *Nature*, 211(5050), 676–681.

Wyborn, L.A.I., Heinrich, C.A., Jaques, A.L. (1994). Australian Proterozoic mineral systems: Essential ingredients and mappable criteria. *The AusIMM Annual Conference*, Darwin.

Zhao, G., Cawood, P.A., Wilde, S.A., Sun, M. (2002). Review of global 2.1–1.8 Ga orogens: Implications for a pre-Rodinia supercontinent. *Earth-Science Reviews*, 59(1–4), 125–162.

Appendix 1

Lithium Mineralization, Contributions of Paleoclimates and Orogens

Éric GLOAGUEN[1,2], Jérémie MELLETON[1], Blandine GOURCEROL[1] and Romain MILLOT[1]

[1] *BRGM, Orléans, France*

[2] *ISTO, CNRS, University of Orléans, France*

Table A1.1 illustrates the wide variety of pegmatites. Two large families have been defined according to Černý and Ercit (2005): NYF, for pegmatites rich in niobium-yttrium-fluorine, and LCT, for pegmatites rich in lithium-cesium-tantalum.

All the pegmatites that make up lithium deposits belong to this second group.

Classes	Subclasses	Types	Subtypes	Families
Abyssal	HREE	–	–	NYF
	LREE	–	–	–
	U	–	–	NYF
	BBe	–	–	LCT
Muscovite	–	–	–	–
Muscovite-rare elements	REE	–	–	NYF
	Li	–	–	LCT
Rare elements	REE	Allanite-monazite	–	NYF
		Euxenite	–	NYF
		Gadolinite	–	NYF
	Li	Beryl	Beryl-columbite	LCT
			Beryl-columbite-phosphate	LCT
		Complex	***Spodumene***	LCT
			Petalite	LCT
			Lepidolite	LCT
			Elbaite	LCT
			Amblygonite	LCT
		Albite-Spodumene	–	LCT
		Albite	–	LCT
Miarolitic	REE	Topaz-beryl	–	NYF
		Gadolinite-fergusonite	–	NYF
	Li	Beryl-topaz	–	LCT
		Spodumene	–	LCT
		Petalite	–	LCT
		Lepidolite	–	LCT

Table A1.1. *Classification of pegmatites according to Černý and Ercit (2005). The pegmatites that make up lithium deposits sensu lato are marked in bold italics*

Appendix 2

Metallogeny of the Abitibi Greenstone Belt, Canada

Patrick MERCIER-LANGEVIN, Benoît DUBÉ, Michel G. HOULÉ, Valérie BÉCU, Anne-Aurélie SAPPIN, Jean-Luc PILOTE and Sébastien CASTONGUAY

Geological Survey of Canada, Natural Resources Canada, Quebec, Canada

Metallic Resources 2,
coordinated by Sophie DECRÉE.

A2.1. Location of the Superior Province and of the Abitibi greenstone belt

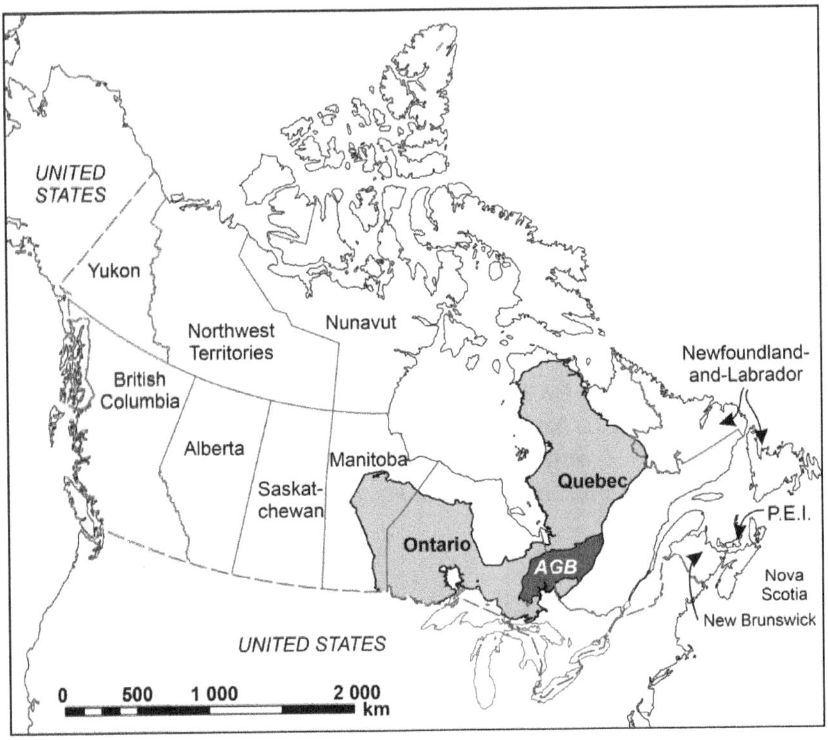

Figure A2.1. *Map of the northern part of North America showing the Canadian provinces and territories, as well as the location of the Superior Province and the Abitibi greenstone belt*

COMMENT ON FIGURE A2.1.– *The limits of the Superior Province correspond to the exposed part of the bedrock in Canada; large parts of the craton are covered with younger rocks to the south and to the north. AGB = Abitibi greenstone belt, PEI = Prince Edward Island.*

A2.2. Gold deposits of the Abitibi greenstone belt[1,2]

Some of the reported data may not be compliant with the National Instrument 43-101 standard. For official information, please refer to the cited literature, websites and industry reports.

Name of deposit or ore deposit/mine (3)	Mining camp	Deposit type/style (4)	Age (5)	Main commodities	Secondary metals	Cumulative production (t Au)	Reserves (6) (t Au)	Resources (7) (t Au)
Cedar Bay	Chibougamau-Chapais	Sulfide-rich Cu-Au-Ag veins	ca. 2,715 Ma	Cu-Au	–	12.5	1.4	
Chevrier	Chibougamau-Chapais	Auriferous ± polymetallic veins hosted in volcanic rocks	<2,705 Ma	Au	–	–	–	13.2
Copper Rand	Chibougamau-Chapais	Sulfide-rich Cu-Au-Ag veins	ca. 2,715 Ma	Cu-Au	–	43.2	3.3	1.8
Lemoine	Chibougamau-Chapais	Auriferous sulfide lenses	2,728–2,727 Ma	Cu-Zn-Au	Ag	3.5	–	–
MOP-II	Chibougamau-Chapais	Sulfide-rich Cu-Au-Ag veins	ca. 2,715 Ma	Au-Ag	Cu	–	–	9.2
Portage-Henderson	Chibougamau-Chapais	Sulfide-rich Cu-Au-Ag veins	ca. 2,715 Ma	Cu-Au	–	30.8	–	–
Principale (Main Mine)	Chibougamau-Chapais	Sulfide-rich Cu-Au-Ag veins	ca. 2,715 Ma	Cu-Au	–	7.4	0.2	–
Bousquet 2; Dumagami	Doyon-Bousquet-LaRonde	Auriferous sulfide lenses	2,698–2,697 Ma	Au	Cu-Zn-Ag	112.6	–	–
Bousquet No. 1; LZ5	Doyon-Bousquet-LaRonde	Auriferous ± polymetallic veins hosted in volcanic rocks	2,698–2,697 Ma	Au	Ag	41.8	21.3	19.4
Doyon; Silverstack	Doyon-Bousquet-LaRonde	*Stockworks*, veinlets and replacement associated with pre-Timiskaming syn-volcanic intrusive rocks	2,698–2,697 Ma	Au	Cu	168.3	–	3.8
LaRonde-Penna	Doyon-Bousquet-LaRonde	Auriferous sulfide lenses	2,698–2,697 Ma	Au	Cu-Zn-Ag-Pb	147.6	89.8	15.2

Mic Mac; Goldhurst	Doyon-Bousquet-LaRonde	Quartz-sulfide veins associated with syn-volcanic intrusive rocks	2,698–2,697 Ma	Au	Cu	3.3	–	–
Mouska	Doyon-Bousquet-LaRonde	Quartz-sulfide veins associated with syn-volcanic intrusive rocks	2,698–2,697 Ma	Au	Cu	28.0	0.3	0.4
Westwood	Doyon-Bousquet-LaRonde	Auriferous sulfide lenses and auriferous ± polymetallic veins hosted in volcanic rocks	2,698–2,697 Ma	Au	Ag	19.0	18.6	47.4
Sleeping Giant	Joutel	Auriferous ± polymetallic veins hosted in volcanic rocks	ca. 2,722 Ma	Au	–	33.4	–	5.5
Horne; Horne 5	Rouyn-Noranda	Auriferous sulfide lenses	2,702–2,700 Ma	Au	Zn-Cu-Ag	325.4	–	141.2
Quémont	Rouyn-Noranda	Auriferous sulfide lenses	2,702–2,700 Ma	Au-Zn	Cu-Ag	76.3	0.8	–
Remnor	Rouyn-Noranda	Auriferous sulfide lenses	2,702–2,700 Ma	Au	Cu	3.2	–	–
Côté Gold	Swayze	*Stockworks*, veinlets and replacement associated with pre-Timiskaming syn-volcanic intrusive rocks	2,740 Ma	Au	–	–	–	317.3
Akasaba; Akasaba West	Val-d'Or	Auriferous ± polymetallic veins hosted in volcanic rocks	2,730–2,698 Ma	Au	Cu-Ag	1.0	4.6	3.0
Kiena	Val-d'Or	*Stockworks*, veinlets and replacement associated with pre-Timiskaming post-volcanic intrusive rocks	<2,690–2,686 Ma	Au	–	55.5	–	25.4
Norlartic; Kierens	Val-d'Or	*Stockworks*, veinlets and replacement associated with pre-Timiskaming post-volcanic intrusive rocks	<2,690–2,686 Ma	Au	–	7.5	–	13.3

Siscoe	Val-d'Or	*Stockworks*, veinlets and replacement associated with pre-Timiskaming post-volcanic intrusive rocks	<2,690–2,686 Ma	Au	–	27.4	–	–
Windfall Lake	Urban-Barry	*Stockworks*, veinlets and replacement associated with pre-Timiskaming post-volcanic intrusive rocks	2,698–2,697 Ma	Au	Ag	0.1	–	37.5
Barry	Urban-Barry	*Stockworks*, veinlets and replacement associated with pre-Timiskaming post-volcanic intrusive rocks	2,698–2,697 Ma	Au	Ag	1.4	–	12.0
Chadbourne	Rouyn-Noranda	Breccia pipe	<2,690–2,686 Ma	Au	–	5.6	–	–
Swanson	Barraute	*Stockworks*, veinlets and replacement associated with syn-Timiskaming intrusive rocks	ca. 2,670–2,660 Ma	Au	–		–	3.2
Bachelor Lake	Chibougamau-Chapais	*Stockworks*, veinlets and replacement associated with syn-Timiskaming intrusive rocks	<2,692 Ma	Au	–	10.7	–	6.6
Moroy	Chibougamau-Chapais	*Stockworks*, veinlets and replacement associated with syn-Timiskaming intrusive rocks	ca. 2,670–2,660 Ma	Au	–	–	–	3.5
Shortt Lake; Gand	Chibougamau-Chapais	*Stockworks*, veinlets and replacement associated with syn-Timiskaming intrusive rocks	<2,688 Ma	Au	–	10.9	–	–
Douay	Douay	*Stockworks*, veinlets and replacement associated with syn-Timiskaming intrusive rocks	<2,676 +6/–5	Au	–		–	14.9

Beattie; Donchester: Duparquet	Duparquet	*Stockworks*, veinlets and replacement associated with syn-Timiskaming intrusive rocks	ca. 2,670–, 660 Ma	Au	Ag	34.7	–	96.8
Garrcon; Jonpol; Newfield	Harker-Holloway	*Stockworks*, veinlets and replacement associated with syn-Timiskaming intrusive rocks	2,683–2,670 Ma	Au	–	0.4	–	48.8
Holloway; Lightning Zone (Holt Complex)	Harker-Holloway	*Stockworks*, veinlets and replacement associated with syn-Timiskaming intrusive rocks	ca. 2,670–2,660 Ma	Au	–	35.3	1.2	6.4
Holt; Holt McDermott (Holt Complex)	Harker-Holloway	*Stockworks*, veinlets and replacement associated with syn-Timiskaming intrusive rocks	2,683–2,670 Ma	Au	–	57.3	16.6	22.8
Eagle; Telbel	Joutel	*Stockworks*, veinlets and replacement associated with syn-Timiskaming intrusive rocks	ca. 2,670–2,660 Ma	Au	–	34.2	1.6	–
Upper Canada	Kirkland Lake	*Stockworks*, veinlets and replacement associated with syn-Timiskaming intrusive rocks	<2,690–2,686 Ma	Au	–	43.5	–	21.6
Upper Beaver; Beaverhouse Lake; Argonaut	Larder Lake	*Stockworks*, veinlets and replacement associated with syn-Timiskaming intrusive rocks	≤2,685–2,680 Ma	Au	Cu	4.4	43.4	12.5
Camflo; Malartic Hygrade	Malartic	*Stockworks*, veinlets and replacement associated with syn-Timiskaming intrusive rocks	<2,690–2,686 Ma	Au	–	58.8	1.1	–
Canadian Malartic; Barnat-Sladen; East Malartic	Malartic	*Stockworks*, veinlets and replacement associated with syn-Timiskaming intrusive rocks	ca. 2,670–2,660 Ma	Au	–	317.0	148.6	52.6

Matachewan Consolidated	Matachewan	*Stockworks*, veinlets and replacement associated with syn-Timiskaming intrusive rocks	ca. 2,670–2,660 Ma	Au	–	12.0	–	–
Young-Davidson	Matachewan	*Stockworks*, veinlets and replacement associated with syn-Timiskaming intrusive rocks	ca. 2,670–2,660 Ma	Au	–	56.5	97.9	37.2
Black Fox; Glimmer	Matheson	*Stockworks*, veinlets and replacement associated with syn-Timiskaming intrusive rocks	ca. 2,670–2,660 Ma	Au	–	30.7	–	15.0
Fenn-Gib	Matheson	*Stockworks*, veinlets and replacement associated with syn-Timiskaming intrusive rocks	ca. 2,670–2,660 Ma	Au	–	–	–	40.5
Hislop East; New Kelore; Goldpost	Matheson	*Stockworks*, veinlets and replacement associated with syn-Timiskaming intrusive rocks	ca. 2,670–2,660 Ma	Au	–	4.0	1.0	4.1
Ross	Matheson	*Stockworks*, veinlets and replacement associated with syn-Timiskaming intrusive rocks	ca. 2,670–2,660 Ma	Au	Ag-Cu	31.0	–	–
Francoeur	Rouyn-Noranda	*Stockworks*, veinlets and replacement associated with syn-Timiskaming intrusive rocks	ca. 2,670–2,660 Ma	Au	–	16.0	–	2.1
Wasamac No. 1	Rouyn-Noranda	*Stockworks*, veinlets and replacement associated with syn-Timiskaming intrusive rocks	ca. 2,670–2,660 Ma	Au	–	7.9	–	80.5
Timmins West	Timmins-Porcupine	*Stockworks*, veinlets and replacement associated with syn-Timiskaming intrusive rocks	ca. 2,670–2,660 Ma	Au	–	32.0	15.8	9.9

Detour Lake	Detour	Orogenic (post-Timiskaming) quartz and carbonate veins	2,660–2,640 ± 10 Ma	Au	–	165.1	575.0	120.7
Lower Detour Zone 58N	Detour	Orogenic (post-Timiskaming) quartz and carbonate veins	2,660–2,640 ± 10 Ma	Au	–	–	–	16.6
McKenzie Break	Barraute	Orogenic (post-Timiskaming) quartz and carbonate veins	2,660–2,640 ± 10 Ma	Au	–	–	–	3.2
Central Cadillac; Wood	Cadillac	Orogenic (post-Timiskaming) quartz and carbonate veins	ca. 2,660–2,640 ± 10 Ma	Au	Ag	3.8	–	–
Lapa; Tonawanda	Cadillac	Orogenic (post-Timiskaming) quartz and carbonate veins	ca. 2,660–2,640 ± 10 Ma	Au	–	25.6	–	–
O'Brien; Darius	Cadillac	Orogenic (post-Timiskaming) quartz and carbonate veins	ca. 2,660–2,640 ± 10 Ma	Au	–	18.9	–	9.0
Casa Berardi; Golden Pond	Casa Berardi	Orogenic (post-Timiskaming) quartz and carbonate veins	ca. 2,660–2,640 ± 10 Ma	Au	–	80.5	59.3	33.1
Cooke	Chapais	Orogenic (post-Timiskaming) quartz and carbonate veins	ca. 2,715	Cu-Au	–	9.9	–	–
Joe Mann; Chibex	Chibougamau	Orogenic (post-Timiskaming) quartz and carbonate veins	ca. 2,660–2,640 ± 10 Ma	Au	Cu	40.0	–	–
Norbeau; McKenzie-07	Chibougamau	Orogenic (post-Timiskaming) quartz and carbonate veins	ca. 2,660–2,640 ± 10 Ma	Au	Ag	5.1	–	–
Destor; Thurbois; Yvan Vezina	Duparquet	Orogenic (post-Timiskaming) quartz and carbonate veins	ca. 2,660–2,640 ± 10 Ma	Au	Ag	3.9	–	–
Duquesne	Duparquet	Orogenic (post-Timiskaming) quartz and carbonate veins	ca. 2,660–2,640 ± 10 Ma	Au	–	2.1	–	6.2
Fayolle	Duparquet	Orogenic (post-Timiskaming) quartz and carbonate veins	ca. 2,660–2,640 ± 10 Ma	Au	–	–	–	3.5

Vezza	Joutel	Orogenic (post-Timiskaming) quartz and carbonate veins	ca. 2,660–2,640 ± 10 Ma	Au	–	0.3	–	8.1
Borden gold	Kapuskasing Structural Zone (Chapleau)	Orogenic (post-Timiskaming) quartz and carbonate veins	ca. 2,660–2,640 ± 10 Ma	Au	–	2.2	29.5	17.4
Amalgamated Kirkland	Kirkland Lake	Orogenic (post-Timiskaming) quartz and carbonate veins	ca. 2,660–2,640 ± 10 Ma	Au	–	–	–	8.2
Bidgood; Moffay-Hall	Kirkland Lake	Orogenic (post-Timiskaming) quartz and carbonate veins	ca. 2,660–2,640 ± 10 Ma	Au	–	5.1	–	3.9
McBean; Anoki; Queenston	Kirkland Lake	Orogenic (post-Timiskaming) quartz and carbonate veins	ca. 2,660–2,640 ± 10 Ma	Au	–	1.4	–	10.0
Omega	Kirkland Lake	Orogenic (post-Timiskaming) quartz and carbonate veins	ca. 2,660–2,640 ± 10 Ma	Au	–	6.7	–	6.8
Kerr Addison; Chesterville	Larder Lake	Orogenic (post-Timiskaming) quartz and carbonate veins	ca. 2,660–2,640 ± 10 Ma	Au	–	336.4	2.6	–
Kirkland Lake; Macassa	Kirkland Lake	Orogenic (post-Timiskaming) quartz and carbonate veins	ca. 2,660–2,640 ± 10 Ma	Au	–	917.8	76.3	22.7
McGarry	Larder Lake	Orogenic (post-Timiskaming) quartz and carbonate veins	ca. 2,660–2,640 ± 10 Ma	Au	–	0.1	–	3.5
East Amphi (Hybrid Zone)	Malartic	Orogenic (post-Timiskaming) quartz and carbonate veins	ca. 2,660–2,640 ± 10 Ma	Au	–	1.9	–	7.7
Malartic Gold Fields; Goldstack	Malartic	Orogenic (post-Timiskaming) quartz and carbonate veins	ca. 2,660–2,640 ± 10 Ma	Au	–	52.9	–	8.8
Marban	Malartic	Orogenic (post-Timiskaming) quartz and carbonate veins	ca. 2,660–2,640 ± 10 Ma	Au	–	10.3	–	36.1
Creek Zone	Matheson	Orogenic (post-Timiskaming) quartz and carbonate veins	ca. 2,660–2,640 ± 10 Ma	Au	–	–	–	3.2

Golden Highway	Matheson	Orogenic (post-Timiskaming) quartz and carbonate veins	ca. 2,660–2,640 ± 10 Ma	Au	Ag	–	–	17.3
Grey Fox	Matheson	Orogenic (post-Timiskaming) quartz and carbonate veins	ca. 2,660–2,640 ± 10 Ma	Au	–	–	–	19.3
Stock; St. Andrew East Zone	Matheson	Orogenic (post-Timiskaming) quartz and carbonate veins	ca. 2,660–2,640 ± 10 Ma	Au	–	4.3	–	3.8
Taylor (Holt Complex)	Matheson	Orogenic (post-Timiskaming) quartz and carbonate veins	ca. 2,660–2,640 ± 10 Ma	Au	–	4.8	4.0	3.4
Astoria; Yorbeau	Rouyn-Noranda	Orogenic (post-Timiskaming) quartz and carbonate veins	ca. 2,660–2,640 ± 10 Ma	Au	–	0.8	–	10.9
Donalda; Kerralda Deka	Rouyn-Noranda	Orogenic (post-Timiskaming) quartz and carbonate veins	ca. 2,660–2,640 ± 10 Ma	Au	–	4.6	0.1	–
Elder; Peel-Elder; Aunore	Rouyn-Noranda	Orogenic (post-Timiskaming) quartz and carbonate veins	ca. 2,660–2,640 ± 10 Ma	Au	–	13.0	–	3.0
Granada	Rouyn-Noranda	Orogenic (post-Timiskaming) quartz and carbonate veins	ca. 2,660–2,640 ± 10 Ma	Au	–	2.2	–	25.1
McWatters	Rouyn-Noranda	Orogenic (post-Timiskaming) quartz and carbonate veins	ca. 2,660–2,640 ± 10 Ma	Au	–	3.9	–	–
Pelletier Lake	Rouyn-Noranda	Orogenic (post-Timiskaming) quartz and carbonate veins	ca. 2,660–2,640 ± 10 Ma	Au	–	0.1	–	2.4
Pierre Beauchemin; Eldrich	Rouyn-Noranda	Orogenic (post-Timiskaming) quartz and carbonate veins	ca. 2,660–2,640 ± 10 Ma	Au	–	5.4	–	–
Powell Rouyn	Rouyn-Noranda	Orogenic (post-Timiskaming) quartz and carbonate veins	ca. 2,660–2,640 ± 10 Ma	Au	Ag	11.9	–	–
Rocmec 1; Russian Kid; El Coco	Rouyn-Noranda	Orogenic (post-Timiskaming) quartz and carbonate veins	ca. 2,660–2,640 ± 10 Ma	Au	–	0.1	–	3.7

Senator Rouyn	Rouyn-Noranda	Orogenic (post-Timiskaming) quartz and carbonate veins	ca. 2,660–2,640 ± 10 Ma	Au	–	7.8	0.7	–
Silidor	Rouyn-Noranda	Orogenic (post-Timiskaming) quartz and carbonate veins	ca. 2,660–2,640 ± 10 Ma	Au	–	15.0	–	–
Stadacona	Rouyn-Noranda	Orogenic (post-Timiskaming) quartz and carbonate veins	ca. 2,660–2,640 ± 10 Ma	Au	Ag	15.0	2.7	–
Juby	Shining Tree	Orogenic (post-Timiskaming) quartz and carbonate veins	ca. 2,660–2,640 ± 10 Ma	Au	–	–	–	32.4
Aquarius	Timmins-Porcupine	Orogenic (post-Timiskaming) quartz and carbonate veins	ca. 2,660–2,640 ± 10 Ma	Au	–	0.8	–	28.8
Aunor; Pamour No. 3	Timmins-Porcupine	Orogenic (post-Timiskaming) quartz and carbonate veins	ca. 2,660–2,640 ± 10 Ma	Au	–	77.8	–	–
Bell Creek; Marlhill	Timmins-Porcupine	Orogenic (post-Timiskaming) quartz and carbonate veins	ca. 2,660–2,640 ± 10 Ma	Au	–	17.5	14.8	15.3
Broulan Reef	Timmins-Porcupine	Orogenic (post-Timiskaming) quartz and carbonate veins	ca. 2,660–2,640 ± 10 Ma	Au	–	19.2	–	4.3
Buffalo Ankerite; Ankerite; March	Timmins-Porcupine	Orogenic (post-Timiskaming) quartz and carbonate veins	ca. 2,660–2,640 ± 10 Ma	Au	Ag	31.7	–	22.8
Clavos	Timmins-Porcupine	Orogenic (post-Timiskaming) quartz and carbonate veins	ca. 2,660–2,640 ± 10 Ma	Au	–	0.8	–	6.1
Delnite	Timmins-Porcupine	Orogenic (post-Timiskaming) quartz and carbonate veins	ca. 2,660–2,640 ± 10 Ma	Au	–	28.7	–	–
Dome; Century	Timmins-Porcupine	Orogenic (post-Timiskaming) quartz and carbonate veins	ca. 2,660–2,640 ± 10 Ma	Au	–	518.0	146.5	–
Fuller-Vedron; Tisdale Ankerite	Timmins-Porcupine	Orogenic (post-Timiskaming) quartz and carbonate veins	ca. 2,660–2,640 ± 10 Ma	Au	–	0.3	–	10.9

Gold River	Timmins-Porcupine	Orogenic (post-Timiskaming) quartz and carbonate veins	ca. 2,660–2,640 ± 10 Ma	Au	–	–	–	3.7
Hallnor; Broulan Porcupine; Pamour West	Timmins-Porcupine	Orogenic (post-Timiskaming) quartz and carbonate veins	ca. 2,660–2,640 ± 10 Ma	Au	–	60.4	–	41.9
Hollinger; McIntyre; Coniaurum	Timmins-Porcupine	Orogenic (post-Timiskaming) quartz and carbonate veins	ca. 2,660–2,640 ± 10 Ma	Au	Ag	998.7	24.1	–
Hoyle Pond	Timmins-Porcupine	Orogenic (post-Timiskaming) quartz and carbonate veins	ca. 2,660–2,640 ± 10 Ma	Au	–	127.6	11.2	0.9
Nighthawk Lake; Porcupine Peninsular	Timmins-Porcupine	Orogenic (post-Timiskaming) quartz and carbonate veins	ca. 2,660–2,640 ± 10 Ma	Au	–	6.3	–	9.3
Owl Creek	Timmins-Porcupine	Orogenic (post-Timiskaming) quartz and carbonate veins	ca. 2,660–2,640 ± 10 Ma	Au	–	7.4	–	5.5
Pamour	Timmins-Porcupine	Orogenic (post-Timiskaming) quartz and carbonate veins	ca. 2,660–2,640 ± 10 Ma	Au	–	173.2	34.2	21.8
Paymaster	Timmins-Porcupine	Orogenic (post-Timiskaming) quartz and carbonate veins	ca. 2,660–2,640 ± 10 Ma	Au	Ag	37.1	–	7.5
Preston; Preston East Dome	Timmins-Porcupine	Orogenic (post-Timiskaming) quartz and carbonate veins	ca. 2,660–2,640 ± 10 Ma	Au	–	47.9	–	–
Vipond	Timmins-Porcupine	Orogenic (post-Timiskaming) quartz and carbonate veins	ca. 2,660–2,640 ± 10 Ma	Au	–	12.9	–	–
Vogel; Schumacker	Timmins-Porcupine	Orogenic (post-Timiskaming) quartz and carbonate veins	ca. 2,660–2,640 ± 10 Ma	Au	–	–	–	3.9
Bradshaw	Timmins-Porcupine	Orogenic (post-Timiskaming) quartz and carbonate veins	ca. 2,660–2,640 ± 10 Ma	Au	–	–	–	13.1
Gladiator; Xemac	Urban-Barry	Orogenic (post-Timiskaming) quartz and carbonate veins	2,660–2,640 ± 10 Ma	Au	–	–	–	6.3
Bevcon; Buffadison; Lencourt	Val-d'Or	Orogenic (post-Timiskaming) quartz and carbonate veins	ca. 2,660–2,640 ± 10 Ma	Au	Ag	12.7	–	–
Chimo	Val-d'Or	Orogenic (post-Timiskaming) quartz and carbonate veins	ca. 2,660–2,640 ± 10 Ma	Au	–	10.9	–	1.1

Croinor	Val-d'Or	Orogenic (post-Timiskaming) quartz and carbonate veins	ca. 2,660–2,640 ± 10 Ma	Au	–	0.5	4.0	7.3
Dumont; Bras d'Or; Payore	Val-d'Or	Orogenic (post-Timiskaming) quartz and carbonate veins	ca. 2,660–2,640 ± 10 Ma	Au	–	8.0	–	–
Ferderber; Belmoral	Val-d'Or	Orogenic (post-Timiskaming) quartz and carbonate veins	ca. 2,660–2,640 ± 10 Ma	Au	–	11.3	–	–
Goldex	Val-d'Or	Orogenic (post-Timiskaming) quartz and carbonate veins	ca. 2,660–2,640 ± 10 Ma	Au	–	39.9	33.8	62.5
Lac Herbin	Val-d'Or	Orogenic (post-Timiskaming) quartz and carbonate veins	ca. 2,660–2,640 ± 10 Ma	Au	–	5.4	–	–
New Pascalis; Lucien C. Beliveau	Val-d'Or	Orogenic (post-Timiskaming) quartz and carbonate veins	ca. 2,660–2,640 ± 10 Ma	Au	–	5.7	–	26.9
Orenada Zone 4; Orenada Zone 2	Val-d'Or	Orogenic (post-Timiskaming) quartz and carbonate veins	ca. 2,660–2,640 ± 10 Ma	Au	Ag	0.0	–	6.1
Perron; Beaufor; North Pascalis	Val-d'Or	Orogenic (post-Timiskaming) quartz and carbonate veins	ca. 2,660–2,640 ± 10 Ma	Au	–	36.7	1.0	2.7
Sigma; Lamaque; Triangle	Val-d'Or	Orogenic (post-Timiskaming) quartz and carbonate veins	ca. 2,660–2,640 ± 10 Ma	Au	–	300.1	–	48.1
Sigma-2; Vicour	Val-d'Or	Orogenic (post-Timiskaming) quartz and carbonate veins	ca. 2,660–2,640 ± 10 Ma	Au	–	5.0	–	–
Sullivan; Sullivan Consolidated	Val-d'Or	Orogenic (post-Timiskaming) quartz and carbonate veins	ca. 2,660–2,640 ± 10 Ma	Au	–	35.3	–	2.4

(1) Only deposits with over 3.11 metric tons (100,000 ounces) of gold are listed here.
(2) Data from Gosselin and Dubé (2005, 2015) and from Bécu et al. (in press).
(3) This list is not necessarily exhaustive.
(4) Classification adapted from Dubé and Mercier-Langevin (2020). See Figure 2.7.
(5) Ages from Dubé and Mercier-Langevin (2020).
(6) Proven and probable reserves.
(7) Measured and indicated resources. Inferred resources are not included.

A2.2.1. *References*

Bécu, V., Dubé, B., Mercier-Langevin, P. (in press). Gold deposits of the Superior Province – Database. Report, Geological Survey of Canada, Ottawa.

A2.3. VMS deposits of the Abitibi greenstone belt[1]

Some of the reported data may not be compliant with the National Instrument 43-101 standard. For official information, please refer to the cited literature, websites and industry reports.

1. Only major deposits with over 0.5 million tonnes of ore are listed here. Note that 44 deposits with a past production and/or reserves and/or resources below 0.5 million tonnes are identified in the Abitibi (Krushnisky et al. in press).

Name of deposits, ore deposit/mine [2]	Mining camp	Ore deposit type/style [3]	Age	Main commodities	Secondary commodities	Cumulative production (Mt)	Reserves [4] (Mt)	Resources [5] (Mt)	Reference(s)
Abcourt-Barvue	Amos-Barraute	Bimodal mafic	–	Zn-Ag	–	5.63	–	8.09	Bérubé 2014
Jay Copper	Amos-Barraute	Bimodal mafic	–	Cu-Ag	Zn-Au	–	–	2.81	Cooke 1992
Vendôme No. 1	Amos-Barraute	Bimodal mafic	–	Zn-Au-Ag	Cu	–	–	0.56	Bérubé 2013
B26	Casa Berardi	Bimodal mafic	2,728–2,727 Ma	Cu-Zn	Au-Ag	–	–	6.97	Fayard et al. 2020a, 2020b
Estrades	Casa Berardi	Bimodal mafic	2,723–2,720 Ma	Cu-Zn-Au-Ag		0.17	–	1.50	Pressacco 2018
Gemini	Casa Berardi	Bimodal felsic	–	Zn-Au-Ag	Cu		–	0.67	Girard 2000
Selbaie	Selbaie	Bimodal felsic	2,728 Ma	Zn-Ag	Cu-Au	56.90	–	–	Faure 2012
Lemoine	Chibougamau-Chapais	Bimodal mafic	2,728–2,727 Ma	Cu-Zn-Au	Ag	0.76	–	–	Mercier-Langevin et al. 2014
Scott Lake	Chibougamau-Chapais	Bimodal mafic	2,727 Ma	Zn	Cu-Au-Ag	–	–	3.57	Lecuyer and Rosco 2017
Bousquet 2-Dumagami	Doyon-Bousquet-LaRonde	Bimodal mafic	2,698–2,697 Ma	Au	Cu-Zn-Ag	15.55	–	2.09	Mercier-Langevin et al. 2017
Ellison	Doyon-Bousquet-LaRonde	Bimodal mafic	2,698–2,697 Ma	Au	–	–	–	0.65	Krushnisky et al. in press

LaRonde Penna	Doyon-Bousquet-LaRonde	Bimodal mafic	2,698–2,697 Ma	Au	Cu-Zn-Ag-Pb	38.99	30.98	13.39	Krushnisky et al. in press
LaRonde Zone 5 (LZ5)	Doyon-Bousquet-LaRonde	Bimodal mafic	2,698–2,697 Ma	Au	–	7.45	–	18.04	Boily-Auclair et al. 2020
Westwood	Doyon-Bousquet-LaRonde	Bimodal mafic	2,698–2,697 Ma	Au	–	3.32	5.27(6)	5.27(5)	Krushnisky et al. in press
Joutel Copper	Joutel	Bimodal mafic	2,728 Ma	Cu-Zn	–	1.66	–	0.06	Lacroix 1998
Poirier	Joutel	Bimodal mafic	2,728 Ma	Cu-Zn	Ag	4.84	–	1.48	Lacroix 1998
Bell Allard	Matagami	Bimodal mafic	2,726–2,724 Ma	Cu-Zn-Ag	Au	3,59	–	–	Adair 2011
Bracemac-McLeod	Matagami	Bimodal mafic	2,726–2,724 Ma	Zn	Cu-Au-Ag	5.80	2.10(6)	3.10	Glencore report(s) 2013, 2019
Caber	Matagami	Bimodal mafic	2,726–2,724 Ma	Cu-Zn	Au-Ag	–	–	1.50	Glencore report 2019
Isle-Dieu	Matagami	Bimodal mafic	2,726–2,724 Ma	Cu-Zn-Ag	Au	3.05	–	–	Adair 2011
Mattagami Lake	Matagami	Bimodal mafic	2,726–2,724 Ma	Zn	Cu-Au-Ag	25.64	–	–	Adair 2011
New Hosco	Matagami	Bimodal mafic	2,726–2,724 Ma	Cu-Zn	Au-Ag	1.83	–	–	Adair 2011
Norita	Matagami	Bimodal mafic	2,726–2,724 Ma	Cu-Zn	Au-Ag	3.89	–	–	Adair 2011
Norita East	Matagami	Bimodal mafic	2,726–2,724 Ma	Zn-Ag	Cu-Au	1.08	–	–	Adair 2011
Orchan	Matagami	Bimodal mafic	2,726–2,724 Ma	Cu-Zn-Ag	Au	4.51	–	–	Adair 2011
PD-1	Matagami	Bimodal mafic	2,726–2,724 Ma	Cu-Zn	Ag	–	–	1.60	Glencore report 2019
Persévérance	Matagami	Bimodal mafic	2,726–2,724 Ma	Cu-Zn	–	4.82	–	–	Krushnisky et al. in press
Aldermac	Noranda	Bimodal mafic	2,701–2,698 Ma	Cu-Zn	Au-Ag	1.87	–	1.05	Cattalani et al. 1995
Amulet A	Noranda	Bimodal mafic	2,700–2,698 Ma	Cu-Zn-Au-Ag	–	5.31	–	–	Mercier-Langevin et al. 2011
Amulet C	Noranda	Bimodal mafic	2,700–2,698 Ma	Cu-Zn-Ag	Au	0.56	–	–	Mercier-Langevin et al. 2011

Ansil	Noranda	Bimodal mafic	2,700–2,698 Ma	Cu-Zn-Au	Ag	1.60	–	–	Mercier-Langevin et al. 2011
Corbet	Noranda	Bimodal mafic	2,700–2,698 Ma	Cu-Zn	Au-Ag	2.65	–	–	Mercier-Langevin et al. 2011
East Waite	Noranda	Bimodal mafic	2,700–2,698 Ma	Cu-Zn-Au-Ag	–	1.50	–	–	Mercier-Langevin et al. 2011
Fabie (New Insco)	Noranda	Bimodal mafic	2,702–2,700 Ma	Cu	Au-Ag	0.55	–	–	Mercier-Langevin et al. 2011
Gallen (West MacDonald)	Noranda	Bimodal mafic	2,700–2,698 Ma	Zn	Cu-Au-Ag	3.54	–	–	Mercier-Langevin et al. 2011
Horne	Noranda	Bimodal mafic	2,702–2,700 Ma	Au-Cu	Zn-Ag	54.34	–	–	Mercier-Langevin et al. 2011
Horne 5	Noranda	Bimodal mafic	2,702–2,700 Ma	Au	Cu-Zn-Ag	0.20	80.90[(6)]	91.20	Hardie et al. 2017
Joliet	Noranda	Bimodal mafic	2,700–2,698 Ma	Cu-Au	–	2.08	–	–	Lacroix 1998
Magusi River	Noranda	Bimodal mafic	2,702–2,700 Ma	Cu-Zn-Au-Ag	–		–	1.31	Salmon and Krutzelmann 2012
Millenbach	Noranda	Bimodal mafic	2,700–2,698 Ma	Cu-Zn-Ag	Au	3.48	–	–	Mercier-Langevin et al. 2011
Mobrun (Bouchard-Hébert)	Noranda	Bimodal mafic	2,697–2,695 Ma	Zn-Au-Ag	Cu	11.24	–	–	Mercier-Langevin et al. 2011
Norbec	Noranda	Bimodal mafic	2,700–2,698 Ma	Cu-Zn-Ag	Au	4.60	–	–	Mercier-Langevin et al. 2011
Old Waite	Noranda	Bimodal mafic	2,700–2,698 Ma	Cu-Zn-Au	Ag	1.12	–	–	Mercier-Langevin et al. 2011
Quemont	Noranda	Bimodal mafic	2,702–2,700 Ma	Cu-Zn-Au-Ag	–	13.91	–	–	Mercier-Langevin et al. 2011
West Ansil	Noranda	Bimodal mafic	2,700–2,698 Ma	Cu	Zn-Au-Ag		–	0.53	Masson 2006
Normétal	Normétal	Bimodal mafic	2,729–2,725 Ma	Cu-Zn-Ag	Au	10.11	–	–	Lacroix 1998
Coniagas	Quévillon-Desmaraisville	Bimodal mafic	2,723–2,720 Ma	Zn-Ag	–	0.72	–	–	Doucet et al. 1995
Langlois	Quévillon-Desmaraisville	Bimodal mafic	2,718 Ma	Zn-Ag	Cu-Au	5.75	0.84[(6)]	2.48	Nyrstar Report(s) 2012, 2018
Shunsby	Swayze	Bimodal mafic	2,732–2,724 Ma	Zn	Cu	–	–	3.71	Sobie 1990; Atkinson et al. 1999

Canadian Jamieson	Kamiskotia	Bimodal mafic	2,701–2,698 Ma	Cu-Zn	Au-Ag	0.83	–	–	Middleton 1975
Kam Kotia	Kamiskotia	Bimodal mafic	2,701–2,698 Ma	Cu-Zn	Au-Ag	5.99	–	–	van Hees et al. 2020
Kidd Creek	Timmins	Bimodal mafic	2,717–2,714 Ma	Cu-Zn-Ag		162.44	4.60(6)	8.60	van Hees et al. 2020; Glencore 2019
Potter	Timmins	Bimodal mafic	2,717–2,716 Ma	Cu-Zn	Au-Ag-Co	0.49	–	3.03	Hannington et al. 1999; Gamble and Bettiol 2008
Tillex	Timmins	Bimodal mafic	2,708–2,704 Ma	Cu	–	–	–	0.91	Bath 1990
Abitibi Copper	Val-d'Or	Bimodal mafic	2,706–2,702 Ma	Cu	Au-Ag	–	–	0.99	Bayne 1969
Dunraine	Val-d'Or	Bimodal mafic	2,704 Ma	Cu	Au-Ag	0.26	–	2.63	Lacroix 1998
East Sullivan	Val-d'Or	Bimodal mafic	2,704 Ma	Cu	Zn-Au-Ag	14.95	–	0.09	Lavergne 1985
Louvem	Val-d'Or	Bimodal mafic	2,704 Ma	Zn	Cu-Au-Ag	1.60	–	0.54	Lacroix 1998
Louvicourt	Val-d'Or	Bimodal mafic	2,704 Ma	Cu-Zn	Au-Ag	15.91	–		MERN, DV 2006-01 2006
Manitou-Barvue	Val-d'Or	Bimodal mafic	2,704 Ma	Zn-Ag	Cu-Au	11.20	–	0.09	Lacroix 1998
Québec Manitou	Val-d'Or	Bimodal mafic	2,704 Ma	Cu	Au-Ag	–	–	0.50	Lacroix 1998

(1) Only major deposits with over 0.5 million tons of ore are listed here. Forty-four additional deposits with a past production and/or reserves and/or resources below 0.5 million tons are identified in the Abitibi (Krushnisky et al. in press).
(2) Deposits whose main commodity is Au are also listed in Appendix 2.2.
(3) Based on the classification by Franklin et al. (2005).
(4) Proven and probable reserves.
(5) Measured and indicated resources. Inferred resources are not taken into consideration.
(6) Reserves are included in resources.

A2.3.1. *References*

Adair, R.N. (2011). Technical report on the resource calculation for the PD1 deposit, Matagami Project, Quebec. Report NI 43-101, Metals Ltd., Zorayda Consulting Ltd.

Atkinson, B.T., Hailstone, M.H., Pressacco, R., Wilson, A.C., Draper, D.M., Hope, P., Morra, P.M., Egerland, D.C. (1999). Report of activities, 1998 Resident Geologist Program, Timmins Regional Resident Geologist Report: Timmins and Sault Ste. Marie Districts. Report, Ontario Geological Survey, 5990.

Bath, A.C. (1990). Mineral occurrences, deposits, and mines of the Black River – Matheson Area. Report, Ontario Geological Survey, 5735.

Bayne, A.S. (1969). Report on Louvicourt Property of Abitibi Copper Mines Limited. Report, Ministère de l'Énergie et des Ressources naturelles du Québec, GM 25155.

Bérubé, J.P. (2013). Resources evaluation report for the Vendôme Property. Report, NI 43-101, Abcourt Mines Inc., Ministère de l'Énergie et des Ressources naturelles du Québec, GM 68379.

Bérubé, J.P. (2014). Mineral Resources Report for the Abcourt-Barvue Property. Report, NI 43-101, Abcourt Mines Inc.

Boily-Auclair, É., Mercier-Langevin, P., Ross, P.S., Pitre, D. (2020). Stratigraphic setting of the LZ5 and Ellison mineralized zones, LaRonde Zone 5 Project, Doyon-Bousquet-LaRonde mining camp, Abitibi, Quebec. In *Targeted Geoscience Initiative 5: Contributions to the Understanding of Canadian Gold Systems*, Mercier-Langevin, P., Lawley, C.J.M., Castonguay, S. (eds). Geological Survey of Canada, Open File 8712, 57–73.

Cattalani, S., Barrett, T.J., Chartrand, F. (1995). Géologie et géochimie du gisement d'Aldermac (région de Rouyn-Noranda). Report, Ministère de l'Énergie et des Ressources naturelles du Québec, ET 90–13.

Cooke, C.M. (1992). A review of the 1990–1991 exploration program and previous work, JonpolAmos Cu-Zn-Ag Property, Project 6801. Report, Ministère de l'Énergie et des Ressources naturelles du Québec, GM 52074.

Doucet, P., Mueller, W., Chartrand, F. (1995). Environnement volcanique, géochimique et minéralisation du gisement de Coniagas, Desmaraisville – District de Chibougamau. Report, Ministère de l'Énergie et des Ressources naturelles du Québec, MB 95–14.

Faure, S. (2012). Réévaluation paléo environnementale du complexe volcanique de Selbaie et de son potentiel métallogénique. Report, Consorem.

Fayard, Q., Mercier-Langevin, P., Wodicka, N., Daigneault, R., Perreault, S. (2020a). The B26 Cu-Zn-Ag-Au project, Brouillan volcanic complex, Abitibi greenstone belt, part 1: Geological setting and geochronology. In *Targeted Geoscience Initiative 5: Contributions to the Understanding of Canadian Gold Systems*, Mercier-Langevin, P., Lawley, C.J.M., Castonguay, S. (eds). Geological Survey of Canada, Open File 8712, 93–107.

Fayard, Q., Mercier-Langevin, P., Daigneault, R., Perreault, S., Wodicka, N. (2020b). The B26 Cu-Zn-Ag-Au project, Brouillan volcanic complex, Abitibi greenstone belt, part 2: Hydrothermal alteration and mineralization. In *Targeted Geoscience Initiative 5: Contributions to the Understanding of Canadian Gold Systems*, Mercier-Langevin, P., Lawley, C.J.M., Castonguay, S. (eds). Geological Survey of Canada, Open File 8712, 109–125.

Gamble, A.D.P. and Bettiol, E.I. (2008). Technical report on the resources at Potter Mine Property, Munro Township – Larder Lake Mining Division. Internal report, Millstream Mines Ltd.

Girard, P. (2000). Rapport d'évaluation, calcul des ressources géologiques, recommandation d'un programme d'exploration, Projet Gemini. Report, Ministère de l'Énergie et des Ressources naturelles du Québec, GM 58239.

GLENCORE (2019). Rapport de ressources et de réserves au 31 décembre 2019. Report, GLENCORE.

Hannington, M.D., Barrie, C.T., Bleeker, W. (1999). The Giant Kidd Creek Volcanogenic Massive Sulfide Deposit, Western Abitibi Subprovince, Canada: Preface and Introduction. In *The Giant Kidd Creek Volcanogenic Massive Sulfide Deposit, Western Abitibi Subprovince, Canada*, Hannington, M.D. and Barrie, C.T. (eds). Society of Economic Geologists, Littleton.

Hardie, C., Lacombe, P., Bertrand, V., Bewick, R., Boulianne, Y., Bratty, M., Drozdiak, J., Kissiova, M., Maillous, M., Ouellet, S. et al. (2017). Feasibility study, Horne 5 gold project, Rouyn-Noranda, Québec, Canada. Report, Falco Resources.

van Hees, E.H., Bousquet, P., Suma-Momoh, J., Daniels, C.M., Hinz, S.L.K., Boucher, C., Sword, P., Wang, L., Fudge, S.P., Millette, A., Patterson, C. (2020). Report of Activities, 2019 Resident Geologist Program, Timmins Regional Resident Geologist Report: Timmins and Sault Ste. Marie Districts. Report, Ontario Geological Survey, 6366.

Kerr, D.J. and Mason, R. (1990). A re-appraisal of the geology and ore deposits of the Horne mine complex at Rouyn-Noranda, Quebec. Report, Canadian Institute of Mining and Metallurgy, 43, 153–165.

Krushnisky, A., Mercier-Langevin, P., Meagher, D. (in press). Volcanogenic massive sulfide deposits of the Superior Province, Canada – Compilation and database. Report, Laurentian University, Sudbury.

Lacroix, S. (1998). Compilation et répartition des gisements polymétalliques à tonnage évalué dans la sous-province de l'Abitibi. Report, Ministère de l'Énergie et des Ressources naturelles du Québec, 98–06.

Lavergne, C. (1985). Gîtes minéraux à tonnage évalué et production minérale du Québec. Report, Ministère de l'Énergie et des Ressources naturelles du Québec, 85–08.

Lecuyer, N.L. and Roscoe, W.E. (2017). Technical report on the preliminary economic assessment for the Scott Lake Project, Northwestern Québec, Canada. Report, Yorbeau Resources Inc.

Masson, M. (2006). West Ansil resource estimate. Report, Falconbridge Limited.

Mercier-Langevin, P., Goutier, J., Ross, P.S., McNicoll, V., Monecke, T., Dion, C., Dubé, B., Thurston, P., Bécu, V., Gibson, H. et al. (2011). The Blake River group of the Abitibi greenstone belt and its unique VMS and gold-rich VMS endowment. In *GAC-MAC-SEG-SGA Joint Annual Meeting 2011, Ottawa, Field Trip 02B Guidebook*. Geological Survey of Canada, Open File 6869.

Mercier-Langevin, P., Lafrance, B., Bécu, V., Dubé, B., Kjarsgaard, I., Guha, J. (2014). The Lemoine auriferous volcanogenic massive sulfide deposit, Chibougamau camp, Abitibi greenstone belt, Quebec, Canada: Geology and genesis. *Economic Geology*, 109(1), 231–269.

Mercier-Langevin, P., Dubé, B., Blanchet, F., Pitre, D., Laberge, A. (2017). The LaRonde Penna Au-rich volcanogenic massive sulfide deposit. In *Archean Base and Precious Metal Deposits, Southern Abitibi Greenstone Belt, Canada*, Monecke, T., Mercier-Langevin, P., Dubé, B. (eds). Society of Economic Geologists, Littleton.

Middleton, R.S. (1975). Geology of Turnbull and Godfrey townships, District of Thunder Bay. Ontario Division of Mines. Report, Geological Branch, 5118.

Ministère de l'Énergie et des Ressources naturelles du Québec (2006). Rapport sur les activités d'exploration minière au Québec en 2005. Report, Ministère de l'Énergie et des Ressources naturelles du Québec, DV 2006-01.

Pressacco, R. (2018). Technical report on the mineral resource estimate for the Estrades Project, Northwestern Québec, Canada. Report, Galway Metals Inc., RPA.

Salmon, B. and Krutzelmann, H. (2012). Technical Report on the mineral resource estimate for the Magusi Project Abitibi region, Canada. Report, Roscoe Postle Associates Inc., Mag Copper Ltd.

Sobie, P.A. (1990). Report on phase 1a exploration program on the Shunsby Property, Cunningham two of Kirkton Resources Corp. Report, Kirkton Resources Corp.

A2.4. Ni-Cu-PGE, Fe-Ti-V, talc, magnesite and chrysotile deposits

Some of the reported data may not be compliant with the National Instrument 43-101 standard. For official information, please refer to the cited literature, websites and industry reports.

Ni-Cu-(PGE)									
Name of deposit, ore deposit/mine (1)	Mining camp	Deposit type/style	Age	Main commodities	Secondary commodities	Cumulative production	Reserves (2)	Resources (3)	Reference(s)
Montcalm [1]	Timmins West	Gabbro	ca. 2,702 Ma	Ni-Cu	PGE	3.9 Mt at 1.1% Ni, 0.6% Cu	2.8 Mt to 1.26% Ni 0.59% Cu	3.0 Mt at 1.26% Ni, 0.59% Cu	Houlé et al. in press
Warren Claim [2]	Timmins West	Gabbro	–	Ni-Cu		–	–	0.4 Mt* at 1% Ni+Cu	OGS 2021
Loveland [3]	Timmins West	Gabbro	–	Ni-Cu	PGE	–	–	0.5 Mt* at 0.7% Ni, 0.5% Cu	Houlé et al. in press
Enid [4]	Timmins West	Gabbro	–	Ni-Cu	PGE	–	–	1.1 Mt* at 0.5% Ni, 0.3% Cu	Houlé et al. in press
Crawford [5]	Timmins West	Komatiite (Style 2)	–	Ni	Cu- PGE	–	–	1,150 Mt at 0.28% Ni	Houlé et al. in press
Alexo [6]	Dundonald	Komatiite (Style 1)	2,720–2,710 Ma	Ni	Cu- PGE	0.09 Mt at 3.63% Ni	–	0,03 Mt at 2.91% Ni	Houlé et al. in press
Kelex [6]	Dundonald	Komatiite (Style 1)	2,720–2,710 Ma	Ni	Cu- PGE	3,900 t at 1.68% Ni	–	0.61 Mt at 0.73% Ni	Houlé et al. in press
Dundeal [6]	Dundonald	Komatiite (Style 1)	2,720–2,710 Ma	Ni	Cu- PGE	–	–	0.4 Mt at 2.0% Ni	Houlé et al. 2017
Dundonald South [6]	Dundonald	Komatiite (Style 1)	2,720–2,710 Ma	Ni	Cu- PGE	–	–	0.12 Mt at 3.15% Ni	Harron 2009
Redstone [7]	Shaw Dome	Komatiite (Style 1)	2,710–2,704 Ma	Ni	Cu- PGE	0.53 Mt at 1.93% Ni	–	1.6 Mt at 1.45% Ni	Houlé et al. in press

McWatters [7]	Shaw Dome	Komatiite (style 1)	2,710–2,704 Ma	Ni	Cu-PGE	0.37 Mt to 0.58% Ni	–	0.8 Mt at 0.55% Ni	Houlé et al. in press
Langmuir #1 [7]	Shaw Dome	Komatiite (style 1)	2,710–2,704 Ma	Ni	Cu- PGE	0.1 Mt to 1.74% Ni	–	1.7 Mt at 0.5% Ni	Houlé et al. in press
Langmuir #2 [7]	Shaw Dome	Komatiite (style 1)	2,710–2,704 Ma	Ni	Cu- PGE	1.1 Mt to 1.45% Ni	–	0.26 Mt at 1.4% Ni	Houlé et al. in press
Langmuir North [7]	Shaw Dome	Komatiite (style 2)	2,710–2,704 Ma	Ni	Cu- PGE	–	–	8.3 Mt at 0.4% Ni	Houlé et al. 2017
Langmuir South [7]	Shaw Dome	Komatiite (style 1)	2,710–2,704 Ma	Ni	Cu- PGE	–	–	0.18 Mt at 1.5% Ni	Houlé et al. 2017
Hart [7]	Shaw Dome	Komatiite (style 1)	2,710–2,704 Ma	Ni	Cu- PGE	–	–	1.87 Mt at 1.4% Ni	Houlé et al. 2017
Langmuir W4 [7]	Shaw Dome	Komatiite (style 1)	2,710–2,704 Ma	Ni	Cu- PGE	–	–	0.85 Mt at 0.97% Ni	Houlé et al. 2017
Texmont [8]	Bartlett Dome	Komatiite (style 2)	2,710–2,704 Ma	Ni	Cu- PGE	0.2 Mt to 0.96% Ni	0.67 Mt to 0.96% Ni	2.9 Mt at 0.92% Ni	Houlé et al. in press
Sothman [9]	Halliday Dome	Komatiite (style 1)	2,720–2,710 Ma	Ni	Cu- PGE	–	–	0.32 Mt at 0.9% Ni	Houlé et al. in press
Shining Tree [10]	–	Gabbro	–	Ni-Cu	–	–	–	2.5 Mt at 0.7% Ni, 0.4% Cu	Houlé et al. in press
R.M. Nickel [11]	–	Gabbro	2,695 ± 1 Ma	Ni-Cu	–	–	–	0.1 Mt* at 0.5% Ni, 0.8% Cu	Honsberger 1965; McNicoll et al. 2014
Quebec Moly [12]	La Motte	–	2,720–2,710 Ma	Ni	–	–	–	0.1 Mt* at 0.7% Ni	Honsberger 1965

Marbridge [12]	La Motte	Komatiite (style 1)	2,714 ± 2 Ma	Ni	Cu-PGE	0.7 Mt at 2.3% Ni	–	–	Pilote et al. 1998
Dumont [13]	Amos	Komatiite (style 2)	2,710–2,704 Ma	Ni	Cu- PGE	–	1,028 Mt at 0.27% Ni	2,165 Mt at 0.26% Ni	Staples et al. 2019
Bilson-Cubric [14]	La Motte	Komatiite (style 1)	2,720–2,710 Ma	Ni	Cu- PGE	–	–	0.2 Mt at 0.8% Ni	Houlé et al. 2017
Vendôme 2 [15]	–	–	–	Ni-Cu	–	–	–	0.3 Mt at 0.8% Ni, 0.7% Cu	SIGÉOM 2021
Commander-Zulapa [16]	–	Gabbro	–	Ni-Cu	–	–	–	1.3 Mt at 0.6% Ni, 0.4% Cu	Chainey 1993
Céré-Villebon [17]	–	Komatiite (style 1)	–	Ni	Cu- PGE	–	–	0.4 Mt* at 0.7% Ni	SIGÉOM 2021
Grasset [18]	Matagami	Komatiite (style 1)	–	Ni	Cu- PGE	–	–	3.5 Mt at 1.6% Ni, 0.2% Cu	Houlé et al. 2017

(1) Numbers between square brackets refer to the numbers of deposits in Figure 2.17.
(2) Proven and probable reserves.
(3) Measured, indicated and inferred resources.
* Uncategorized resources.

Fe-Ti-V									
Name of deposit or ore deposit/mine (1)	Mining camp	Deposit type/style	Age	Main commodities	Secondary commodities	Cumulative production	Reserves (2)	Resources (3)	Reference(s)
Iron-T [19]	Mattagami	Stratiform magnetitite	ca. 2,725 Ma	Fe-Ti-V	–	–	–	14.4 Mt at 27% Fe, 6.6% TiO_2, 0.42% V_2O_5	Dupéré 2011
Doré Lake -North-West/Armitage-Southwest [20]	Chibougamau	Stratiform magnetitite	ca. 2,728 Ma	Fe-Ti-V	–	–	–	942.0 Mt at 24.4-27.4% Fe, 6.7-7.2% TiO_2, 0.42% V_2O_5	Houlé et al. in press
Ile Portage [21]	Chibougamau	Stratiform magnetitite	ca. 2,728 Ma	Fe-Ti	V	–	–	0.5 Mt at 1.3% TiO_2	Houlé et al. in press
Mont Sorcier [21]	Chibougamau	Stratiform magnetitite	ca. 2,728 Ma	Fe-Ti-V	–	–	–	634.1 Mt at 24.9% Fe, 0.7-1.2% TiO_2, 0.25-0.5% V_2O_5	Houlé et al. in press

(1) Numbers between square brackets refer to the numbers of deposits in Figure 2.17.
(2) Proven and probable reserves.
(3) Measured and indicated resources. Inferred resources are not taken into consideration. In some cases, data precede the establishment of standard NI 43-101.

Talc, magnesite									
Name of deposit or ore deposit/mine [(1)]	Mining camp	Deposit type/style	Age	Main commodities	Secondary commodities	Cumulative production	Reserves [(2)]	Resources [(3)]	Reference(s)
Penhorwood [22]	Swayze	Invasive alteration	2,710–2,704 Ma	Talc	–	6.5 Mt with 2.1 Mt produced talc			OGS 2021
Timmins Talc-Magnesite [23]	Shaw Dome	Invasive alteration	2,710–2,704 Ma	Magnesite	Talc	–	31.5 Mt at 52.7% Mgs, 33.2% Tlc	5.0 Mt at 34.2% Mgs, 33.4% Tlc	Hennessey et al. 2012
Allerston-North Zone [24]	Shaw Dome	Invasive alteration	2,710–2,704 Ma	Talc	Magnesite	–	–	5.7 Mt* at 46% Tlc, 33.3% Mgs	OGS 2021
Allerston-South Zone [24]	Shaw Dome	Invasive alteration	2,710–2,704 Ma	Talc	Magnesite	–	54,076,357 at 46.4% Tlc, 34.4% Mgs	43.0 Mt at 47% Tlc, 34% Mgs	OGS 2021

(1) Numbers between square brackets refer to the numbers of deposits in Figure 2.17.
(2) Measured and indicated resources.
(3) Presumed resources.
* Uncategorized resources.
Abbreviations: Tlc = talc, Mgs = magnesite.

Chrysotile									
Name of deposit or ore deposit/mine (1)	Mining camp	Deposit type/style	Age	Main commodities	Secondary commodities	Cumulative production	Reserves (2)	Resources (3)	Reference(s)
Reeves [22]	Swayze	Veins	2,720–2,704 Ma	Chrysotile	–	6.1 Mt of chrysotile	–	20 Mt at 3.0% recoverable fibers	Vos 1971; OGS 2021
Daffodil [23]	Shaw Dome	Veins	2,720–2,704 Ma	Chrysotile	–	6,581 t produced chrysotile	–	18.1 Mt at 3.2% recoverable fibers	Vos 1971; Kretschmar and Kretschmar 1986
United Asbestos – Midlothian [25]	Halliday	Veins	2,720–2,704 Ma	Chrysotile	–	1.4 Mt at ~1.8% recoverable fibers	–	22.4 Mt at 4.7% recoverable fibers	Kretschmar and Kretschmar 1986
Munro [26]	Matheson	Veins	2,720–2,704 Ma	Chrysotile	–	7.0 Mt between 3.9 and 6.4% recoverable fibers	–	–	Bath 1990
Bird – Garrisson [27]	Matheson	Veins	2,720–2,704 Ma	Chrysotile	–	–	–	29.9 Mt at 9.8% recoverable fibers	Bath 1990
Abitibi Asbestos [28]	Lebel-sur-Quévillon	Veins	–	Chrysotile	–	–	–	92.1 Mt at 3.5% recoverable fibers	SIGÉOM 2021
Canadian Bolduc [29]	Amos	Veins	–	Chrysotile	–	0.7 Mt at 2.0% recoverable fibers	–	–	SIGÉOM 2021

Courvan-Amiante [30]	Val d'Or	Veins	–	Chrysotile	–	–	–	3.7 Mt at 2.5% recoverable fibers	Germain 1984
Ile de l'Amiante [31]	Chibougamau	Veins	–	Chrysotile	–	–	–	11.3 Mt at 6.4% recoverable fibers	Tiphane 1973
Lac Roberge [32]	Chibougamau	Veins	–	Chrysotile	–	–	–	150.4 Mt at 4.0% recoverable fibers	SIGÉOM 2021
Lac Roberge-Zone B [32]	Chibougamau	Veins	–	Chrysotile	–	–	–	40.3 Mt at 3.5% recoverable fibers	SIGÉOM 2021

(1) Numbers between square brackets refer to the numbers of deposits in Figure 2.17.
(2) Proven and probable reserves.
(3) Uncategorized resources.

A2.5. Fe, Cu-Mo, Mo-Bi, aluminosilicates, rare metals and Li deposits

Some of the reported data may not be compliant with the National Instrument 43-101 standard. For official information, please refer to the cited literature, websites and industry reports.

Fe									
Name of ore deposit/mine	Mining camp	Ore deposit type/style	Age	Main commodities	Secondary commodities	Cumulative production	Reserves	Resources	References
Adams	Kirkland Lake	Algoma	ca. 2,724 Ma >2,680Ma	Fe	–	3.8 Mt of ore, 1.1 Mt of iron ore pellets	21.4 Mt of ore at 26% Fe(3)		Basa 1991
Sherman	Kirkland Lake	Algoma	ca. 2,724 Ma >2,680 Ma	Fe	–	86 Mt of ore, 25.4 Mt of Fe concentrate			Basa 1991
Rivière Adam	Matagami	Algoma	<2,696 Ma	Fe	–			244.4 Mt at 24.6% Fe(2)	Grandillo and Bisaillon 2015
Iron Hills	Matagami	Algoma	<2,696 Ma	Fe	–		173.7 Mt at 23.8% Fe(1)	1,263.5 Mt at 20.7% Fe(2)	Grandillo and Bisaillon 2015

(1) Proven and probable reserves.

(2) Measured and indicated resources. Inferred resources are not taken into consideration.

(3) Not consistent with NI 43-101 standard.

Cu-Mo and Mo-Bi									
Name of ore deposit/mine	Mining camp	Ore deposit type/style	Age	Main commodities	Secondary commodities	Cumulative production	Reserves [1]	Resources [2]	Reference(s)
Don Rouyn	Noranda	Magmatic-hydrothermal/veins	2,700–2,690 Ma	Cu, Au, Mo	–	4.8 Mt of ore, 4,589 t Cu	36 Mt at 0.15% Cu[3]	40 Mt at 0.15% Cu[3]	Couture 1996; Goldie et al. 1979
R-2	Lac Doré, Chibougamau	Magmatic-hydrothermal/veins	~2,715 Ma	Cu, Au, Mo	–	–	–	10.9 Mt at 0.36% Cu and 0.015% Mo[3]	Kirkham et al. 1982
Clark Lake	Lac Doré, Chibougamau	Magmatic-hydrothermal/veins	~2,715 Ma	Cu, Au, Mo	–	–	–	–	–
Grandroy	Lac Doré, Chibougamau	Magmatic-hydrothermal/veins	~2,715 Ma	Cu, Au, Mo	–	349,000 t at 1.24% Cu and 0.022 g/t Au	123,600 t at 1.68% Cu and 0.51 g/t Au; Mo estimated at 0.1% [3]	44,000 t at 2.65% and 1.8 g/t Au [3]	Kirkham et al. 1982; Pilote 1995
Queylus (Talbot)	Lac Doré, Chibougamau	Magmatic-hydrothermal/veins	~2,715 Ma	Cu, Au, Mo	–	–	–	–	–
Preissac Moly	Preissac-La Corne	Magmatic-hydrothermal/veins	2,670–2,640 Ma	Mo, Bi	Cu	2.24 Mt at 0.161% Mo and 0.023% Bi	1.0 Mt at 0.186% Mo and 0.024% Bi [3]	–	Kirkham et al. 1982

Cadillac	Preissac-La Corne	Magmatic-hydrothermal/veins	2,670–2,640 Ma	Mo, Bi	Ag, Cu	2.0 Mt at 0.36% Mo	507,000 t at 0.19% Mo and 0.030% Bi (3)	–	Exploratech ltée 1979; Kirkham et al. 1982
La Corne	Preissac-La Corne	Magmatic-hydrothermal/veins	2,670–2,640 Ma	Mo, Bi	Be, Cu	3.86 Mt at 0.21% Mo and 0.03% Bi	200,000 t at 0.07% Mo and 0.019% Bi (3)	–	Kirkham et al. 1982; Trudel 1980
Moly Hill	Preissac-La Corne	Magmatic-hydrothermal/veins	2,670–2,640 Ma	Mo, Bi	Cu, Ta	–	272,155 t at 0.12% Mo and 0.061% Bi (3)	–	Boily 1995; Kirkham et al. 1982
Height of Land	Preissac-La Corne	Magmatic-hydrothermal/veins	2,670–2,640 Ma	Mo, Bi	Be, Cu	–	218,520 t at 0.12 Mo and 0.04% Bi (3)	–	Exploratech ltée 1979
Ryan Lake	Matachewan	Magmatic-hydrothermal/veins	2,680–2,675 Ma	Cu, Mo	Au, Ag	1.81 Mt at 0.25% Cu and 0.192% Mo	3,5 Mt at 0.4% Cu and 0.09% MoS2 (3)	–	Sinclair 1979; Kirkham et al. 1982

(1) Proven and probable reserves.
(2) Measured and indicated resources. Inferred resources are not taken into consideration.
(3) Prior to the establishment of standard NI 43-101.

Aluminosilicates									
Name of ore deposit/mine	**Mining camp**	**Ore deposit type/style**	**Age**	**Main commodities**	**Secondary commodities**	**Cumulative production**	**Reserves and resources**	**Mineral inventory (1)**	**Reference(s)**
Carpentier	Amos-Barraute	Aluminosilicates (pyrophyllite)	–	Pyrophyllite	Andalusite, silica	–	–	≥2.4 Mt at 60% pyrophyllite	Assad 1990

(1) Considered as reserves in Assad (1990).

Rare metals and lithium									
Name of ore deposit/mine	**Mining camp**	**Ore deposit type/style**	**Age**	**Main commodities**	**Secondary commodities**	**Cumulative production**	**Reserves (1)**	**Resources (2)**	**Reference(s)**
North American Lithium (Québec Lithium)	Amos-Barraute	Granitic pegmatite	2,641–2,639 Ma	Li	Be-Mo-Ta-Bi	0.91 Mt at 1.40% Li_2O	30 Mt at 1.12% Li_2O	–	Boily et al. 1989; Stone and Selway 2009
Authier (Colombe)	Amos-Barraute	Granitic pegmatite	2,641–2,639 Ma	Li	Be-Mo	–	12.1 Mt at 1.00% Li_2O	17.2 Mt at 1.01% Li_2O	Sayona Mining 2021
Montviel	Quévillon-Desmaraisville	Carbonatite	1,894 Ma	REE-Nb	–	–	–	82.4 Mt at 1.51% total rare earth oxides	David et al. 2006; Belzile et al. 2015

(1) Proven and probable reserves.

(2) Measured and indicated resources. Inferred resources are not taken into consideration.

A2.5.1. *References*

Assad, J.R. (1990). Le gîte de pyrophyllite de Carpentier. Report, Ministère de l'Énergie et des Ressources naturelles, GM 50145.

Basa, A. (1991). Cobalt resident geologist's district 1990: Report of activities 1990. Report, Ontario Geological Survey, MP152, 261–279.

Belzile, E., Marchand, R., Bouajila, A. (2015). NI 43-101 technical report – Montviel rare earth project, Québec, Canada. Report, Géoméga Resources Inc. [Online]. Available at: https://geomega.ca/fr/montviel-ree-nb-2/ [Accessed 8 January 2021].

Boily, M. (1995). Pétrogénèse du batholite de Preissac-Lacorne : implications pour la métallogénie des gisements de métaux rares. Report, Ministère de l'Énergie et des Ressources, ET93-05.

Boily, M., Pilote, P., Raillon, H. (1989). La métallogénie des métaux de haute technologie en Abitibi-Témiscamingue. Report, Ministère de l'Énergie et des Ressources, MB89-29.

Couture, J.F. (1996). Les gisements métalliques du district de Rouyn-Noranda. Métallogénie et évolution tectonique de la région de Rouyn-Noranda. Report, Ministère de l'Énergie et des Ressources, MB96-06, 11–22.

David, J., Dion, C., Goutier, J., Roy, P., Bandyayera, D., Legault, M., Rhéaume, P. (2006). Datations U-Pb effectuées dans la Sous-province de l'Abitibi à la suite des travaux de 2004–2005. Report, Ministère des Ressources naturelles et de la Faune, RP2006-04.

Exploratech Ltée. (1979). Évaluation du potentiel en molybdène au Québec ; évaluation sommaire. Report, Ministère de l'Énergie et des Ressources, DPV-619.

Goldie, R.J., Kotila, B., Seward, D. (1979). The Don Rouyn Mine: An Archean porphyry copper deposit near Noranda, Québec. *Economic Geology*, 74(7), 1680–1684.

Grandillo, A. and Bisaillon, C. (2015). Barlow Mine Inc.: Iron Hills, Adam River, and Orvilliers projects, West-Abitibi, Quebec, Canada. Report, NI43-101.

Kirkham, R.V., McCann, C., Prasad, A.E., Soregaroli, F.M., Vokes, F.M., Wine, G. (1982). Molybdenum in Canada, Part 2: MOLYFILE – An index-level computer file of molybdenum deposits and occurrences in Canada. Report, Geological Survey of Canada, 33.

Pilote, P. (1995). Metallogenic evolution and geology of the Chibougamau area – From porphyry Cu-Au-Mo to mesothermal lode gold deposits, field trip guidebook. Report, Geological Survey of Canada, 3143.

Sayona Mining (2021). Sayona mining [Online]. Available at: https://sayonamining.com.au/authier-project/ [Accessed 8 January 2021].

Sinclair, W.D. (1979). Copper-molybdenum occurrences of the Matachewan area, Ontario. Report, Geological Survey of Canada, Open File 79-1A.

Stone, M. and Selway, J. (2009). Independent technical report, Quebec Lithium Property. Report, Ministère de l'Énergie et des Ressources naturelles, GM65011.

Trudel, P. (1980) Métallogénie et prospection du molybdène au Québec. Report, Ministère de l'Énergie et des Ressources du Québec, DPV733.

List of Authors

Anne-Sylvie ANDRÉ-MAYER
GeoRessources
University of Lorraine
Vandœuvre-lès-Nancy
France

Valérie BÉCU
Geological Survey of Canada
Natural Resources Canada
Quebec
Canada

Antonio BENEDICTO
GEOPS
CNRS
University of Paris-Saclay
Orsay
France

Mohammed BOUABDELLAH
Laboratoire des gîtes minéraux
University Mohammed Premier
Oujda
Morocco

Sébastien CASTONGUAY
Geological Survey of Canada
Natural Resources Canada
Quebec
Canada

Guoxiang CHI
University of Regina
Canada

Sophie DECRÉE
Royal Belgian Institute of
Natural Sciences
Geological Survey of Belgium
Brussels
Belgium

Benoît DUBÉ
Geological Survey of Canada
Natural Resources Canada
Quebec
Canada

Aurélien EGLINGER
GeoRessources
University of Lorraine
Vandœuvre-lès-Nancy
France

Éric GLOAGUEN
BRGM
and
ISTO
CNRS
University of Orléans
France

Blandine GOURCEROL
BRGM
Orléans
France

Michel G. HOULÉ
Geological Survey of Canada
Natural Resources Canada
Quebec
Canada

Charles KHAIRALLAH
Orano Canada
Saskatoon
Canada

Patrick LEDRU
GeoRessources
University of Lorraine
Vandœuvre-lès-Nancy
France

Quentin MASUREL
University of Western Australia
Crawley
Australia

Jérémie MELLETON
BRGM
Orléans
France

Julien MERCADIER
GeoRessources
University of Lorraine
Vandœuvre-lès-Nancy
France

Patrick MERCIER-LANGEVIN
Geological Survey of Canada
Natural Resources Canada
Quebec
Canada

Romain MILLOT
BRGM
Orléans
France

Jean-Luc PILOTE
Geological Survey of Canada
Natural Resources Canada
Quebec
Canada

Jonathan POH
Géosciences Rennes
CNRS
University of Rennes
France

John ROBBINS
Orano Canada
Saskatoon
Canada

Anne-Aurélie SAPPIN
Geological Survey of Canada
Natural Resources Canada
Quebec
Canada

Nicolas THÉBAUD
University of Western Australia
Crawley
Australia

Index

A, B

C, D

E, F

G, H

I, J

K, L

M, N

O, P, Q

R, S

T, U

W, Z

Summary of Volume 1

Printed and bound by CPI Group (UK) Ltd, Croydon, CR0 4YY
13/09/2026
14961068-0005